微课版

OpenStack私有云基础架构与运维

（openEuler版）

沈建国 代丽 ◉ 主编
张云 张婷婷 龙全波 ◉ 副主编

人 民 邮 电 出 版 社
北 京

图书在版编目（CIP）数据

OpenStack私有云基础架构与运维 : openEuler版 : 微课版 / 沈建国，代丽主编. -- 北京 : 人民邮电出版社，2024.7
工业和信息化精品系列教材. 云计算技术
ISBN 978-7-115-64427-5

Ⅰ. ①O… Ⅱ. ①沈… ②代… Ⅲ. ①云计算－教材
Ⅳ. ①TP393.027

中国国家版本馆CIP数据核字(2024)第097057号

内容提要

本书基于国产操作系统 openEuler 和云基础架构平台 OpenStack 编写，主要介绍云基础架构平台的部署、运维、管理与应用。全书共 6 个项目，包括 OpenStack 云基础架构平台技术概述、私有云基础架构、云基础架构平台部署、云基础架构平台运维、云基础架构平台管理、云基础架构平台应用。每个项目均包含相关实际案例的实施过程，项目末尾还设置项目小结、拓展知识、知识巩固和拓展任务，帮助读者巩固所学内容，强化相关技能。

本书可以作为高校云计算技术应用专业及其他相关专业的教材，也可以作为云计算技术培训班的培训教材，还可以作为从事云计算运维工作的专业人员和广大云计算爱好者的自学参考书。

◆ 主　　编　沈建国　代　丽
　副 主 编　张　云　张婷婷　龙全波
　责任编辑　顾梦宇
　责任印制　王　郁　焦志炜

◆ 人民邮电出版社出版发行　北京市丰台区成寿寺路 11 号
　邮编　100164　电子邮件　315@ptpress.com.cn
　网址　https://www.ptpress.com.cn
　固安县铭成印刷有限公司印刷

◆ 开本：787×1092　1/16
　印张：14.75　　2024 年 7 月第 1 版
　字数：435 千字　　2025 年 3 月河北第 3 次印刷

定价：59.80 元

读者服务热线：(010)81055256　印装质量热线：(010)81055316
反盗版热线：(010)81055315

前言

在数字化和云原生时代，云计算作为一种灵活且高效的技术架构，已经在全球范围内得到广泛的认可和应用。作为开源云基础架构平台中的佼佼者，OpenStack 已经成为私有云市场的主流选择之一，并且正在逐渐改变着传统基础设施的运作方式。

本书模拟企业真实生产环境，详细介绍 OpenStack 云基础架构平台的部署、运维、管理与应用，不仅注重理论知识的讲解，还注重实践操作的介绍，以及知识的巩固和拓展。本书通过一系列实际案例，为读者深入讲解 OpenStack 云基础架构平台的相关知识，帮助读者提升实操经验。这些案例包括 WordPress 部署、云应用系统迁移、高可用架构应用、主从数据库部署及 HAProxy 负载均衡方案实施等。通过这些案例，读者能够更好地构建和管理私有云，从而实现稳定、可靠的云计算应用。

党的二十大报告中提出，教育、科技、人才是全面建设社会主义现代化国家的基础性、战略性支撑。必须坚持科技是第一生产力、人才是第一资源、创新是第一动力，深入实施科教兴国战略、人才强国战略、创新驱动发展战略，开辟发展新领域新赛道，不断塑造发展新动能新优势。在国家推动国产软件自主可控的大背景下，本书选择 openEuler 22.09 进行案例实施，以展现国产自主可控的操作系统在私有云领域的卓越表现。openEuler 作为一个开源的、自主可控的操作系统，提供了强大的功能和稳定的性能，适用于多种云计算应用场景。

本书共 6 个项目，建议教师采用理论实践一体化的教学模式，参考学时为 64 学时，其中项目讲解为 60 学时，课程考评为 4 学时，具体的学时分配可参见以下学时分配表。

学时分配表

项目	项目内容	学时
项目 1	OpenStack 云基础架构平台技术概述	4
项目 2	私有云基础架构	8
项目 3	云基础架构平台部署	12
项目 4	云基础架构平台运维	12
项目 5	云基础架构平台管理	12
项目 6	云基础架构平台应用	12
	课程考评	4
学时总计		64

本书由沈建国、代丽任主编并统稿，由张云、张婷婷、龙全波任副主编。本书的实验案例由江苏一道云科技发展有限公司提供，杜纪魁负责全书的技术审核，慕旭东、应天龙、袁婉敏参与全书案例的验证与审阅，在此表示诚挚的感谢！

由于搭建环境复杂及编者水平有限，书中难免存在不妥之处，敬请读者批评指正，编者邮箱：shenjianguo@wxic.edu.cn。

主 编

2024 年 2 月

目录

项目 1

项目 2

项目 3

项目 4

项目 5

项目 6

项目1 OpenStack云基础架构平台技术概述

01

学习目标

【知识目标】

① 理解云计算的基本概念及其核心技术。
② 了解常用的云基础架构平台。
③ 学习 OpenStack 的整体平台架构。

【技能目标】

① 具备运用常见云基础架构平台的能力。
② 具备安装及部署虚拟机的能力。
③ 掌握 OpenStack 云基础架构平台 Dashboard 的使用方法。

【素养目标】

① 培养结构化思维。
② 培养自主思考、独立分析问题与解决问题的能力。
③ 具备良好的职业道德和职业素养。

项目概述

小张即将从学校毕业，已被某公司聘为云计算助理工程师。公司现准备将原有的计算机服务器改造成云计算服务平台，为此，小张必须学习云计算的基础概念及其核心技术的相关知识，并对常用的云基础架构平台进行对比和了解，以便制订详细的改造方案。

知识准备

1.1 初识云计算

云计算是公用计算与网格计算理念的革新成果，历经演进现已成为信息时代的重要计算技术。本节将精练概述云计算的起源与发展历程，解析其基本特征，并介绍不同部署方式的特点与应用。

1.1.1 云计算的起源与发展历程

早在 2006 年 3 月，亚马逊（Amazon）首先提出弹性计算云服务，2006 年 8 月，谷歌（Google）

前首席执行官埃里克·施密特（Eric Schmidt）在搜索引擎大会（SES San Jose 2006）首次提出“云计算”（Cloud Computing）的概念。从那时候起，云计算开始受到关注，这是云计算较准确的诞生记录。

云计算作为一种计算技术和服务理念，有着极其深厚的技术背景，谷歌作为头部搜索引擎公司，首创这一概念，有着很大的必然性。随着众多互联网公司的崛起，各家互联网公司对云计算的投入不断增多、对云计算的研发不断加深，陆续形成完整的云计算技术架构、硬件网络，逐步向数据中心、全球网络连接、软件系统等方面发展，完善了操作系统、文件系统、并行计算架构、并行计算数据库、开发工具等云计算系统关键部件。

云计算经历了从集中式时代到网络时代，最终到分布式时代的转变，并在分布式时代的基础之上形成了云时代，如图 1-1 所示。

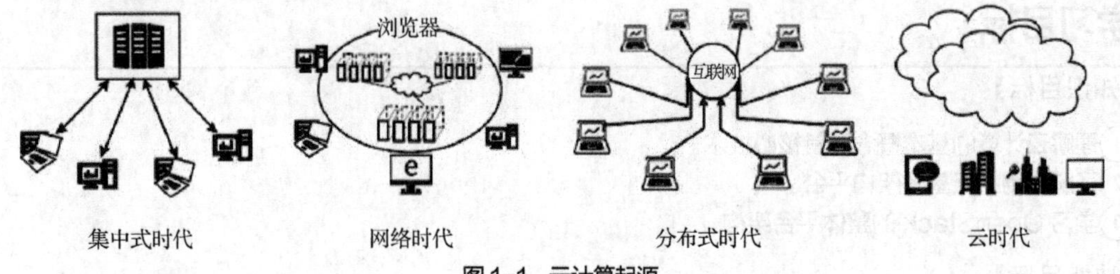

图 1-1　云计算起源

云计算的最终目标是将计算、服务和应用作为一种公共设施提供给公众，使人们能够像使用水、电、煤气一样便捷地使用计算资源。2010 年 7 月，美国国家航空航天局（National Aeronautics and Space Administration，NASA）和 Rackspace、超威半导体（AMD）、英特尔（Intel）、戴尔（Dell）等支持厂商共同宣布“OpenStack”开源计划。微软（Microsoft）在 2010 年 10 月表示支持 OpenStack 与 Windows Server 2008 R2 的集成；Ubuntu 操作系统也已把 OpenStack 加至其 11.04 版本中。2011 年 2 月，思科操作系统正式加入 OpenStack，重点研制 OpenStack 的网络服务。在这些相关厂商发布的操作系统的支持下，云计算服务发展速度变得更快，OpenStack 项目也得到了空前的发展，迎来了良好的发展时机。

1.1.2　云计算基本特征

云计算具有按需自助服务、通过互联网获取资源、资源池化、快速伸缩和可计量这 5 个基本特征。

1. 按需自助服务

用户可以单方面部署资源，如服务器、网络存储等，资源是按需自动部署的，不需要与服务供应商进行人工交互。

2. 通过互联网获取资源

资源可以通过互联网获取，并可以通过标准方式访问。例如，用户通过瘦客户端（没有硬盘的轻量计算机）或胖客户端（移动电话、笔记本电脑、工作站等）获取资源。

3. 资源池化

服务供应商的资源被池化，以便以多用户租用模式被不同用户使用。例如，不同的物理和虚拟资源可根据用户需求动态分配和重新分配，这些分配通常与地域无关，这些资源包括存储、处理器、内存、网络带宽等。

4. 快速伸缩

资源可以弹性地或者自动化地部署和释放，以便能够迅速地按需扩大或缩小规模。

5. 可计量

云计算系统能够自动控制和优化资源的使用，它通过使用一些与服务种类对应的抽象信息（存储、计算、带宽、激活的用户账号）来提供计量能力（通常在此基础上实现按使用量收费）。

1.1.3 云计算部署方式

云计算主要有私有云、社区云、公有云和混合云这 4 种部署方式。

（1）私有云：云基础设施由一个单一的组织部署和使用，可由该组织、第三方或两者的组合进行管理和运营。

（2）社区云：云基础设施由一些具有共同关注点的组织形成的社区中的用户部署和使用，可由一个或多个社区中的组织、第三方或两者的组合进行管理和运营。

（3）公有云：云基础设施被开放部署，公众可以使用。它可由一个商业组织、研究机构、政府机构或者多方的混合所拥有、管理和运营，或者被一个销售云计算服务的组织所拥有，该组织将云计算服务销售至广泛的工业群体。

（4）混合云：云基础设施由两种或两种以上的云（私有云、社区云或公有云）组成，每种云仍然保持独立，可以用标准的或专有的技术将它们组合，混合云使得数据和应用程序具备可移植性。

云计算部署方式与云计算服务模式之间存在密切的关系，它们描述了云计算提供的服务及服务的交付方式。云计算部署方式通常指的是云基础设施的物理或逻辑位置，云计算服务模式则涉及云计算服务的不同层次，由三大服务组成，即基础设施即服务（Infrastructure as a Service，IaaS）、平台即服务（Platform as a Service，PaaS）、软件即服务（Software as a Service，SaaS），又称为云计算 SPI（Saas，Paas，Iaas）模型，如图 1-2 所示，对 IaaS、PaaS、SaaS 的介绍如下。

（1）SaaS：SaaS 为用户提供了一种完善的产品，其运行和管理皆由应用服务供应商负责。用户使用应用程序并按使用量付费，但并不需要掌握操作系统、硬件的运行原理或网络的基础架构。

（2）PaaS：PaaS 提供了更高级别的平台和工具，用于支持应用程序的开发、测试、部署和管理。PaaS 包括了开发框架、数据库、中间件和其他工具，使开发人员能够专注于应用程序逻辑，而不必担心底层的基础设施。

（3）IaaS：IaaS 提供了基础设施层面的计算资源，如虚拟机、存储、网络和基本的操作系统。用户可以在这些虚拟化的基础设施上构建、部署和管理自己的应用程序及操作系统。IaaS 通常允许用户根据需要进行扩展或缩减资源，以满足不同的业务需求。

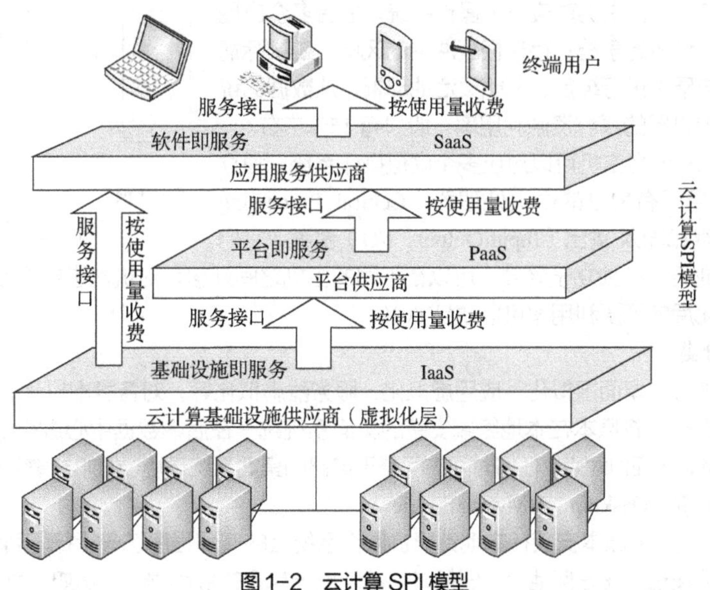

图 1-2　云计算 SPI 模型

在图 1-2 所示的模型中，IaaS 主要对应基础设施，实现底层资源虚拟化，最后部署实际云应用平台，这个过程是一个网络架构由规划架构到最终的物理实现的过程；PaaS 基于 IaaS 技术和平台，部署终端用户使用的应用程序，提供对外的服务接口或者服务产品，最终实现对整个平台的管理和平台的可伸缩化；SaaS 是基于现成的 PaaS、终端用户最后接触的产品，SaaS 对现有资源进行对外服务，并完成服务的租赁化。

1.2 云计算核心技术

云计算的目标是以低成本的方式提供高可靠、高可用、规模可伸缩的个性化服务。为了达到这个目标，需要虚拟化、分布式存储、数据管理、海量数据处理、资源管理与调度、安全与隐私保护等若干关键技术加以支持。

1.2.1 虚拟化技术

在图 1-2 所示的模型中，IaaS 是基础设施平台，主要利用虚拟化技术对计算机硬件资源进行集中管理，高效利用云上硬件资源。谈到云计算就离不开虚拟化的内容，因为虚拟化是云计算重要的支撑技术之一。

在计算机科学领域中，虚拟化代表着对计算资源的抽象，而不仅仅局限于虚拟机（Virtual Machine，VM）的概念。例如，对物理内存的抽象，产生了虚拟内存技术，使得应用程序认为其自身拥有连续可用的地址空间（Address Space），而实际上，应用程序的代码和数据可能被分隔成多个碎片页或段，甚至被交换到磁盘、闪存等外部存储器上，即使物理内存不足，应用程序也能顺利执行。

那么到底什么是虚拟化呢?

1. 虚拟化定义

虚拟化是一个广义的术语，是指计算元件在虚拟而不是真实的基础上运行，是一种为了简化管理、优化资源而产生的解决方案。在计算机运算中，虚拟化通常扮演硬件平台、操作系统、存储设备或者网络资源等角色。

图 1-3 所示为虚拟化示意，从以下几个方面对其进行简单说明。

（1）虚拟化前：一台主机对应一个操作系统。后台多个应用程序会对特定的资源进行争抢，存在相互冲突的风险；在实际情况中，业务系统与硬件进行绑定，不能灵活地部署；从数据的统计来说，虚拟化前机器的系统资源利用率一般只有 15%左右。

（2）虚拟化后：一台主机可以对应多个虚拟操作系统。独立的操作系统和应用拥有独立的中央处理器（Central Processing Unit，CPU）、内存和输入/输出（Input/Output，I/O）资源，它们之间相互隔离；业务系统独立于硬件，可以在不同的主机之间迁移；对机器进行虚拟化能够充分利用系统资源，虚拟化后的资源利用率可以达到 60%。

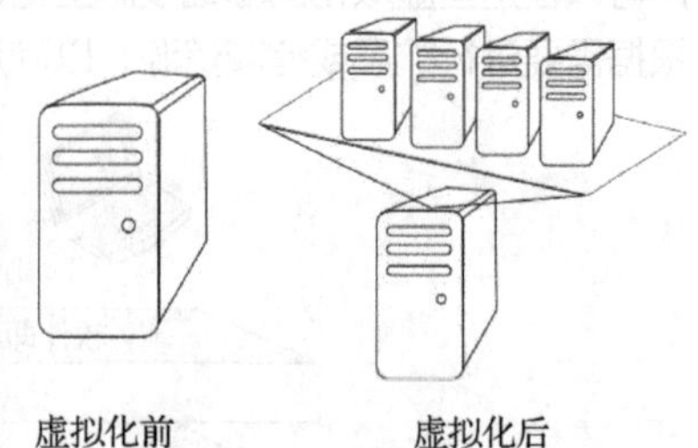

图 1-3 虚拟化示意

2. 虚拟化分类

虚拟化可以被分为桌面虚拟化、应用虚拟化、服务器虚拟化等，对各类虚拟化的介绍如下。

（1）桌面虚拟化：将原本在本地终端安装的桌面系统统一在后端数据中心进行部署和管理；用户可以通过任何设备，在任何地点、任何时间访问属于自己的桌面系统环境。例如，微软的 Remote Desktop Services、Citrix 的 XenDesktop、VMware 的 View。

（2）应用虚拟化：将原本安装在本地计算机操作系统上的应用程序统一运行于后台终端服务器上。用户可以通过任何设备，在任何地点、任何时间访问属于自己的应用程序。例如，微软的 WTS、Citrix 的 XenApp、VMware 的 ThinApp。

（3）服务器虚拟化：将服务器物理资源（如 CPU、内存、磁盘、I/O 等）抽象成逻辑资源，形成动态管理的“资源池”，并创建合适的虚拟服务器，实现服务器资源整合，提升资源利用率，最终更好地适应信息技术（Information Technology，IT）业务的变化。例如，微软的 Hyper-V、Citrix 的 XenServer、VMware 的 ESXi。

1.2.2 分布式存储技术

云计算的另一大优势就是能够快速、高效地处理海量数据。在数据爆炸的今天，这一点至关重要。为了保证数据的高可靠性，云计算通常会采用分布式存储技术，将数据存储在不同的物理设备中。这种模式不仅摆脱了硬件设备的限制，还使得可扩展性更好，能够快速响应用户需求。

分布式存储与传统的网络存储并不完全一样，传统的网络存储系统采用集中的存储服务器存放所有数据，存储服务器成为系统性能的瓶颈，不能满足大规模存储应用的需要；分布式存储系统采用可扩展的系统结构，利用多台存储服务器分担存储负载，同时利用位置服务器定位存储信息，不但提高了系统的可靠性、可用性和存取效率，还易于扩展。

分布式存储技术通过将数据存储在不同的物理设备中，实现动态负载均衡、故障节点自动接管，具有高可靠性、高可用性、高可扩展性。这是因为在多节点的并发执行环境中，各个节点的状态需要同步，并且在单个节点出现故障时，系统需要有效的机制保证其他节点不受影响。该技术在摆脱硬件设备限制的同时提供了更好的可扩展性，并且能够快速响应用户需求。

在当前的云计算领域，谷歌开发的谷歌文件系统（Google File System，GFS）和 Hadoop 开发的开源系统 Hadoop 分布式文件系统（Hadoop Distributed File System，HDFS）是比较流行的两种云计算分布式存储技术。

1. GFS 技术

基于谷歌非开源的 GFS 技术搭建的云计算平台可满足大量用户的需求，并行地为大量用户提供服务，使得云计算的数据存储技术具有了高吞吐率和高传输率的特点。

2. HDFS 技术

大部分信息与通讯技术（Information and Communication Technology，ICT）厂商，包括雅虎、英特尔的“云”计划采用的都是基于 HDFS 的数据存储技术。HDFS 未来的发展将集中在超大规模的数据存储、数据加密和安全性保证，以及继续提高 I/O 速率等方面。

1.2.3 数据管理技术

处理海量数据是云计算的一大优势，高效的数据管理技术也是云计算不可或缺的核心技术之一。云计算不仅要保证数据的存储和访问，还要能够对海量数据进行特定的检索和分析。由于云计算需要对海量的分布式数据进行处理、分析，因此，数据管理技术必须能够高效地管理大量的数据。

谷歌的 Bigtable 数据库和 Hadoop 团队开发的开源数据库 HBase 是业界比较典型的大规模数据管理技术的数据库应用。

1. Bigtable

Bigtable 是非关系型数据库，是一个分布式的、持久化存储的多维有序映射表。Bigtable 建立在 GFS、Scheduler、LockService 和 MapReduce 之上，与传统的关系数据库不同，它把所有数据都作为对象来处理，形成一个巨大的表格来分布式存储大规模结构化数据。Bigtable 的设计目标是能够可靠地处理 PB 级别的数据，并且能够将服务部署到上千台机器上。

2. HBase

HBase 是 Apache 的 Hadoop 项目的子项目，它是一个分布式的、面向列的开源数据库。HBase 不同于一般的关系数据库，它是一个适用于非结构化数据存储的数据库，且它基于列模式存储数据。利用 HBase 数据库可在廉价的个人计算机服务器上搭建起大规模结构化存储集群。

1.3 常用云基础架构平台

在完成前两小节相关知识的学习后，需要进一步了解常用的云基础架构平台，从而在搭建公司的私有云过程中选择最适合的云基础架构平台。这里主要介绍3种云基础架构平台：OpenStack、CloudStack、Eucalyptus。

1.3.1 OpenStack

图1-4所示为OpenStack示意，OpenStack是一个开源的云基础架构平台，由几个主要的组件组合起来完成具体工作。OpenStack支持几乎所有类型的云环境，其目标是提供实施简单、可大规模扩展、丰富、标准统一的云计算管理平台。OpenStack通过各种互补的服务提供了IaaS的解决方案，每个服务都提供应用程序接口（Application Program Interface，API）以进行集成。

图1-4 OpenStack示意

目前OpenStack平台或其演变版本正被广泛应用在各行各业，它的用户包括思科、华为、英特尔、IBM、99Cloud、希捷等。OpenStack采用Python语言开发，支持基于内核的虚拟机（Kernel-based Virtual Machine，KVM）、Xen、Docker等虚拟化软件，通过调用这些虚拟化软件来构建虚拟机（也称为云主机）为用户服务。

1. 起源与发展

OpenStack最早由NASA研发的Nova和Rackspace研发的Swift组成。它后来以Apache许可证授权，旨在为公有及私有云平台的建设与管理提供帮助。OpenStack主要用来为企业内部实现类似于亚马逊EC2和S3的云基础架构服务。OpenStack每6个月更新一次，基本与Ubuntu同步，以A～Z作为首字母来命名。

OpenStack是云基础架构平台中的佼佼者，在云基础架构平台研发方面，国外有IBM、微软、谷歌及亚马逊等厂商，国内则有优刻得、海云捷迅、UnitedStack、易捷行云、金山云、阿里云等厂商。现在比较流行的云基础架构相关平台有CloudStack、Eucalyptus、vCloud Director和OpenStack。OpenStack在市场中占据了绝对的份额优势。

OpenStack云基础架构平台作为开源IaaS平台，适用于部署裸金属、虚拟机、图形处理单元（Graphics Processing Unit，GPU）及容器等多种环境。OpenStack在实际生产中部署广泛，整体部署规模庞大，全球多个公有云数据中心都基于OpenStack运行。OpenStack社区也正在不断演进，广泛集成了Ceph、Kubernetes、TensorFlow等新兴技术，社区聚集着一批有实力的厂商和研发公司，这些厂商和研发公司把代码贡献给社区，不断完善OpenStack技术和推动其发展，拓展新的功能。随着软件可用性的进一步提升，OpenStack的部署场景更加丰富，部署规模也更大。

2. OpenStack平台架构

OpenStack提供了一个部署云的操作平台或工具集，其宗旨在于帮助组织运行实现虚拟计算或存储服务的云，为公有云、私有云提供可扩展的、灵活的云计算。

整个OpenStack由不同功能的节点（Node）组成，包括控制节点、计算节点、网络节点、存储节点（这4种节点也可以安装在一台机器上，此时为单机部署）。其中，控制节点负责对其余节点进行控制，包含虚拟机的建立、迁移、网络分配、存储分配等；计算节点负责虚拟机的运行；网络节点负责外网与内网之间的通信；存储节点负责对虚拟机的额外存储管理等。

OpenStack的主要目标是管理数据中心的资源，简化资源分配。它用于管理3部分资源，分别是计算资源、存储资源和网络资源，对这3部分资源的说明如下。

（1）计算资源：OpenStack可以规划并管理大量虚拟机，从而允许企业或服务供应商按需提供计算资源。开发人员可以通过API访问计算资源从而创建云应用，管理员与用户则可以通过Web访问这些资源。

（2）存储资源：OpenStack 可以为云服务或云应用提供所需的对象及块存储资源。因为对性能及价格有需求，传统的企业级存储技术无法满足很多组织的存储需求，而 OpenStack 可以根据用户需要提供可配置的对象存储或块存储功能。

（3）网络资源：如今的数据中心存在大量的配置工作，如服务器、网络设备、存储设备、安全设备等均需要配置，而它们还将被划分成更多的虚拟设备或虚拟网络，这会导致 IP 地址的数量、路由配置、安全规则的爆炸式增长。传统的网络管理技术无法真正地实现高扩展、高自动化地管理下一代网络，而 OpenStack 提供了插件式、可扩展、API 驱动型的网络及 IP 地址管理功能。OpenStack 为了实现云计算的各项功能，将存储、计算、监控、网络服务划分为几个项目来进行开发，每个项目对应 OpenStack 中的一个或多个组件。图 1-5 所示为 OpenStack 的整体架构。

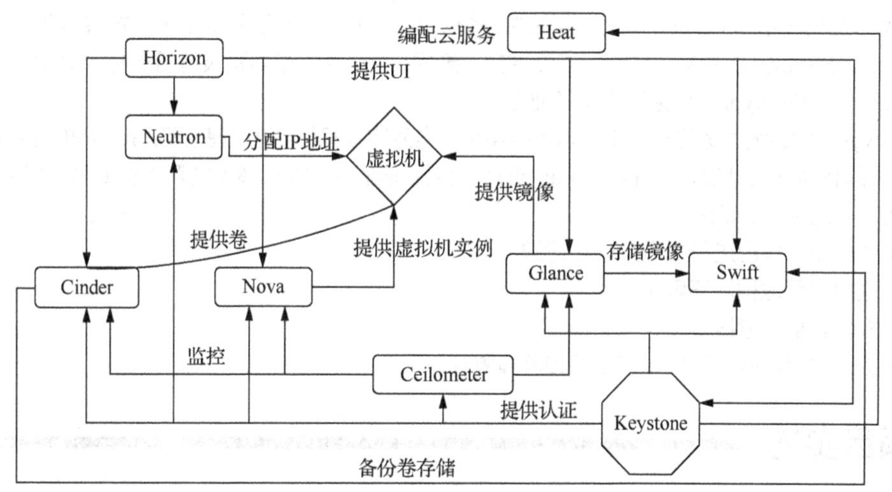

图 1-5　OpenStack 的整体架构

OpenStack 各个组件之间是松耦合的，其中，Keystone 是各个组件之间的通信核心，它依赖自身 RESTful API（基于 Identity API）为所有的 OpenStack 组件提供认证和访问策略服务。云平台用户在经过 Keystone 服务认证授权后，通过 Horizon 模式创建虚拟机服务，创建过程中包括利用 Nova 服务创建虚拟机实例，虚拟机实例通过 Glance 提供镜像服务，然后使用 Neutron 为新建的虚拟机分配 IP 地址，并将其纳入虚拟网络中，之后通过 Cinder 创建的卷为虚拟机挂载存储块，整个过程都在 Ceilometer 模块资源的监控下进行，Cinder 产生的卷（Volume）和 Glance 提供的镜像（Image）可以通过 Swift 的对象存储机制进行保存。此外，Heat 也是一个重要的服务，它提供了编排功能，可以帮助用户快速地创建和管理复杂的云服务应用。

1.3.2 CloudStack

CloudStack 开源云基础架构平台于 2012 年加入 Apache 基金会，CloudStack 的开发语言为 Java。

CloudStack 是一种开源的、具有高可用性及高扩展性的云基础架构平台，同时是一种开源的云计算解决方案。一些知名的信息驱动公司，如 Zynga 等已经使用 CloudStack 进行云部署。CloudStack 平台除了有自己的 API 外，还支持 CloudBridge、亚马逊 EC2，它可以把亚马逊 API 转换成自己的 API。

CloudStack 平台可以帮助用户快速地进行公有云和私有云的部署、管理、配置。另外，CloudStack 兼容亚马逊 API，允许跨 CloudStack 和亚马逊平台实现负载兼容。使用 CloudStack 作为基础，数据中心操作者可以快速、方便地通过现存基础架构创建云服务。它包含如下 7 个主要特性。

（1）虚拟机管理程序（如 KVM、Xen、ESXi、Oracle VM 和 BareMetal 等）不可知。

（2）根据不同角色进行权限分配和管理。

（3）提供支持虚拟局域网（Virtual Local Area Network，VLAN）的虚拟网络。

（4）提供资源池。

（5）提供快照和卷。

（6）提供虚拟路由器（包含一个防火墙和一个负载均衡器）。

（7）实现带有主机维护的实时迁移。

1.3.3 Eucalyptus

Eucalyptus（Elastic Utility Computing Architecture for Linking Your Programs To Useful Systems）是一种开源的软件基础平台，它通过计算集群或工作站群实现弹性的、实用的云计算。它最初是美国加利福尼亚大学圣巴巴拉分校计算机科学学院的一个研究项目，现在已经商业化，发展成为 Eucalyptus Systems Inc。Eucalyptus 是又一种流行的云平台。索尼、NASA、趋势科技等机构和公司都选择用它来部署其私有云。Eucalyptus 有免费版和商业版。

Eucalyptus 的最大优势之一是 Eucalyptus API 完全兼容亚马逊 API，基于亚马逊 API 的所有脚本和软件产品都可轻易用于私有云。Eucalyptus 支持 3 种管理程序：Xen、KVM 和 ESXi（只对商业版的用户开放）。其主要特性如下。

（1）根据不同角色进行权限分配和管理。

（2）虚拟机管理程序不可知。

（3）提供集群和分区。

（4）提供灵活的网络管理、安全组和流量隔离。

项目实施

任务 1.1 参观并分析学校信息化中心机房

在学习云计算的相关基本概念后，通过参观具体的系统环境，可帮助读者认识设备及其功能，直观了解云计算相关技术的实际应用，从而达到更好的学习效果。

任务 1.1.1 参观信息化中心机房

参观学校信息化中心机房的配套设施并与相关老师进行交流。读者需要了解的主要内容如下。

（1）学校的网络使用了哪种拓扑结构？由哪几个层次组成？

（2）系统采用的是真实的云计算平台还是虚拟化平台？搭建平台使用的具体技术是什么？

（3）系统有几台物理服务器？服务器的性能指标如何？

（4）系统中运行的应用项目有哪些？

任务 1.1.2 分析与调研

针对以上问题进行分析，并撰写调研报告。

任务 1.2 安装和部署虚拟化环境

通过安装 openEuler 22.09 操作系统来熟悉虚拟机的安装方法，在操作过程中熟悉计算机虚拟化资源的分配管理。

微课 1.1 安装和部署虚拟化环境

任务 1.2.1 VMware Workstation Pro 17 的安装

在个人计算机上安装 VMware 公司的虚拟机软件 VMware Workstation Pro 17，详细的安装教程可自行上网查询。

安装完成后，在桌面上会生成 VMware Workstation Pro 图标，双击该图标即可进入虚拟机软件工作界面，如图 1-6 所示。

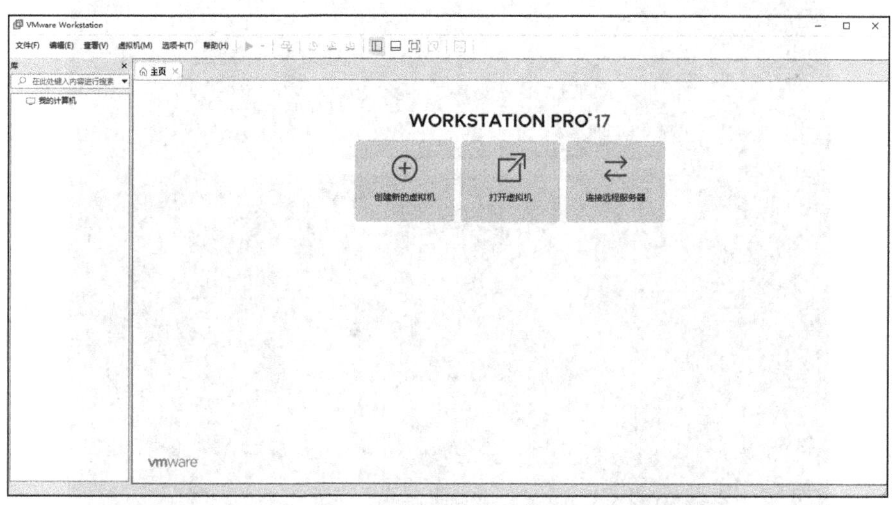

图 1-6 虚拟机软件工作界面

任务 1.2.2 安装虚拟机

本书选择 openEuler 操作系统替换经典的 CentOS 作为实验虚拟化环境。

启动 VMware Workstation Pro 虚拟机软件，安装一台使用 openEuler 22.09 操作系统的虚拟机，节点类型为 2 个 vCPU/2GB 内存/100GB 硬盘，虚拟机的安装结果如图 1-7 所示。

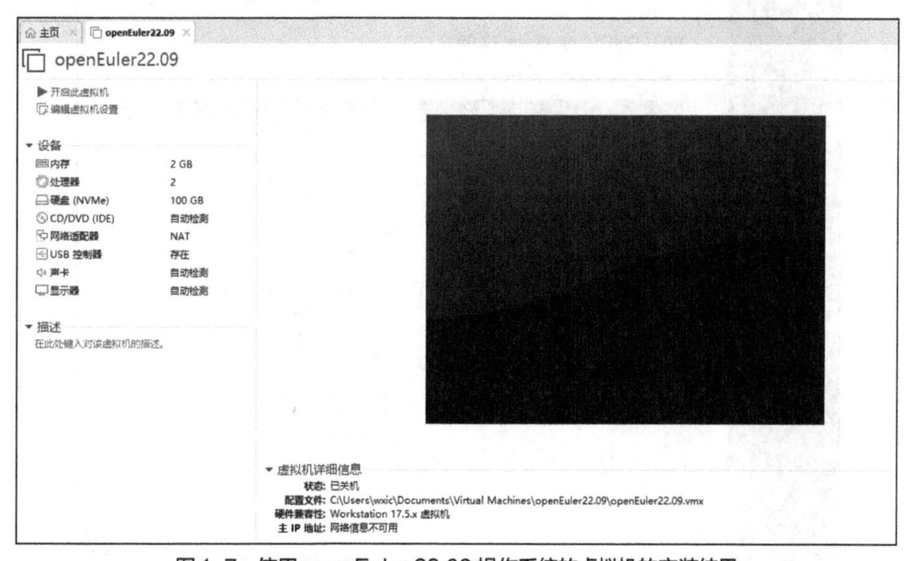

图 1-7 使用 openEuler 22.09 操作系统的虚拟机的安装结果

接下来，在“openEuler 22.09”界面中单击“开启此虚拟机”按钮，启动虚拟机。

成功引导系统后，会进入图 1-8 所示的界面，该界面的菜单中有以下 3 个选项。

（1）Install openEuler 22.09：安装 openEuler 22.09。

（2）Test this media & install openEuler 22.09：测试并安装 openEuler 22.09。

（3）Troubleshooting：故障排除。

这里使用键盘上的上下方向键选择“Install openEuler 22.09”选项并按 Enter 键。

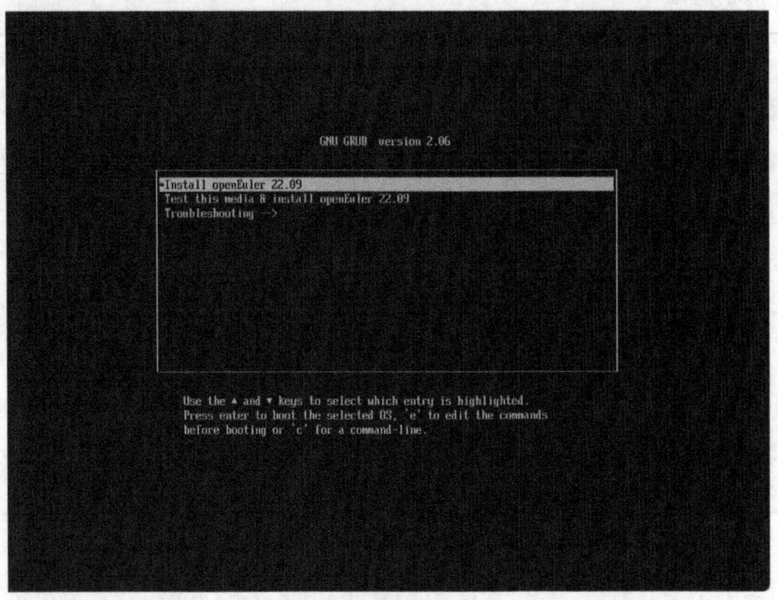

图 1-8　系统安装界面

等待片刻后，系统自动进入系统语言选择界面，如图 1-9 所示，保持默认选项，单击“Continue”按钮，进入系统安装向导界面，如图 1-10 所示。

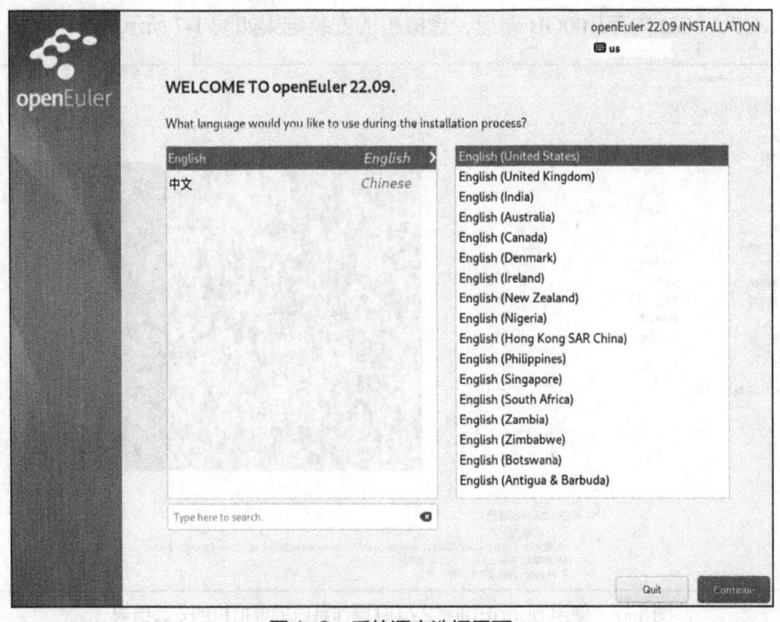

图 1-9　系统语言选择界面

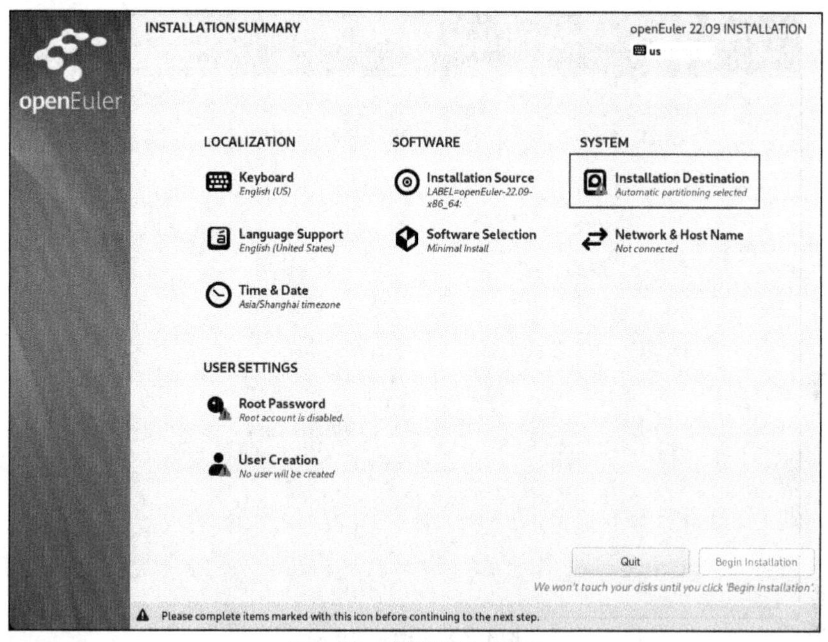

图1-10　系统安装向导界面

在系统安装向导界面中单击“Installation Destination”图标，进入磁盘分区界面，如图 1-11 所示。进入磁盘分区界面后，有两个单选按钮：Automatic（自动配置分区）及 Custom（手动配置分区）。分区应该按照实际服务器用途而定，这里选中“Automatic”单选按钮。单击该界面左上角的“Done”按钮完成设置，分区完成。

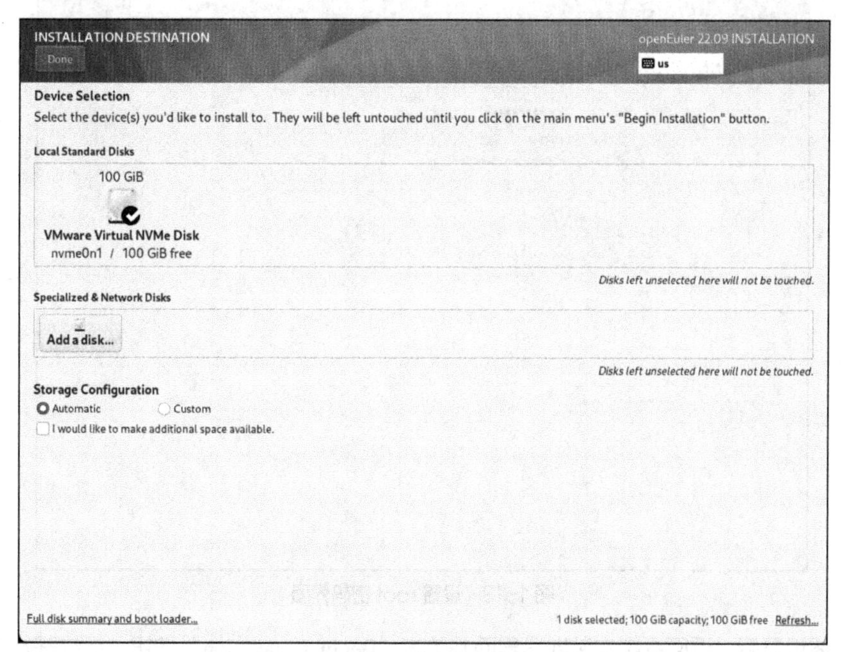

图1-11　磁盘分区界面

返回系统安装向导界面并单击“Time & Date”图标，进入设置时区界面，用户可选择所在时区，如“Asia/Shanghai”时区，并设置“24-hour”小时制，如图 1-12 所示。

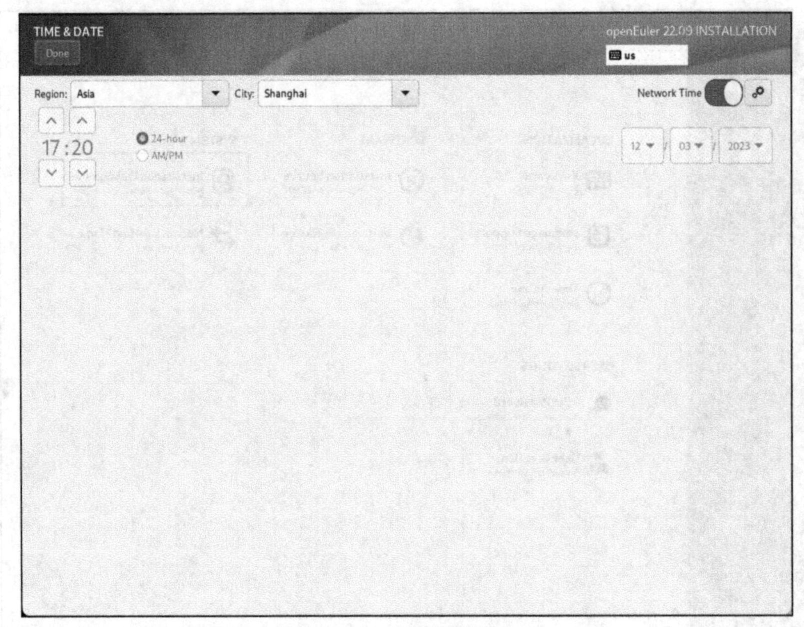

图 1-12　设置时区界面

返回系统安装向导界面并单击“Root Password”图标，进入设置 root 密码界面，设置 root 用户密码并选中“Use SM3 to encrypt the password”复选框，使用国密算法 SM3 加密用户密码，如图 1-13 所示。如果设置的密码过于简单，则需单击两次“Done”按钮完成确认。

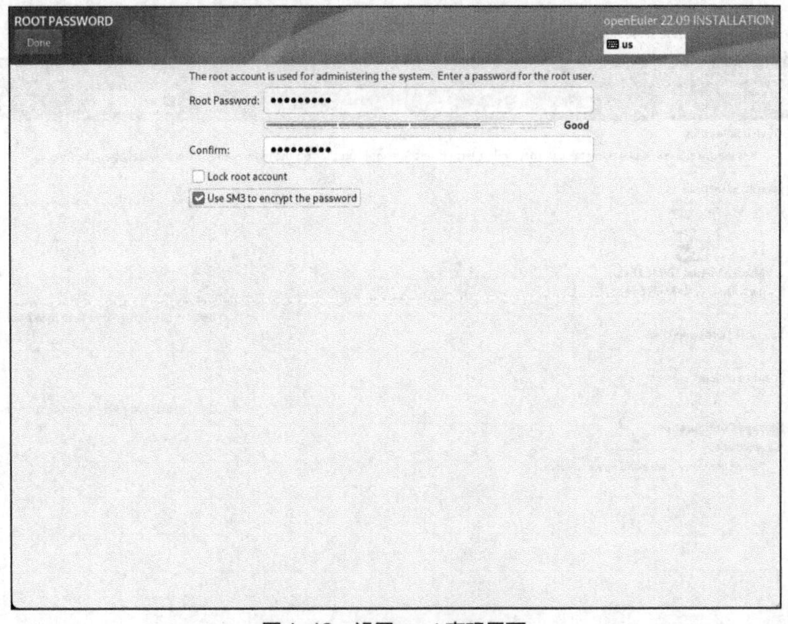

图 1-13　设置 root 密码界面

完成上述设置后，返回系统安装向导界面并单击“Begin Installation”按钮，系统开始安装，安装完成后，进入安装完成界面，如图 1-14 所示，单击“Reboot System”按钮即可重启系统。

等待片刻后，进入系统登录成功界面，如图 1-15 所示。在这里输入 root 用户名和设置的 root 用户的密码，即可以管理员身份成功进入 openEuler 22.09 操作系统。

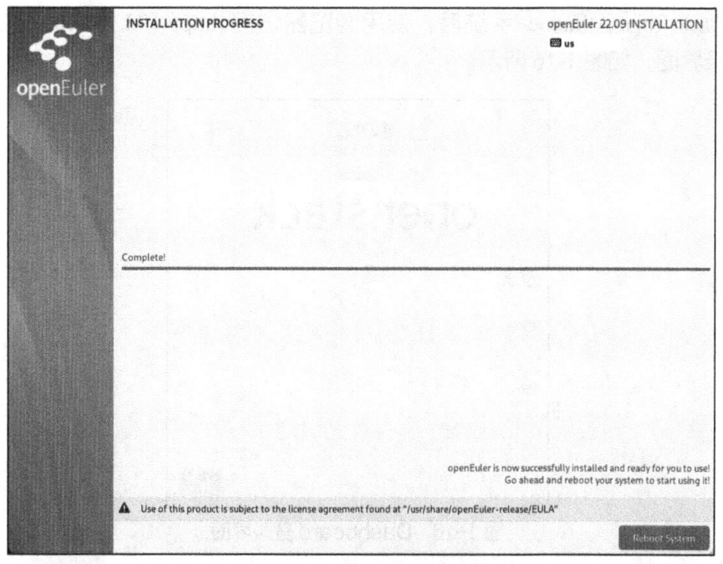

图 1-14　安装完成界面

```
Authorized users only. All activities may be monitored and reported.
localhost login: root
Password:
Last failed login: Sun Dec  3 17:40:33 CST 2023 on tty1
There were 2 failed login attempts since the last successful login.

Authorized users only. All activities may be monitored and reported.


Welcome to 5.10.0-106.18.0.68.oe2209.x86_64

System information as of time:  Sun Dec  3 05:40:47 PM CST 2023

System load:    1.05
Processes:      233
Memory used:    3.9%
Swap used:      0.0%
Usage On:       2%
IP address:     192.168.100.135
Users online:   1


[root@localhost ~]#
```

图 1-15　系统登录成功界面

至此，openEuler 22.09 操作系统安装成功。

任务 1.3　OpenStack Yoga 初体验

微课 1.2
OpenStack Yoga
初体验

在任务 1.2 中，我们已经完成了虚拟化环境的部署，为了让读者快速体验私有云平台，并进一步加深对 OpenStack 各组件的认识与理解，本任务介绍如何部署一台 OpenStack-Yoga-Allinone 系统。

任务 1.3.1　任务准备

单节点 OpenStack-Yoga-Allinone 系统已经提前安装好 Yoga 版本的 OpenStack，使用 openEuler 22.09 操作系统，配置为双网卡，云主机类型使用 4 个 vCPU/12GB 内存/100GB 硬盘。节点网络规划如表 1-1 所示。

表 1-1　节点网络规划

网卡名	网段	网关	网卡模式
eth0	192.168.100.0/24	无	仅主机
eth1	任意	动态主机配置协议（Dynamic Host Configuration Protocol，DHCP）	网络地址转换（Network Address Translation，NAT）

启动 OpenStack-Yoga-Allinone 系统后，使用浏览器访问 http://192.168.100.10/即可进入仪表板（Dashboard）登录界面，如图 1-16 所示。

图 1-16　Dashboard 登录界面

输入用户名“admin”，密码“000000”，单击“登入”按钮，登录并进入 Dashboard 操作界面，如图 1-17 所示。

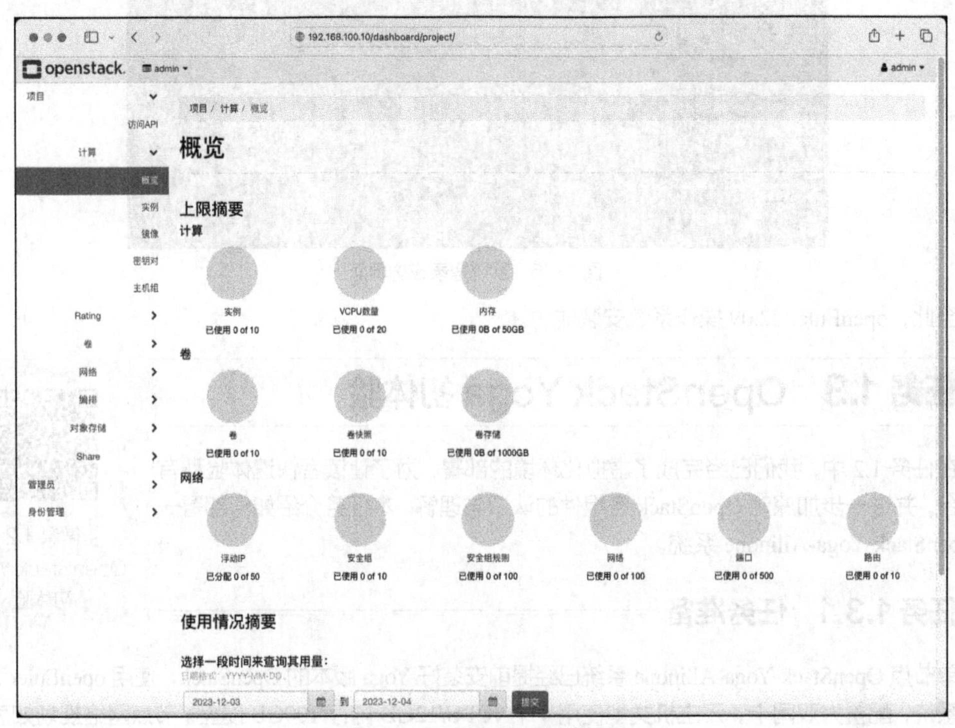

图 1-17　Dashboard 操作界面

任务 1.3.2　Dashboard 核心组件验证

1. 账户管理模块

在 Dashboard 操作界面中选择“身份管理→用户”选项，单击“创建用户”按钮，进入创建用户界面，如图 1-18 所示。在输入对应参数之后，单击“创建用户”按钮，即可创建用户。

图 1-18　创建用户界面

返回主界面，在 Dashboard 操作界面的用户列表中可以查看创建成功的用户，如图 1-19 所示。

openstack　admin ▾　admin ▾

☐	nova	-	dd762a54918c48d29f1f2f246917425d	True	Default	编辑 ▾
☐	neutron	-	12bbbf697be74bc4bc4838542cbe6aac	True	Default	编辑 ▾
☐	cinder	-	3f62614fe51e49ecbc0b64dcefd9125b	True	Default	编辑 ▾
☐	swift	-	4eeccacc614143848e18c3b35047e788	True	Default	编辑 ▾
☐	manila	-	b9789a706fe54e72a43a7569222f088a	True	Default	编辑 ▾
☐	heat	-	376cb80339d54fd69452836e3c30ac5b	True	Default	编辑 ▾
☐	heat_domain_admin	-	afebecadd3f24092be7363e0e10d1e2a	True	-	编辑 ▾
☐	aodh	-	cf18676db6394d33acf7cec24403bab2	True	Default	编辑 ▾
☐	gnocchi	-	eff406f9498a40a2860323a69cbefb3f	True	Default	编辑 ▾
☐	ceilometer	-	102a244794be4f2b9d0e0f7e2154d3da	True	Default	编辑 ▾
☐	cloudkitty	-	c8fb1f6f63ea44c69271ccfa1c9c28e5	True	Default	编辑 ▾
☐	barbican	-	1f6be432323748859697be03436e0aa2	True	Default	编辑 ▾
☐	skyline	-	1f020487eeb74b869e55f502058a04df	True	Default	编辑 ▾
☐	openstack-test	-	4b4459b655744a018c9d076e056543c9	True	Default	编辑 ▾

图 1-19　用户列表

使用 SecureCRT 工具连接 controller 节点可以查看创建的用户列表，命令如下。

```
[root@controller ~]# openstack user list | grep openstack-test
| 4b4459b655744a018c9d076e056543c9   | openstack-test     |
```

使用 openstack user show 命令可以查询 openstack-test 用户的详细信息，命令如下。

```
[root@controller ~]# openstack user show openstack-test
```

2. 镜像模块

在 Dashboard 操作界面中选择“管理员→镜像→创建镜像”选项，进入创建镜像界面。在创建镜像界面中，可以自定义镜像名称，添加本地镜像文件（cirros-0.6.1-x86_64-disk.img），在设置对应的镜像格式后，可以根据其他要求进行相应配置，最后单击“创建镜像”按钮即可完成镜像的创建。

3. 网络模块

在 Dashboard 操作界面中选择“网络”选项，根据要求创建相应的网络“testnet”，在其下拉列表中选择“admin”选项，提供商网络类型选择“Flat”，物理网络为“provider”，选中“共享的”及“外部网络”复选框，使云主机能够连通外网。单击“下一步”按钮，进入创建子网界面，填写子网名称为“testsubnet”，网络地址为“192.168.200.0/24”，网关 IP 地址为“192.168.200.2”，单击“下一步”按钮，进入确认界面，单击“创建网络”按钮。

4. 云主机模块

为了顺利创建实例，还需要提前创建实例类型。在 Dashboard 操作界面中选择“管理员→计算→实例类型”选项，单击“创建实例类型”按钮，在弹出的窗口中输入相应的属性参数，名称为“test_flavor”，vCPU 数量为 1，内存为 2048MB，根磁盘为 10GB，最后单击该界面右下方的“创建实例类型”即可。

在以上几个模块都完成设置之后，就可以创建并使用实例了。如果缺少了上述任何一个操作，则都可能使实例创建失败。

在 Dashboard 操作界面中选择“计算→实例”选项，单击右侧的“创建实例”按钮，进入创建实例界面，输入实例名称“openstack-test”，如图 1-20 所示。

图 1-20　创建实例界面

接下来依次选择上述模块创建的“源*”“实例类型*”“网络”，单击“创建实例”按钮，进行实例的创建。

创建完成后，等待片刻，即可在云主机列表中看到实例“openstack-test”正在运行中，实例创建成功，此时实例列表如图 1-21 所示。

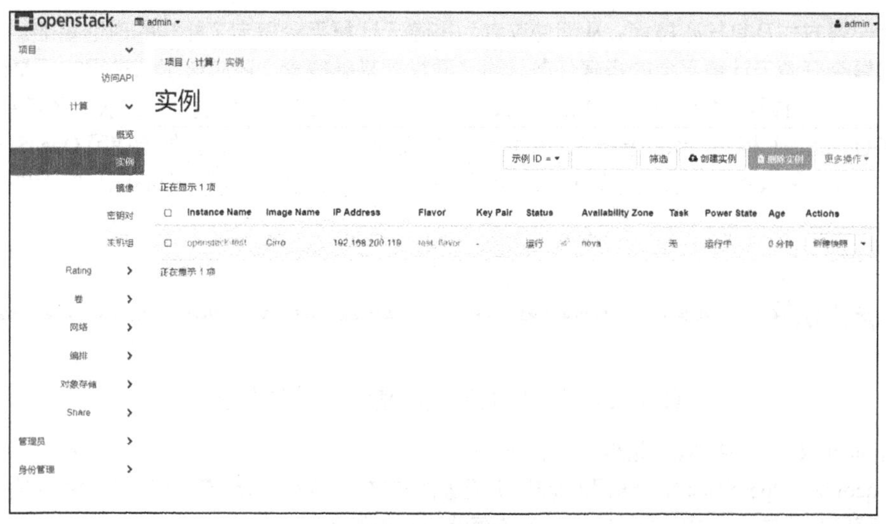

图 1-21　实例列表

选择当前实例“Actions”下拉列表中的“控制台”选项，进入云主机控制台界面，按照提示输入正确的登录名及密码，即可成功登录云主机，如图 1-22 所示。

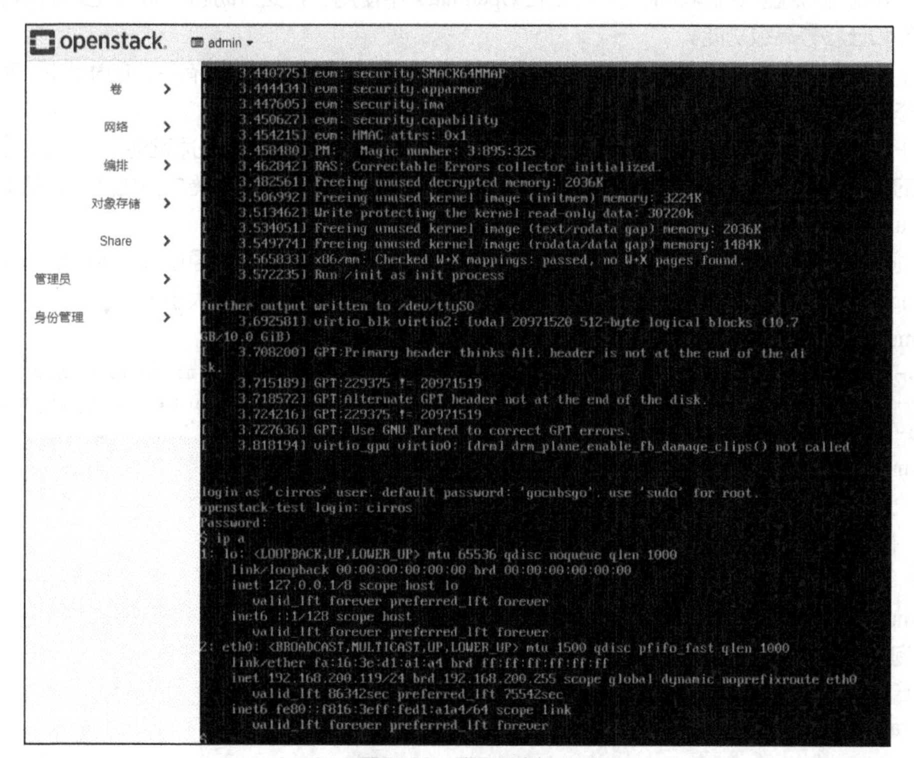

图 1-22　登录云主机

项目小结

本项目首先介绍了云计算的起源与发展历程，然后对云计算的特征进行了分析，并介绍了云计算的几大部署方式及其核心技术，从而使读者为搭建云计算平台奠定了扎实的理论基础。对于一个系统化的复杂开源云计算平台的搭建任务，除了要掌握基础概念，还应该了解常用的云基础架构平台。其中，本项目重点讲解了开源的 OpenStack 平台，介绍它的起源与发展，并认识了其平台架构。在介绍了 OpenStack 相关知识的基础上，本项目通过安装和部署虚拟化环境、体验使用 OpenStack-Yoga-Allinone 系统，使读者快速体验了私有云平台，并进一步加深了读者对 OpenStack 各组件的认识与理解。

拓展知识

OpenStack 平台支持的虚拟机镜像格式

OpenStack 支持多种虚拟机镜像格式，包括以下几种。

（1）qcow2：OpenStack 中最常用的虚拟机镜像格式之一。它支持动态增加和减少存储容量，支持快照和迁移等高级功能，同时能够实现高效的存储管理和备份。

（2）raw：原始的虚拟机镜像格式，不带任何压缩和编码。它具有较高的读写性能和可靠性，但不支持快照和在线备份等高级功能。

（3）vhd：微软虚拟机镜像格式，可以在 OpenStack 中使用。它支持动态增加和减少存储容量，实现了快照和迁移等高级功能。

（4）vmdk：VMware 虚拟机镜像格式，可以在 OpenStack 中使用。它具有高度的兼容性和可靠性，支持动态增加和减少存储容量，同时能够实现快照和迁移等高级功能。

（5）ploop：针对容器虚拟化设计的镜像格式，它支持高效的增量块备份和还原，以及支持容器文件系统的高性能快照和恢复。在 OpenStack 中，ploop 镜像通常用于支持容器虚拟化技术，如 Linux 容器（Linux Container，LXC）和 Docker 等。

不同虚拟机镜像格式（raw、qcow2、vmdk、vhd 等）之间的转换可以通过 qemu-img 命令完成，qemu-img 命令还可以用来管理虚拟磁盘，如创建、查看、调整虚拟磁盘的大小等。

qemu-img 部分命令格式如下所示。

```
# qemu-img create [-6] [-e] [-b base_image] [-f format] [-p] filename [size]
# qemu-img convert [-c] [-e] [-f format] filename [-O output_format] output_filename
# qemu-img info [-f format] filename
```

qemu-img 命令部分参数说明如下。

-p：显示镜像转换进度。

-f：指定原镜像格式。

-O：指定需要转换的格式。

qemu-img 命令的部分使用案例如下。

查询镜像文件的详细信息，命令如下。

```
# qemu-img info image.qcow2
```

将 raw 格式转换为 qcow2 格式，命令如下。

```
# qemu-img convert -p -f raw -O qcow2 image.img image.qcow2
```

将 vmdk 格式转换为 raw 格式，命令如下。

```
# qemu-img convert -p -f vmdk -O img image.vmdk image.img
```

将 vhd 格式转换为 qcow2 格式，命令如下。

```
# qemu-img convert -p -f vpc -O qcow2 image.vhd image.qcow2
```

调整虚拟磁盘空间大小，命令如下。

```
# qemu-img resize image.img 20G
```

知识巩固

1. 单选

（1）云计算资源可以弹性地或自动化地部署和释放，以便能够迅速地按需扩大或缩小规模。这说明云计算具有（　　）的特征。

A. 按需自助服务　B. 快速伸缩　C. 资源池化　D. 可计量

（2）云计算 SPI 模型的服务大致分为 3 类，（　　）不属于这 3 类服务。

A. IaaS　B. PaaS　C. DaaS　D. SaaS

（3）云计算部署方式包括（　　）。

A. 公有云、私有云、社区云和应用云　B. 基础设施云、平台云、应用云和混合云
C. 公有云、私有云、社区云和混合云　D. 基础设施云、平台云、私有云和应用云

（4）下列关于 OpenStack 的描述错误的是（　　）。

A. OpenStack 是一款开源软件平台
B. OpenStack 是硬件之上提供的基础设施服务
C. OpenStack 是 SaaS 组件，可建立和提供云端运算服务
D. OpenStack 具有功能丰富、可扩展等特性

2. 填空

（1）SaaS 是________的简称。

（2）将基础设施作为服务的云计算 SPI 模型的服务种类是________。

3. 简答

（1）简述什么是虚拟化技术。

（2）云计算有哪几种部署方式？请分别介绍。

拓展任务

对比 3 种云基础架构平台

分别访问 OpenStack、CloudStack 和 Eucalyptus 的官网，阅读官方文档，学习相关使用手册，从系统功能的合理性、安装及配置的难易程度、用户界面的友好度等方面对 3 种云基础架构平台进行对比。

项目2 私有云基础架构

02

学习目标

【知识目标】

1. 了解传统架构及其存在的问题。
2. 学习集群架构及其优势和局限。
3. 学习私有云基础架构的构成要素及其与集群架构、传统架构的区别。
4. 了解 IT 基础架构的发展趋势。

【技能目标】

1. 掌握集群架构下系统部署的方法。
2. 具备私有云基础架构下部署环境的能力，从而进一步了解不同 IT 架构的区别。

【素养目标】

1. 培养读者的逻辑思维能力。
2. 培养读者的方案设计能力。
3. 培养读者的团队合作意识。

项目概述

小张经过云计算基础知识及核心技术的学习，希望进一步了解 IT 基础架构的演变过程，并通过学习传统架构、集群架构及私有云基础架构的相关知识，认识到企业从 IT 基础架构向私有云基础架构转型的必要性。

知识准备

2.1 传统架构

IT 基础架构由企业运营所需的所有核心技术组件组成，其中硬件、软件和网络通常都被认为是 IT 基础架构的支柱。然而，由于传统架构的局限性，当前传统企业都在进行数字化转型，云原生、微服务是不可逆的技术发展趋势。

2.1.1 IT 基础架构

IT 基础架构是指企业通过规划、建设和管理现有技术，为员工、客户和合作伙伴提供 IT 服务及解决方案的框架与基础设施。IT 基础架构为组织和企业提供了必要的技术支持，确保其业务运作顺畅、安全可靠。

IT 基础架构的 3 个基本要素包括硬件、软件和网络。

1. 硬件

硬件包括计算机、Web 服务器、数据中心等设备，以及数据中心的容纳设施、冷却设施和供电设施等配套设备。

2. 软件

软件指企业内外所使用的向用户提供服务的应用，包括 Web 服务、企业资源规划和 SaaS 应用等，以及管理系统资源和硬件的操作系统（Operating System，OS）。

3. 网络

网络是支持企业内外部系统和设备之间连接、通信及操作的组件，主要包括 Internet 访问、防火墙、安全监控、配置管理和设备访问管理，以及路由器、交换机和因特网服务提供商（Internet Service Provider，ISP）。

2.1.2 传统架构的特点

传统架构又称集中式架构或单体式架构，指企业在自有设施内管理所有组件。传统架构是“烟囱式”的，或者叫作“专机专用”系统，如图 2-1 所示。

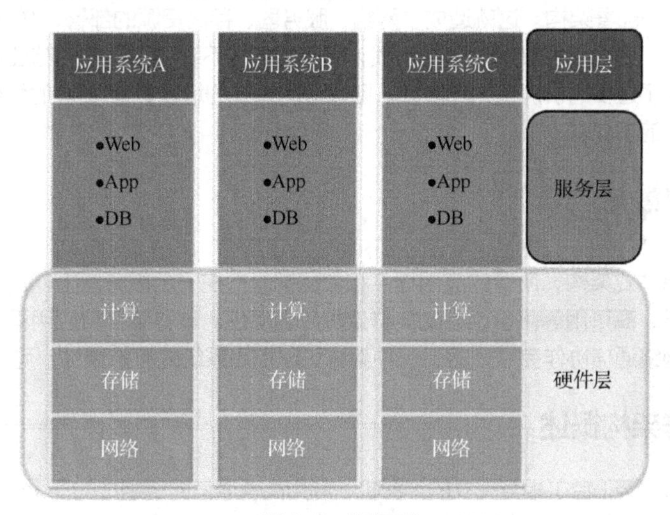

图 2-1 传统架构

传统架构由硬件层、服务层和应用层构成。

（1）硬件层：硬件层包括提供计算、存储、网络等资源的各类硬件设备，企业可根据需求和规模选择合适的硬件设备，并对其进行合理配置和布局，以确保系统的稳定运行。

（2）服务层：服务层包括数据库（Database，DB）、应用（Application，App）、Web 等，企业可以对其进行合理集成和配置，以满足企业的业务需求。

（3）应用层：应用层包含各类应用系统，企业可根据自身业务需求进行安装。

传统架构的主要优点是易于开发、部署简单和易于伸缩。

（1）易于开发：当前开发工具的功能目标是支持单片应用程序的开发，这对于一个传统的单体应用来说非常容易实现。

（2）部署简单：只需将应用部署为简单的 Web 应用程序归档（Web Application ARchive，WAR）

文件即可，且只需要部署一个单体应用即可。

（3）易于伸缩：可以通过在负载均衡器后运行应用程序的多个副本来伸缩应用程序。随着用户人数的增加，一台机器已经满足不了系统的负载，此时就会考虑系统的水平扩展。

2.1.3 传统架构存在的问题

在传统架构中，新的应用系统上线的时候需要分析该应用系统的资源需求，确定基础架构所需的计算、存储、网络等资源的规格和数量。这种架构存在的主要问题如下。

1. 硬件高配低用

考虑到应用系统未来 3～5 年的业务发展，以及业务突发的需求，为满足应用系统的性能、容量承载需求，往往在选择计算、存储和网络等硬件设备的配置时会留有一定比例的余量。但硬件资源上线后，应用系统在一定时间内的负载并不会太高，使得较高配置的硬件设备使用率不高。

2. 整合困难

用户在实际使用中也注意到了硬件设备使用率不高的问题，因此，当需要上线新的应用系统时，会优先考虑将其部署在既有的基础架构上。但因为不同的应用系统所需的运行环境、对资源的抢占情况有很大的差异，更重要的是考虑到可靠性、稳定性和运维管理的难易程度问题，将新、旧应用系统整合在一套基础架构上的难度非常大，更多的用户往往选择新增与应用系统配套的计算、存储和网络等硬件设备。

总而言之，传统架构造成了每套硬件设备与所承载应用系统的“专机专用”，多套硬件设备和应用系统构成了“烟囱式”部署架构。这就造成了网络、服务器、存储资源的部署、联调周期长、难度大、投资大，且结构复杂、建设过程复杂、运维烦琐，横向扩展性不够高，架构伸缩性不强，使得整体资源利用率不高，占用了过多的机房空间和能源。随着应用系统的增多，IT 资源的利用率、可扩展性、可管理性都面临很大的挑战。

2.2 集群架构

传统架构都是单体式架构，所有的应用程序和数据都运行在一台服务器上。这种架构简单易用，但是无法满足高并发、高可用等需求，因此集群架构应运而生。集群架构将应用程序和数据分散到多台服务器上，通过网络通信协作完成任务。这种架构可以提高系统的可扩展性、可靠性和性能。

2.2.1 集群架构概述

集群就是指一组（若干台）相互独立的计算机，利用高速通信网络组成的一个较大的计算机服务系统，每个集群节点（即集群中的每台计算机）都是运行各自服务的独立服务器。这些服务器之间可以彼此通信，协同向用户提供应用程序、系统资源和数据，并以单一系统的模式加以管理。当用户请求集群系统时，集群给用户的感觉就是一台单一且独立的服务器，而实际上用户请求的是一组集群服务器。

集群架构是一种分布式系统的基础架构，其通过将多个计算机节点连接在一起形成一个统一的整体，能够提高应用程序的性能和可用性，扩展应用程序的能力和容量。单一应用服务器能够处理的请求连接有限，因此，在网站访问高峰期，应用服务器的负载压力成为整个网站的瓶颈。

应用服务器集群能够改善网站的并发处理能力。通过负载均衡调度服务器，可将来自用户浏览器的访问请求分发到应用服务器集群中的任何一台服务器上，如果有更多的用户，则可在集群中加入更多的应用服务器，使应用服务器的负载压力不再成为整个网站的瓶颈。

集群架构将多个计算机节点组合在一起以完成共同的任务。这种架构可以提供更高的可用性和可扩展性，可以被用于各种应用场景，如大型 Web 应用、数据库管理系统、云计算平台和大数据处理系统等。集群架构的常用应用场景如图 2-2 所示。

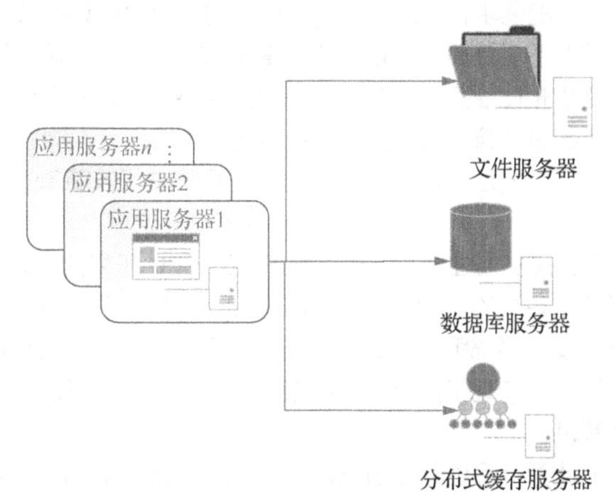

图 2-2 集群架构的常用应用场景

1. 应用服务器

应用程序是最常见的集群应用程序之一。在集群应用程序中，多个应用服务器节点被连接到一起，共同处理用户请求，通过负载均衡和故障转移等技术来提高可用性及可靠性。

2. 文件服务器

在集群架构中，文件服务器负责进行中央存储和数据文件管理，集群中的其他计算机可以访问文件服务器中的文件。

3. 数据库服务器

集群架构可以帮助数据库实现可扩展性和高可用性。基于集群的数据库服务器可以将数据分布在多个节点上，以提供更好的性能和可靠性，并能够支持大规模的数据存储和查询操作。

4. 分布式缓存服务器

分布式缓存服务器是典型的集群应用场景的应用程序之一。大规模的数据处理和分析任务可通过使用分布式缓存服务器，将计算资源和存储资源分散在多个节点上来实现，从而提高计算速度和扩展能力。

2.2.2 集群架构的优势

集群中每台服务器都是这个集群的一个“节点”，所有节点构成了一个集群。每个节点都提供相同的服务，这样集群的处理能力就相当于提升了许多倍（有几个节点就相当于提升了几倍）。所以集群架构在一定程度上解决了传统架构存在的问题。

1. 传统架构的局限性

传统架构在不采用极端优化方案时，是不存在一致性问题的，只有可用性问题和性能问题，主要体现在单点故障和性能瓶颈这两点。

（1）单节点故障。单节点部署很容易出现服务崩溃之后，没有备用节点的情况，从而影响用户使用。单机对外提供服务风险很大，服务器出现任何故障都可能造成整个服务的不可用。

（2）性能瓶颈。当单节点的资源无法满足支持更大用户量的需求时，若想要性能有较大的提升，一般采取的措施是优化业务和增加服务器配置。然而，这样只是杯水车薪，成本高昂且效果非常有限。

2. 集群架构的主要优势

相较于传统架构，集群架构的主要优势体现在以下几点。

（1）计算能力强：因为集群架构中的计算机通过高速通信网络连接实现了使用多台不同计算机的

计算性能，合作处理一个共同的需求，所以集群计算机相对于单台计算机来说计算能力更强。

（2）性价比高：通常情况下，计算机集群架构只需要数台或数十台服务器主机即可。其与动辄价值上百万元的专用超级计算机相比便宜了很多。在满足同样性能需求的条件下，采用计算机集群架构比采用同等计算能力的专用超级计算机具有更高的性价比。

（3）可伸缩性强：当服务负载压力增大时，针对集群系统进行较简单的扩展即可满足需求，且不会降低服务质量。

通常情况下，硬件设备要想扩展性能，不得不增加新的CPU和存储设备，甚至购买更高性能的服务器，但可以增加的设备总是有限的。如果采用集群架构，则将新的单台服务器加入现有集群架构中即可。从访问的用户角度来看，系统服务无论是在连续性还是计算性能上都几乎没有变化，系统经过升级后提高了访问能力，轻松地实现了扩展。集群架构中的节点数目可以增长到上千乃至上万个，其可伸缩性远超单台超级计算机。

（4）可用性高：单一的计算机系统总会面临设备损坏的问题，如CPU、内存、主板、电源、硬盘等损坏，只要有一台设备损坏，整个计算机系统就可能会宕机，无法正常提供服务。在集群系统中，尽管部分硬件和软件还是会发生故障，但整个系统的服务是随时可用的。集群架构技术保证了系统在若干硬件设备发生故障时仍可以继续工作，这样就将系统的停机时间缩到最短。集群架构在提高系统可用性的同时，也大大减少了系统故障带来的业务损失。

（5）高度透明：多台独立计算机组成的松耦合集群系统构成了一个虚拟服务器。用户或客户端程序访问集群系统时，就像访问一台高性能、高可用的服务器一样，集群中一部分服务器的上线、下线不会中断整个系统服务，这对用户也是透明的。

（6）易于管理：整个系统可能在物理上很大，但其实容易管理，用户对其进行管理就像管理一个单一映像系统一样。在理想状况下，软硬件模块的插入能做到即插即用。

（7）可编程：在集群系统中，容易开发及修改各类应用程序。

2.2.3 集群架构的局限性

集群架构主要存在如下局限性。

1. 成本更高

与传统架构相比，集群架构需要更多的计算机硬件及相关设备，这也意味着成本更高。

2. 管理复杂

管理集群架构的成本较高，且需要更多的经验和技能，如负载均衡和资源分配等技能。

3. 存在安全漏洞

集群架构中的安全漏洞也是一个问题，如果一个节点被攻击，则所有节点都可能会受到影响。

综合来看，集群架构与传统架构有各自的优势和局限性，用户需要针对不同的场景进行权衡，从而选择适合的架构。同时，在生产环境下，为了提供更高的计算性、可靠性和可扩展性，可以通过添加更多的节点来提高集群的工作负载和响应能力，使其更好地服务于多种应用场景。

2.3 私有云基础架构

为突破传统架构和集群架构的局限性，私有云基础架构应运而生。

2.3.1 私有云基础架构构成要素

随着越来越多的企业设定了构建内部云服务的目标，规划和构建企业内部云服务平台就成为IT部门的职责。通过部署合适的基础设施，企业能够较容易地完成从传统架构到私有云基础架构的转变。

私有云基础架构至少包括以下 6 个构成要素。

1. 虚拟存储

共享存储是移动性负载均衡的一个重要推动因素。它允许多台服务器和一个公用存储对象关联，以便可执行文件可以根据需要进行移动，从而实现负载均衡，维护可扩展性，维护窗口功能和进行性能调试。然而，并不是所有的共享存储都是真正虚拟化的。目前市场上很多的存储区域网（Storage Area Network，SAN）和网络附属存储（Network Attached Storage，NAS）解决方案实际上使用的还是传统架构，并不是真正的从物理硬盘驱动器和固态设备上虚拟出来的逻辑存储对象。真正的虚拟存储平台会将逻辑存储模块映射到物理存储设备上面，这种方式允许非破坏性的数据物理位置的实时移动，而不会影响可执行文件对数据的访问。

2. 扁平的虚拟化网络

在传统架构中，网络仍然是以硬件为基础设计的，即从客户端到服务器，而它所需要的现代应用程序不同，优化服务器到服务器的流量需要设计扁平的虚拟化网络和孤立的 I/O 以减少每个数据包的跳出数量。当正在运行的虚拟机在服务器之间进行迁移时，随着虚拟机的存储通过网络复制到目标服务器，从服务器到服务器的流量就产生了。在高负载的虚拟机上执行这项操作甚至能承载传输速率为 10Gbit/s 的专用链路。和传统的核心交换解决方案相比，在服务器集群和扁平数据中心网络中的 I/O 能缩短约 30%的虚拟机迁移时间和减少约 60%的网络延迟。

3. 自动预警监控

维护服务可用性的关键之一是有监控整个 IT 系统运行状态的强大的解决方案，对重要的事件进行预警，并自动运行包括主动切换在内的故障主动修复操作。因此，部署私有云基础架构时可以选择内置跟踪并预警关键运行参数的组件级智能基础设施。

4. 业务连续性管理

用户可以通过设计、安排以 SAN 为基础的全天候故障恢复点来大大缩短故障的恢复时间目标（Recovery Time Objective，RTO），以及增强程序和数据的恢复点目标（Recovery Point Objective，RPO）。无论是同步还是异步，利用存储端来确保网站发生故障并造成服务中断时的业务连续性，可以将不利影响降到最小。

5. 集成的基础设施管理

对基础设施进行集成管理使用户能够通过基础设施推进重复的进程，并以此来简化 IT 服务交付和确保具备一定服务水平实施的一致性。程序管理、操作系统管理和集成基础设施管理能减少管理员与分隔区域的交互，减少培训人力，以及管理云环境过程中出现的错误。

6. 端到端工作负载提供

工作负载提供包括查询、存档，以及工作负载准备资源的配置。工作负载准备资源又包括基础设施、应用程序，以及具体的服务水平资源包，它使用户能够通过定义了服务水平范围的共享资源池来管理集群和虚拟机生命周期。

2.3.2 私有云基础架构的优势

私有云基础架构的引入有效解决了传统架构的问题，私有云基础架构如图 2-3 所示。

私有云基础架构在传统架构的应用、服务、硬件层的基础上，增加了虚拟化层和云层，具体说明如下。

（1）虚拟化层

大多数私有云基础架构都采用了虚拟化技术，包括计算虚拟化、存储虚拟化、网络虚拟化等。虚拟化层屏蔽了硬件层自身的差异和复杂度，向上呈现为标准化、可灵活扩展、可收缩和具有弹性的虚拟化资源池。

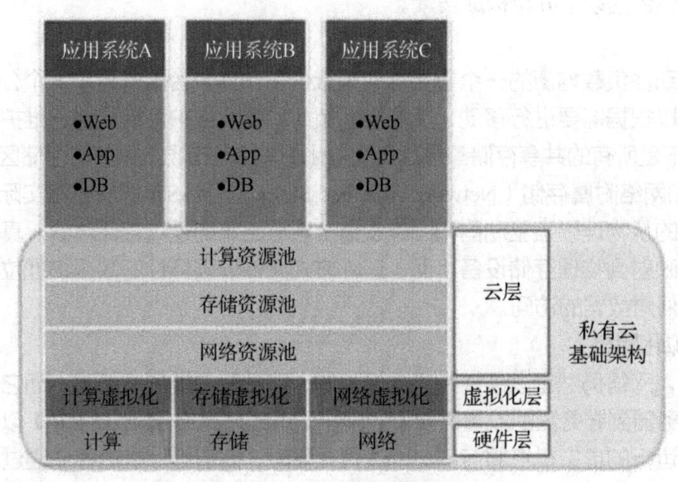

图 2-3 私有云基础架构

（2）云层

该层负责对资源池进行调配、组合，根据应用系统的需要自动生成、扩展所需的硬件资源，对更多的应用系统进行流程化、自动化部署和管理，提升效率。

相对于传统架构的应用系统，通过虚拟化整合与自动化，私有云基础架构的应用系统能够共享基础架构资源池，实现高利用率、高可用性、低成本和低能耗，并且通过对云层的自动化管理，应用系统具备快速部署、易于扩展、智能管理的优点，能够帮助用户构建 IaaS 云业务模式。

与传统架构和集群架构相比，私有云基础架构具有如下不同。

1．与传统架构相比

与传统架构相比，私有云基础架构具有如下不同。

（1）规模扩展能力不同，资源复用性不同：私有云基础架构的搭建方式由垂直扩展变为横向扩展，资源可复用。

（2）硬件依赖性不同，生态链、开放性不同：私有云基础架构由硬件定义变为软件定义，屏蔽了硬件差异性。在同一个硬件平台上，可以运行来自多个不同厂家的软件和操作系统。新时代的 IT 生态链更加繁荣，开放性更好。

（3）可靠性实现方式不同：私有云基础架构由单机硬件器件级的冗余实现可靠性，发展为依赖分布式软件和故障处理自动化实现可靠性。

（4）资源接入方式不同：私有云基础架构的资源由专用发展为开放式接入。

2．与集群架构相比

私有云基础架构是集群架构的进一步应用。在集群架构中，虽然对多台机器进行了联合，但是某项具体的任务在执行的时候还是会被转发到某台服务器上，在私有云基础架构中则可以认为任务被分割成多个进程，在多台服务器上进行并行计算。集群是一种服务器的应用方式，云计算是一种或多种业务的应用方式。集群可以理解为云计算的一个组成部分，云计算必然包括集群，但集群服务器未必构成云。

2.3.3 IT 基础架构发展趋势

IT 基础架构由传统架构向私有云基础架构的转变，极大地提升了基础架构融合的必要性和可行性。IT 基础架构发展历史如图 2-4 所示。

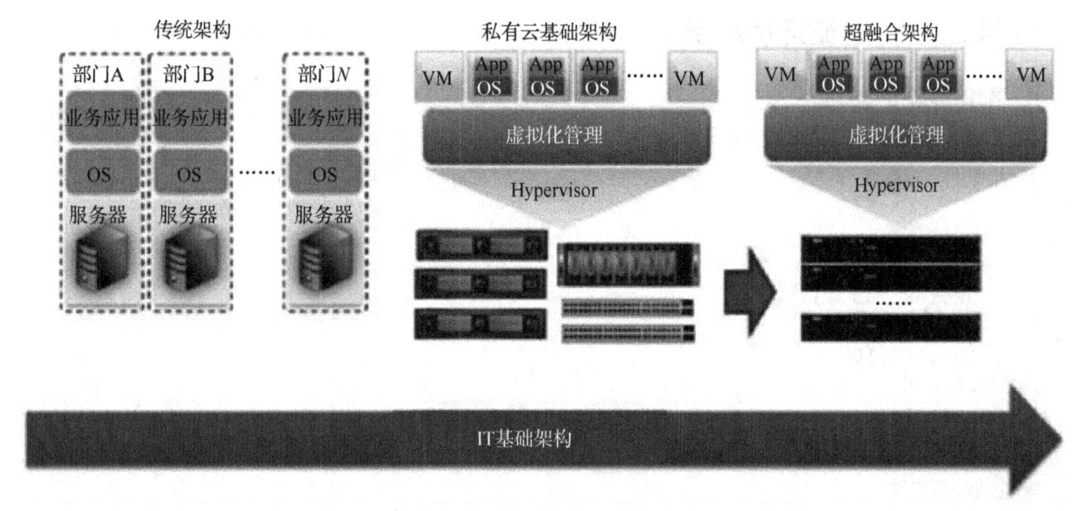

图 2-4　IT 基础架构发展历史

IT 基础架构经历了物理机时代、虚拟机时代，现在已经进入云原生时代，并正朝着超融合架构发展。当前，以微服务、DevOps、容器、多云业务管理为代表的云原生技术已经被广泛、成熟应用，成为加速企业数字化业务高效创新、实现企业数字化转型的最佳技术支撑之一。

微服务架构作为云原生落地的重要技术架构之一，实现了应用软件的模块化、组件化、共享化，实现了开发团队的独立化、小型化和协同化，为数字化应用研发创新更敏捷、更高效打下了坚实的基础。在云原生浪潮下，微服务架构不仅成为主流的架构模式，还在企业的应用中占据了很高的地位，根据相关调查，未来 80%以上的工作负载会转移到微服务架构上。

传统架构向超融合架构过渡是一个必然的趋势，但是并不是一蹴而就的，这是一个长期的过程。随着云原生技术的不断成熟和应用的不断深入，在未来，企业与社会可能会加速进行数字化发展。

值得注意的是，云原生并不仅仅是一项新技术、新工具，也是一种新的服务模式和思维。正确认识云原生，根据企业自身的实际情况来逐步推动云的应用，这样才能更好地服务企业，使企业获得高质高效的发展。

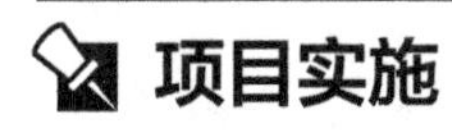

项目实施

任务 2.1　传统架构下的应用部署

微课 2.1　传统架构下的应用部署

本任务的目标为在传统架构下搭建 LAMP（Linux+Apache+MySQL/MariaDB+PHP）环境并部署 WordPress 服务，为了方便演示，本书在项目 1 安装好的单台 openEuler 22.09 操作系统虚拟机中进行模拟。节点基础配置如表 2-1 所示。

表 2-1　节点基础配置

虚拟机版本	主机名	IP 地址	安装服务
openEuler 22.09	web01	192.168.100.100	Apache、页面超文本处理器（Page Hypertext Preprocessor，PHP）、MariaDB

任务 2.1.1 基础环境准备

1. 配置静态 IP 地址

修改网卡配置文件，示例修改内容如下。

```
[root@web01 ~]# vi /etc/sysconfig/network-scripts/ifcfg-ens160
……
BOOTPROTO=none
ONBOOT=yes
IPADDR=192.168.100.100
PREFIX=24
GATEWAY=192.168.100.2
DNS1=223.6.6.6
DNS2=119.29.29.29
```

修改完成后，使配置生效，执行命令如下。

```
# 重新加载配置文件
[root@web01 ~]# nmcli c reload
# 激活配置文件
[root@web01 ~]# nmcli c up ens160
```

2. 配置本地 Yum 源

在/opt 目录下创建 openEuler 目录，将默认的 Yum 源文件移动到/media 目录下，将 openEuler 22.09 镜像上传到/mnt 目录下，并配置本地 Yum 源，具体操作如下。

```
[root@web01 ~]# mkdir /opt/openEuler
[root@web01 ~]# mv /etc/yum.repos.d/* /media
[root@web01 ~]# mount openEuler-22.09-x86_64-dvd.iso /mnt/
[root@web01 ~]# cp -va /mnt/{Packages,repodata} /opt/openEuler/
[root@web01 ~]# cat << WXIC > /etc/yum.repos.d/openEuler.repo
[openEuler22.09]
name=openEuler
baseurl=file:///opt/openEuler/
gpgcheck=0
WXIC
```

3. 关闭防火墙

命令如下。

```
[root@web01 ~]# systemctl disable --now firewalld
```

4. 关闭 SELinux

命令如下。

```
[root@web01 ~]# vi /etc/selinux/config
# 将 SELINUX=enforcing 改为 SELINUX=disabled
[root@web01 ~]# setenforce 0
```

任务 2.1.2 安装 Apache 服务

具体操作如下。

```
[root@web01 ~]# dnf install -y httpd
[root@web01 ~]# systemctl enable --now httpd
[root@web01 ~]# apachectl -v
Server version: Apache/2.4.51 (Unix)
Server built:   Sep  7 2022 00:00:00
```

在浏览器的地址栏中输入虚拟机 IP 地址进行 Apache 访问测试，其测试成功界面如图 2-5 所示。

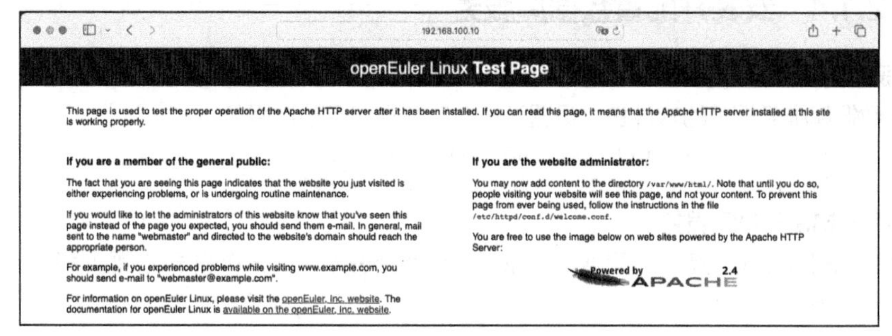

图 2-5　Apache 测试成功界面

任务 2.1.3　安装 PHP 服务

1. 安装 PHP 及其模块

命令如下。

```
[root@web01 ~]# dnf -y install php php-common php-cli php-gd \
php-pdo php-devel php-xml php-mysqlnd
```

2. 编写测试界面文件

命令如下。

```
[root@web01 ~]# vi /var/www/html/php-test.php
<?php
phpinfo();
?>
```

重启 Apache 服务，并在浏览器中访问 http://192.168.100.100/php-test.php，PHP 测试成功界面如图 2-6 所示。

```
[root@web01 ~]# systemctl restart httpd
```

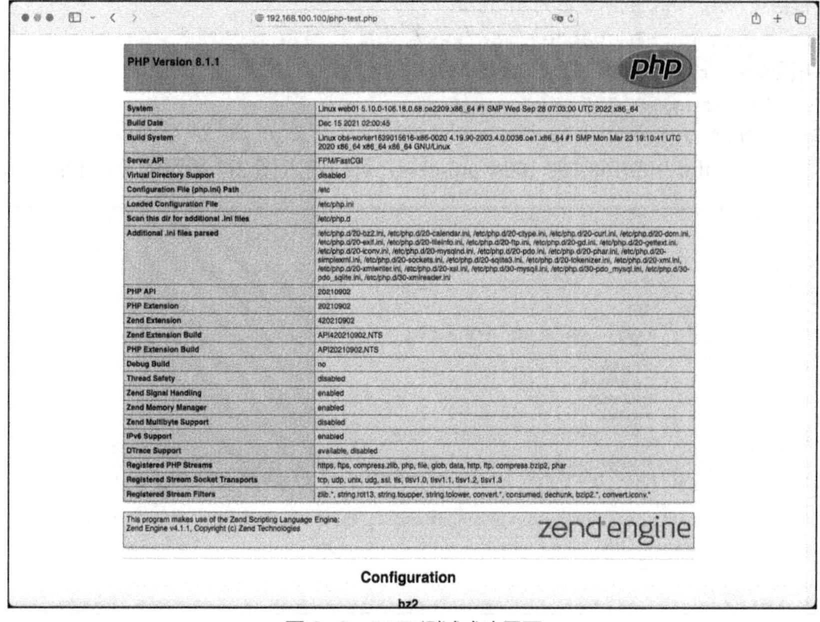

System	Linux web01 5.10.0-106.18.0.68.oe2209.x86_64 #1 SMP Wed Sep 28 07:03:00 UTC 2022 x86_64
Build Date	Dec 15 2021 02:00:45
Build System	Linux obs-worker1639015616-x86-0020 4.19.90-2003.4.0.0036.oe1.x86_64 #1 SMP Mon Mar 23 19:10:41 UTC 2020 x86_64 x86_64 x86_64 GNU/Linux
Server API	FPM/FastCGI
Virtual Directory Support	disabled
Configuration File (php.ini) Path	/etc
Loaded Configuration File	/etc/php.ini
Scan this dir for additional .ini files	/etc/php.d
Additional .ini files parsed	/etc/php.d/20-bz2.ini, /etc/php.d/20-calendar.ini, /etc/php.d/20-ctype.ini, /etc/php.d/20-curl.ini, /etc/php.d/20-dom.ini, /etc/php.d/20-exif.ini, /etc/php.d/20-fileinfo.ini, /etc/php.d/20-ftp.ini, /etc/php.d/20-gd.ini, /etc/php.d/20-gettext.ini, /etc/php.d/20-iconv.ini, /etc/php.d/20-mysqlnd.ini, /etc/php.d/20-pdo.ini, /etc/php.d/20-phar.ini, /etc/php.d/20-simplexml.ini, /etc/php.d/20-sockets.ini, /etc/php.d/20-sqlite3.ini, /etc/php.d/20-tokenizer.ini, /etc/php.d/20-xml.ini, /etc/php.d/20-xmlwriter.ini, /etc/php.d/20-xsl.ini, /etc/php.d/30-mysqli.ini, /etc/php.d/30-pdo_mysql.ini, /etc/php.d/30-pdo_sqlite.ini, /etc/php.d/30-xmlreader.ini
PHP API	20210902
PHP Extension	20210902
Zend Extension	420210902
Zend Extension Build	API420210902,NTS
PHP Extension Build	API20210902,NTS
Debug Build	no
Thread Safety	disabled
Zend Signal Handling	enabled
Zend Memory Manager	enabled
Zend Multibyte Support	disabled
IPv6 Support	enabled
DTrace Support	available, disabled
Registered PHP Streams	https, ftps, compress.zlib, php, file, glob, data, http, ftp, compress.bzip2, phar
Registered Stream Socket Transports	tcp, udp, unix, udg, ssl, tls, tlsv1.0, tlsv1.1, tlsv1.2, tlsv1.3
Registered Stream Filters	zlib.*, string.rot13, string.toupper, string.tolower, convert.*, consumed, dechunk, bzip2.*, convert.iconv.*

图 2-6　PHP 测试成功界面

任务 2.1.4　安装并配置数据库服务

1. 安装 MariaDB 服务

安装与数据库服务相关的软件包，设置开机自启并立即启动该服务。

```
[root@web01 ~]# dnf -y install mariadb mariadb-server
[root@web01 ~]# systemctl enable --now mariadb.service
```

2. 初始化 MariaDB 服务

MariaDB 服务在本地连接时不需要用户密码，可免密进入，第一次登录时需要修改 root 用户的密码，具体操作如下。

```
[root@localhost ~]# mysql
MariaDB [(none)]> set password = password("wxic@2024");
Query OK, 0 rows affected (0.001 sec)
```

3. 创建数据库

创建 wordpress 数据库并开启 root 用户远程访问的权限。

```
MariaDB [(none)]> create database wordpress;
Query OK, 1 row affected (0.00 sec)
MariaDB [(none)]> grant all privileges on wordpress.* to root@'%' identified by '';
Query OK, 0 rows affected (0.00 sec)
```

任务 2.1.5　安装 WordPress

将从官方网站下载的 wordpress-6.4.1-zh_CN.tar.gz 文件存放在/root 目录下，并将压缩包解压到 Apache 网页文件夹中。

```
[root@web01 ~]# tar xvf wordpress-6.4.1-zh_CN.tar.gz -C /var/www/html/
```

为解压后的 wordpress 文件夹赋予权限。

```
# 设置 http 根目录/var/www 的所有组为 apache
[root@web01 ~]# chown -R :apache /var/www//
# 设置 http 根目录/var/www 的所有者为 apache
[root@web01 ~]# chown -R apache /var/www/
# 设置 http 根目录/var/www 的组下的所有用户具有读写权限
[root@web01 ~]# chmod -R 775 /var/www/
[root@web01 html]# systemctl restart httpd
```

在浏览器中访问 http://192.168.100.100/wordpress，可以进入图 2-7 所示的 WordPress 测试界面。

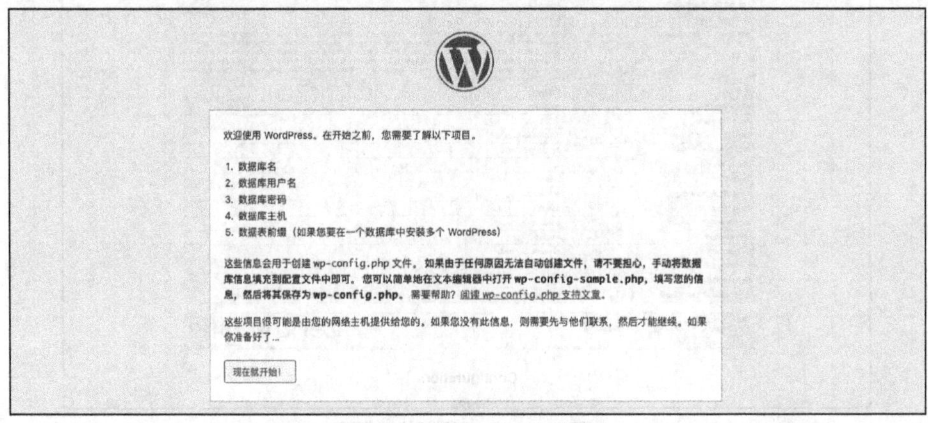

图 2-7　WordPress 测试界面

单击“现在就开始！”按钮，在进入的界面中输入数据库相关配置信息并单击“提交”按钮，即可完成数据库连接，如图 2-8 所示。

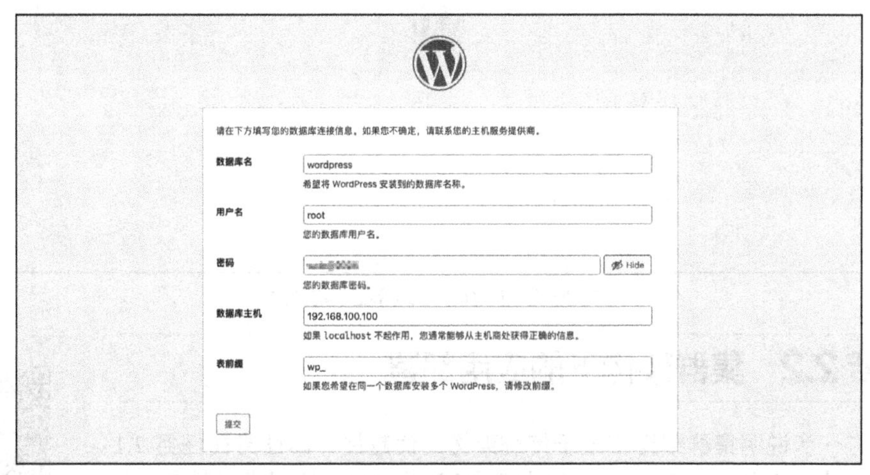

图 2-8　数据库连接

数据库连接成功后，在进入的界面中单击“运行安装程序”按钮继续安装，如图 2-9 所示。

图 2-9　数据库连接成功

自定义与站点相关的表单，配置相应内容，如用户名及密码等，如图 2-10 所示。

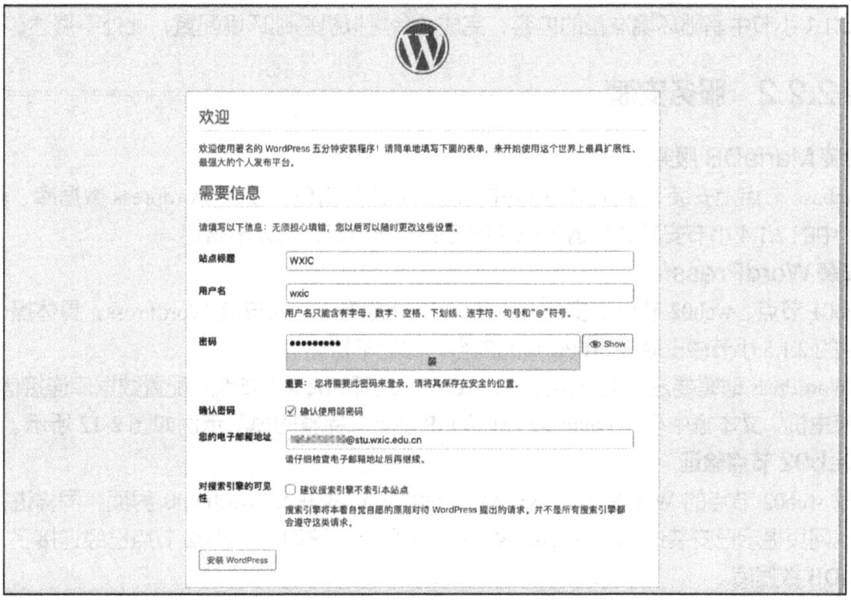

图 2-10　WordPress 站点配置

至此，WordPress 部署成功，其界面如图 2-11 所示。

图 2-11　WordPress 部署成功界面

任务 2.2　集群架构下的应用部署

微课 2.2　集群架构下的应用部署

为了进一步说明集群架构相较于传统架构的优越性，本任务在任务 2.1 的基础上，安装 3 台 openEuler 22.09 操作系统的虚拟机，拓展集群架构，完成 WordPress 服务部署。每台虚拟机所安装的服务及节点基础配置情况如表 2-2 所示。

表 2-2　每台虚拟机所安装的服务及节点基础配置情况

虚拟机版本	主机名	IP 地址	安装服务
openEuler 22.09	web01	192.168.100.100	Apache、PHP
openEuler 22.09	database	192.168.100.101	MariaDB
openEuler 22.09	web02	192.168.100.102	Apache、PHP

任务 2.2.1　基础环境准备

参照 2.1.1 小节中基础环境准备的内容，完成 3 台虚拟机基础环境配置，此处不赘述。

任务 2.2.2　服务安装

1. 安装 MariaDB 服务

在 database 节点上安装 MariaDB 服务并设置 root 用户密码，创建 wordpress 数据库，具体操作参考任务 2.1 中的 2.1.4 小节安装并配置数据库服务的操作步骤，此处不赘述。

2. 安装 WordPress

在 web01 节点、web02 节点上安装 Apache、PHP 服务，然后安装 WordPress，具体操作步骤参考任务 2.1 中的 2.1.5 小节中安装 WordPress 的内容，此处不赘述。

安装 WordPress 时需要注意的一点是，在 web01 节点和 web02 节点的配置数据库连接信息界面中，在“数据库主机”文本框中填写 database 节点的 IP 地址，完整的填写示例如图 2-12 所示。

3. web02 节点验证

在完成 web02 节点的 WordPress 的安装后，使用浏览器进入 WordPress 界面，同样进行数据库连接，会发现网页提示已安装过 WordPress，如图 2-13 所示，这说明 web02 节点已经连接了 database 节点的 MariaDB 数据库。

web02 节点验证成功，直接单击“登录”按钮便可以正常访问站点，其验证成功界面如图 2-14 所示。

请在下方填写您的数据库连接信息。如果您不确定，请联系您的主机服务提供商。

数据库名 wordpress
希望将 WordPress 安装到的数据库名称。

用户名 root
您的数据库用户名。

密码 Hide
您的数据库密码。

数据库主机 192.168.100.101 此处填写的是database节点的IP地址
如果 localhost 不起作用，您通常能够从主机商处获得正确的信息。

表前缀 wp_
如果您希望在同一个数据库安装多个 WordPress，请修改前缀。

提交

图 2-12　配置数据库连接信息界面

图 2-13　提示界面

图 2-14　web02 节点验证成功界面

即使把 web01 节点的 Apache 服务关闭，web02 节点的 WordPress 仍然正常工作，其界面如图 2-15 所示。

```
[root@web01 ~]# systemctl stop httpd
```

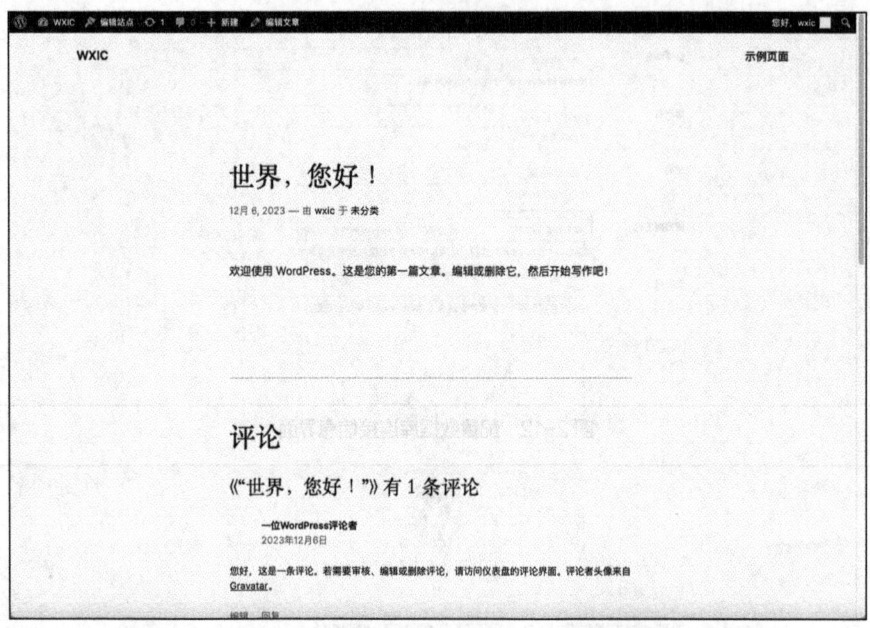

图 2-15　关闭 web01 节点的 Apache 服务后 web02 节点的测试界面

4. database 节点验证

在 database 节点上登录 MariaDB 数据库，查看数据库列表信息，命令和结果如下。

```
[root@database ~]# mysql -uroot -pwxic@2024
MariaDB [(none)]> show databases;
+--------------------+
| Database           |
+--------------------+
| information_schema |
| mysql              |
| performance_schema |
| wordpress          |
+--------------------+
4 rows in set (0.000 sec)
```

进入 wordpress 数据库查看表单详情，结果如下，从中可以发现 web01 和 web02 的用户数据已经录入数据库中，说明在集群架构下部署 WordPress 应用是有效的。

```
MariaDB [wordpress]> select * from wp_users;
+----+------------+------------------------------------+---------------+---------
------------+-----------------------------------+---------------------+---------+
------------+------------+--
| ID | user_login | user_pass                          | user_nicename | user_email
| user_url                          | user_registered     | user_activation_key |
user_status | display_name |
+----+------------+------------------------------------+---------------+--------
------------+-----------------------------------+---------------------+---------
-----------+------------+--------------+
|  1 | wxic       | $P$BwLE9mzSx2l2Dvw1hKRU1o1y4YFcNw1 | wxic          |  |
http://192.168.100.100/wordpress | 2023-12-05 16:41:36 |                     |
0 | wxic       |
```

```
+----+------------+---------------------------------+------------------+---------
------------+---------------------------------+-------------------+----------
----------+-------------+--------------+
1 row in set (0.000 sec)
```

任务 2.3 私有云基础架构下的应用部署

在知识准备 2.3 中，读者已经了解了私有云基础架构相较于传统架构和集群架构的不同，本任务利用私有云基础架构创建云主机，在云主机中实现更灵活的应用部署。

具体操作将在项目 6 的任务 6.1 云应用系统部署中展开。

项目小结

本项目首先介绍了传统架构及其存在的问题，从而引出集群架构和私有云基础架构，讲解了其与传统架构的区别，并带领读者进一步对私有云基础架构的构成要素及关键技术进行学习，从而使读者对 IT 基础架构形成更为清晰、全面的认识。

本项目在完成传统架构、集群架构和私有云基础架构相关知识的讲解后，通过在 3 种架构下部署开源应用进一步介绍了这 3 种架构的区别及私有云基础架构的优势。本项目可帮助读者认识私有云基础架构产生的必要性，加深读者对 IT 基础架构的理解，增强学习效果。

拓展知识

分布式缓存架构

分布式缓存架构是一种将缓存分布在多个计算节点上，从而提高数据访问速度和可扩展性的架构。它可以支持大量并发用户的要求，并减轻数据中心的压力。其主要实现方式包括以下几种。

（1）缓存分区：将缓存数据分布在多个节点分区中，每个分区包含一定数量的缓存数据。这样可以缓解单个节点可能面临的缓存容量和访问并发数量方面的问题。

（2）数据复制：在集群中的每个节点上，缓存数据复制可以确保集群的高可用性。通过将缓存数据在多个节点之间复制，即使某个节点出现故障，数据仍可在整个集群中保持可用。

（3）均衡负载：在多个节点之间均衡分配请求，防止其他节点工作不足或工作过度的情况发生。这可以通过使用负载均衡器、动态路由和自适应算法等技术来实现。

（4）数据一致性：在进行缓存数据复制时，需要确保数据在集群的所有节点上保持一致。其中，通常使用一致性哈希算法或基于 Paxos 算法的一致性协议来实现。

（5）监控和报告：维护关于集群节点和缓存数据使用情况的监控及报告可确保集群管理的最佳状态，并识别潜在的性能瓶颈和故障点。

知识巩固

1. 单选

（1）传统架构是（　　）的部署架构。

A. 塔式　　B. 机架式　　C. 刀片式　　D. 烟囱式

（2）（　　）是私有云基础架构的基石。

A. 虚拟化　　　B. 分布式　　　C. 并行　　　D. 集中式

2. 填空

（1）IT 基础架构的 3 个基本要素包括______、______和______。

（2）________就是指一组（若干台）相互独立的计算机，利用高速通信网络组成的一个较大的计算机服务系统。

（3）私有云基础架构在传统架构的计算、存储、网络硬件层的基础上，增加了_____和______。

3. 简答

（1）简述私有云基础架构与传统架构的主要区别。

（2）为什么要实现传统架构到私有云基础架构的转变？未来 IT 基础架构的发展趋势如何？

拓展任务

部署 MariaDB 主从数据库集群服务

在任务 2.2 的基础上，拓展 database 节点，部署 MariaDB 主从数据库集群服务，验证主从数据库的同步功能。

任务步骤

具体任务步骤参见任务 2.2.1 和任务 2.2.2。

项目3
云基础架构平台部署

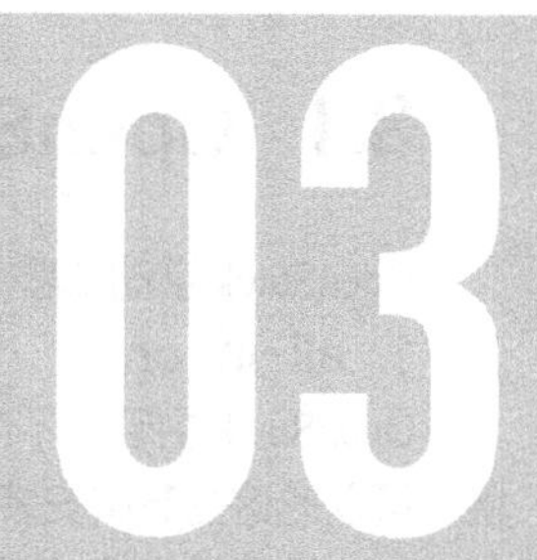

学习目标

【知识目标】

①学习 OpenStack 的多种部署模式。
②学习 Kolla-ansible 概念。
③学习 Ansible 概念。

【技能目标】

①掌握利用 Kolla-ansible 部署模式搭建云平台的技能。
②具备 OpenStack 云平台部署的能力。
③掌握云平台扩容的方法。

【素养目标】

①培养开拓创新的开放思维。
②培养诚实守信的团队精神。
③培养细致严谨的工作态度。

项目概述

小张完成了对 OpenStack 的基础学习，了解了其架构原理，并成功部署了一台 OpenStack-Yoga-Allinone 系统，通过创建网络、云主机类型、安全策略和云主机，对 OpenStack 云平台的使用有了初步的认识。公司需要小张指导技术部门的员工利用 Kolla-ansible 部署模式搭建 OpenStack 云平台。因此接下来小张需要对 OpenStack 进行更深入的学习和实践，并进一步掌握云平台的扩容方法。

知识准备

如今 OpenStack 日渐崛起，有很多企业采用了 OpenStack，OpenStack 开发人员和用户遍及全球。国内 OpenStack 产业生态也正在形成中，三大电信运营商纷纷有所行动，开始引入 OpenStack 架构技术。

为了更好地满足国内 OpenStack 产业应用发展的需求，读者需要更深入地学习和实践 OpenStack 相关技术。接下来本项目将详细介绍 OpenStack 的手动部署模式和自动化部署模式，便于读者根据实际应用需求选择合适的部署模式。

3.1 OpenStack 手动部署

OpenStack 手动部署支持两种部署模式，即 all-in-one 和 multi-node。

1. all-in-one

OpenStack 由若干不同功能的节点（控制节点、网络节点、存储节点和计算节点）组成，在实际环境中可以把这些节点安装在不同的主机（Host）上，而 all-in-one 部署模式可将上述所有的节点都安装在一台主机上进行统一部署。

2. multi-node

multi-node 部署模式可将控制节点和计算节点分离，将这些节点安装在不同的主机上并分开部署。

3.2 OpenStack 自动化部署

事实上，手动部署和自动化部署在命令上没有区别。自动化部署的原理如下：将安装命令脚本化，由主机自动执行安装脚本，自动生成配置文件并自动解决软件包和服务的依赖关系。

1. Ansible 脚本自动化部署

Ansible 是一种自动化部署配置管理工具，已被红帽（Red Hat）公司收购。它基于 Python 语言开发，集合了众多运维工具（Puppet、Chef、SaltStack 等）的优点，实现了批量系统配置、批量程序部署、批量命令运行等功能。

2. DevStack 环境部署

DevStack 不依赖于任何自动化部署工具，通过纯 Bash 脚本实现，开发人员只需要简单地编辑配置文件，然后运行脚本即可一键部署 OpenStack 环境。利用 DevStack 基本上可以部署所有的 OpenStack 组件，但并不是所有的开发人员都需要部署所有的组件，如 Nova 开发人员可能只需要部署核心组件，其他组件（如 Swift、Heat 等）并不需要部署。

3. Puppet

Puppet 由 Ruby 语言编写。Puppet 是 OpenStack 自动化部署早期的一个项目。目前，它的活跃开发群体是红帽、Mirantis、UnitedStack 等。Mirantis 出品的 Fuel 部署工具中大量的模块代码使用的是 Puppet。

如上所述，OpenStack 的部署模式多种多样。容器和 OpenStack 的结合是大势所趋，因此无论是生产环境还是开发环境，容器化部署带来的优势都是吸引人的。针对目前行业实际情况，较为推荐的是容器化部署工具 Kolla-ansible 结合自动化部署的模式，该模式可以让使用者充分体验容器（如 Docker）和 OpenStack 的有机结合。

项目实施

任务 3.1 云基础环境构建

本任务将逐步演示如何利用 Kolla-ansible 工具从零开始搭建单节点 OpenStack 环境，为实际私有云平台建设提供实践参考。

微课 3.1　云基础环境构建 1

任务 3.1.1　规划节点

单节点部署 OpenStack 云平台，各节点主机名和 IP 地址规划如表 3-1 所示。

表 3-1　各节点主机名和 IP 地址规划（单节点部署模式下）

节点类型	主机名	IP 地址规划	
		内部管理	实例通信
all-in-one 节点	controller	192.168.100.11	192.168.200.11

任务 3.1.2　环境准备

在物理机上申请一台安装 openEuler 22.09 操作系统的虚拟机作为 OpenStack all-in-one 节点，在设置处理器处勾选“虚拟化 IntelVT-x/EPT 或 AMD-V/RVI(V)”复选框。all-in-one 节点类型为 4 个 vCPU/8GB 内存/120GB 硬盘。需要给虚拟机设置两个网络接口，网络接口 1 设置为内部网络，其网卡使用仅主机模式，负责该节点的通信和管理；网络接口 2 设置为外部网络，其网卡使用 NAT 模式，主要作为该节点的数据网络，在集群部署完成后，创建的云主机使用的是网络接口 2 的网卡。all-in-one 节点配置如图 3-1 所示。

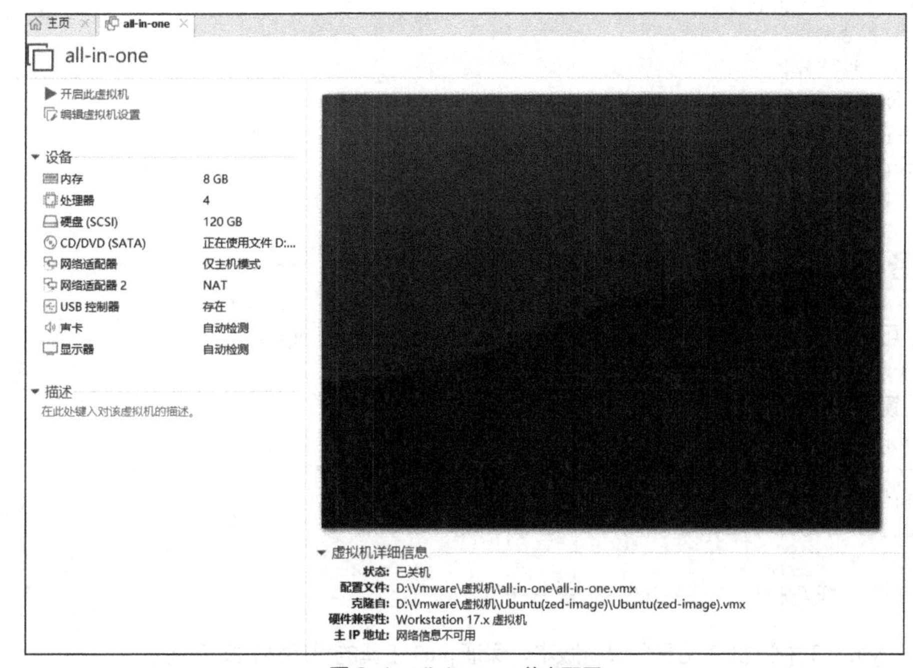

图 3-1　all-in-one 节点配置

任务 3.1.3　系统基本环境设置

1. 修改主机名

修改主机名的命令如下。

```
[root@localhost ~]# hostnamectl set-hostname controller
[root@localhost ~]# exec bash
[root@controller ~]#
```

2. 更新系统软件包

all-in-one节点更新系统软件包，以获得最新的功能支持。

```
[root@controller ~]# dnf -y update && dnf -y upgrade
[root@controller ~]# reboot
```

3. 修改网卡地址

按照表3-1配置all-in-one节点的静态IP地址，修改以下示例配置中的参数。

```
[root@controller ~]# cat \
/etc/sysconfig/network-scripts/ifcfg-ens160
……
BOOTPROTO=none
ONBOOT=yes
IPADDR=192.168.100.11
PREFIX=24
[root@controller ~]# cat \
/etc/sysconfig/network-scripts/ifcfg-ens224
……
BOOTPROTO=none
NBOOT=yes
IPADDR=192.168.200.11
PREFIX=24
GATEWAY=192.168.200.2
DNS1=223.6.6.6
DNS2=119.29.29.29
```

载入网卡配置并启动相应的网卡。

```
[root@controller ~]# nmcli c reload
[root@controller ~]# nmcli c up ens160
[root@controller ~]# nmcli c up ens224
```

任务3.1.4 安装Ansible和Kolla-ansible

为了可以使用pip3安装和管理Python 3软件包，现需要安装pip3。

```
[root@controller ~]# dnf -y install python3-pip
```

使用镜像源来加快pip3安装包的下载速度。

```
[root@controller ~]# mkdir .pip
[root@controller ~]# cat << WXIC > .pip/pip.conf
[global]
index-url = https://pypi.tuna.tsinghua.edu.cn/simple
[install]
trusted-host=pypi.tuna.tsinghua.edu.cn
WXIC
```

更新Python 3中的pip3工具到最新版本，保证pip3工具的可用性和安全性。

```
[root@controller ~]# pip3 install --ignore-installed --upgrade pip
```

使用以下命令安装Ansible，并查看所安装的版本号。

```
[root@controller ~]# pip3 install -U 'ansible>=4,<6'
[root@controller ~]# ansible --version
```

安装Kolla-ansible和Kolla-ansible环境必需的依赖项。

```
[root@controller ~]# dnf -y install \
git python3-devel libffi-devel gcc openssl-devel python3-libselinux
[root@controller ~]# dnf -y install openstack-kolla-ansible
[root@controller ~]# kolla-ansible --version
14.2.0
```

创建 Kolla-ansible 配置文件目录。

```
[root@controller ~]# mkdir -p /etc/kolla/{globals.d,config}
[root@controller ~]# chown $USER:$USER /etc/kolla
```

将 inventory 文件复制到/etc/ansible 目录下。

```
[root@controller ~]# mkdir /etc/ansible
[root@controller ~]# cp \
/usr/share/kolla-ansible/ansible/inventory/* /etc/ansible
```

随着 Kolla-ansible 版本的更迭，从 Yoga 版本开始需要安装 Ansible Galaxy 的依赖项，使用以下命令安装其依赖项。

```
[root@controller ~]# pip3 install cryptography==38.0.4
[root@controller ~]# tar -xvf kolla-ansible-deps.tar.gz -C /root/.ansible
```

任务 3.1.5　Ansible 运行配置优化

在使用 Kolla-ansible 部署 OpenStack 时，其会执行大量的命令和任务，因此对 Ansible 运行配置进行调优可以加快执行速度，具体的优化过程如下。

```
[root@controller ~]# cat << MXD > /etc/ansible/ansible.cfg
[defaults]
#SSH 服务关闭密钥检测
host_key_checking=False
#如果不使用 sudo，则建议将其启用
pipelining=True
#执行任务的并发数
forks=100
timeout=800
#禁用警告#
devel_warning = False
deprecation_warnings=False
#显示每个任务花费的时间
callback_whitelist = profile_tasks
#记录 Ansible 的输出，通过相对路径表示
log_path= wxic_cloud.log
#主机清单文件，通过相对路径表示
inventory = yoga_all-in-one
#命令执行环境，也可更改为/bin/bash
executable = /bin/sh
remote_port = 22
remote_user = root
#默认输出的详细程度
#可选值为 0、1、2、3、4 等
#值越大表示输出越详细
verbosity = 0
show_custom_stats = True
interpreter_python = auto_legacy_silent
```

```
[colors]
#成功的任务使用绿色显示
ok = green
#跳过的任务使用亮灰色显示
skip = bright gray
#警告使用亮紫色显示
warn = bright purple
[privilege_escalation]
become_user = root
[galaxy]
display_progress = True
MXD
```

修改完成以后，可以使用 ansible-config view 命令对其进行查看。

```
[root@controller ~]# ansible-config view
[defaults]
#SSH 服务关闭密钥检测
host_key_checking=False
......
```

任务 3.1.6 Kolla-ansible 环境初始配置

1. 修改主机清单文件

进入/etc/ansible 目录，过滤提供的主机清单 all-in-one 文件的注释和空行，并将其内容覆盖到 yoga_all-in-one 文件。

微课 3.2 云基础环境构建 2

```
[root@controller ~]# cd /etc/ansible/
[root@controller ansible]# awk '!/^#/ && !/^$/' all-in-one > yoga_all-in-one
```

2. 检查主机清单文件配置是否正确

使用以下命令测试主机能否连通。

```
[root@controller ~]# ansible -m ping all
localhost | SUCCESS => {
  "ansible_facts": {
    "discovered_interpreter_python": "/usr/bin/python3.10"
  },
  "changed": false,
  "ping": "pong"
}
```

3. 配置 OpenStack 各服务组件的密码

在使用 Kolla-ansible 部署 OpenStack Yoga 平台时，各个服务组件的密码存储在/etc/kolla/passwords.yml 文件中，此文件默认所有的密码都是空白的，必须手动或者通过运行随机密码生成器来填写这些密码。在部署时建议使用随机密码生成器来生成各个服务组件的密码，命令如下。

```
[root@controller ~]# kolla-genpwd
# 修改 Horizon 界面登录密码为 wxic@2024
[root@controller ~]# sed -i \
's/keystone_admin_password: .*/keystone_admin_password: \
wxic@2024/g' /etc/kolla/passwords.yml
# 验证修改结果
[root@controller ~]# grep keystone_admin /etc/kolla/passwords.yml
keystone_admin_password: wxic@2024
```

4. 编辑 globals.yml 文件

此次部署 all-in-one 时只安装了 OpenStack 的核心组件，在 globals.yml 中指定参数 enable_openstack_core: "yes"，并安装 Glance、Keystone、Neutron、Nova、Heat 和 Horizon 组件。其中，需要注意 kolla_internal_vip_address 的地址，因为此次部署时 HAProxy 和 Keepalived 都未启用，故该地址为内部网卡的地址（此次部署时使用 IP 地址 192.168.100.11），部署完成后使用该 IP 地址登录 Horizon。

```
[root@controller ~]# cd /etc/kolla/
[root@controller kolla]# cp globals.yml{,_bak}
[root@controller kolla]# cat << MXD > globals.yml
---
kolla_base_distro: "ubuntu"
kolla_install_type: "source"
openstack_release: "yoga"
kolla_internal_vip_address: "192.168.100.11"
kolla_sysctl_conf_path: /etc/sysctl.conf
docker_client_timeout: 120
network_interface: "ens160"
network_address_family: "ipv4"
neutron_external_interface: "ens224"
neutron_plugin_agent: "openvswitch"
neutron_ipam_driver: "internal"
openstack_region_name: "RegionWxic"
openstack_logging_debug: "False"
enable_openstack_core: "yes"
glance_backend_file: "yes"
nova_compute_virt_type: "kvm"
nova_console: "novnc"
enable_haproxy: "no"
enable_keepalived: "no"
MXD
```

在/etc/kolla/config 目录下自定义 Neutron 服务的一些配置，这将在部署集群时使用自定义的配置覆盖默认的配置。

```
[root@controller ~]# cd /etc/kolla/config/
[root@controller config]# mkdir neutron
[root@controller config]# cat << MXD > neutron/dhcp_agent.ini
[DEFAULT]
dnsmasq_dns_servers = 8.8.8.8,223.6.6.6,119.29.29.29
MXD
[root@controller config]# cat << MXD > neutron/ml2_conf.ini
[ml2]
tenant_network_types = flat,vxlan,vlan
[ml2_type_vlan]
network_vlan_ranges = provider:10:1000
[ml2_type_flat]
flat_networks = provider
MXD
[root@controller config]# cat << MXD > neutron/openvswitch_agent.ini
[securitygroup]
firewall_driver = openvswitch
[ovs]
bridge_mappings = provider:br-ex
MXD
```

任务 3.1.7 部署集群环境

微课 3.3 云基础环境构建 3

在 all-in-one 节点上安装 OpenStack 命令行界面（Command Line Interface，CLI）客户端。

```
[root@controller ~]# dnf -y install python3-openstackclient
```

为了使部署的 all-in-one 节点网络路由正常工作，需要在 openEuler 22.09 操作系统中启用 IP 地址转发功能，修改所有节点的/etc/sysctl.conf 文件，并配置在系统启动时自动加载 br_netfilter 模块，具体操作如下。

```
[root@controller ~]# cat << MXD >> /etc/sysctl.conf
net.ipv4.ip_forward=1
net.bridge.bridge-nf-call-ip6tables=1
net.bridge.bridge-nf-call-iptables=1
MXD
# 临时加载模块，重启后模块失效
[root@controller ~]# modprobe br_netfilter
# 重新加载配置
[root@controller ~]# sysctl -p /etc/sysctl.conf
# 创建 yoga.service 文件，设置系统启动时自动加载 br_netfilter 模块
[root@controller ~]# cat << MXD > /usr/lib/systemd/system/yoga.service
[Unit]
Description=Load br_netfilter and sysctl settings for OpenStack

[Service]
Type=oneshot
RemainAfterExit=yes
ExecStart=/sbin/modprobe br_netfilter
ExecStart=/usr/sbin/sysctl -p

[Install]
WantedBy=multi-user.target
MXD
[root@controller ~]# systemctl enable --now yoga.service
Created symlink /etc/systemd/system/multi-user.target.wants/yoga.service →
/lib/systemd/system/yoga.service.
```

在 all-in-onc 节点上使用以下命令安装 OpenStack 集群所需要的基础依赖项并修改配置文件，如安装 Docker 和修改 Hosts 文件等，执行结果和用时如图 3-2 所示。

```
[root@controller ~]# kolla-ansible bootstrap-servers
```

```
PLAY RECAP **********************************************************************************
localhost      : ok=26   changed=12   unreachable=0    failed=0    skipped=22   rescued=0    ignored=0

Monday 17 April 2023  13:31:27 +0800 (0:00:00.084)       0:01:22.808 **********
===============================================================================
openstack.kolla.docker : Install packages --------------------------------------------- 68.65s
openstack.kolla.docker_sdk : Install docker SDK for python ---------------------------- 1.98s
Gather facts -------------------------------------------------------------------------- 1.31s
openstack.kolla.docker : Install docker rpm gpg key ----------------------------------- 1.17s
openstack.kolla.packages : Install packages ------------------------------------------- 0.80s
openstack.kolla.baremetal : Disable firewalld ----------------------------------------- 0.78s
openstack.kolla.docker : Restart docker ----------------------------------------------- 0.77s
openstack.kolla.baremetal : Change state of selinux ----------------------------------- 0.69s
openstack.kolla.packages : Remove packages -------------------------------------------- 0.56s
openstack.kolla.docker_sdk : Install packages ----------------------------------------- 0.53s
openstack.kolla.docker : Write docker config ------------------------------------------ 0.51s
openstack.kolla.docker : Start and enable docker -------------------------------------- 0.49s
openstack.kolla.baremetal : Generate /etc/hosts for all of the nodes ------------------ 0.46s
openstack.kolla.baremetal : Ensure unprivileged users can use ping -------------------- 0.30s
openstack.kolla.docker : Enable docker yum repository --------------------------------- 0.30s
openstack.kolla.docker : Ensure yum repos directory exists ---------------------------- 0.29s
openstack.kolla.baremetal : Check if firewalld is installed --------------------------- 0.29s
openstack.kolla.baremetal : Ensure /etc/cloud/cloud.cfg exists ------------------------ 0.27s
openstack.kolla.baremetal : Ensure localhost in /etc/hosts ---------------------------- 0.25s
openstack.kolla.docker : Ensure docker config directory exists ------------------------ 0.20s
[root@controller ~]#
```

图 3-2 执行结果和用时（单节点部署模式下使用命令安装 OpenStack 集群所需要的基础依赖项并修改配置文件）

Docker 默认的镜像拉取地址在国外，在国内拉取镜像的速度比较慢，可以将其修改为国内镜像拉取地址来加速镜像的拉取。编辑 all-in-one 节点的/etc/docker/daemon.json 文件，添加 registry-mirrors 部分的内容。下面给出 all-in-one 节点的修改示例。

```
[root@controller ~]# cat /etc/docker/daemon.json
{
  "bridge": "none",
  "default-ulimits": {
    "nofile": {
      "hard": 1048576,
      "name": "nofile",
      "soft": 1048576
    }
  },
  "ip-forward": false,
  "iptables": false,
  "registry-mirrors": [
       "https://registry.docker-cn.com",
       "http://hub-mirror.c.163.com",
       "https://docker.mirrors.ustc.edu.cn",
       "https://mirror.ccs.tencentyun.com"
  ]
}
[root@controller ~]# systemctl daemon-reload
[root@controller ~]# systemctl restart docker
```

在 all-in-one 节点上进行部署前检查，在 openEuler 22.09 操作系统执行过程中，如果出现“openEuler release NA version 22.09 is not supported. Supported releases are: 20.03”提示，则不必理会，继续进行接下来的操作即可，它不会影响后面的部署，执行结果和用时如图 3-3 所示。

```
[root@controller ~]# kolla-ansible prechecks
```

```
PLAY RECAP ********************************************************************************
localhost        : ok=65    changed=0    unreachable=0    failed=0    skipped=58    rescued=0    ignored=0

Monday 17 April 2023  13:33:53 +0800 (0:00:00.510)       0:00:25.518 **********
===============================================================================
Gather facts ------------------------------------------------------------------------ 1.05s
glance : Checking free port for Glance API ------------------------------------------ 0.93s
glance : Get container facts -------------------------------------------------------- 0.75s
heat : Checking free port for Heat API CFN ------------------------------------------ 0.58s
horizon : Checking free port for Horizon -------------------------------------------- 0.51s
nova-cell : Checking free port for Nova Libvirt ------------------------------------- 0.51s
service-precheck : glance | Validate inventory groups ------------------------------- 0.50s
Group hosts based on enabled services ----------------------------------------------- 0.49s
nova-cell : Checking free port for Nova SSH (API interface) ------------------------- 0.48s
nova-cell : Checking free port for Nova NoVNC Proxy --------------------------------- 0.48s
prechecks : Checking Docker version ------------------------------------------------- 0.48s
nova : Checking free port for Nova Metadata ----------------------------------------- 0.47s
nova-cell : Checking free port for Nova Spice HTML5 Proxy --------------------------- 0.46s
nova-cell : Checking that host libvirt is not running ------------------------------- 0.46s
placement : Checking free port for Placement API ------------------------------------ 0.45s
mariadb : Get container facts ------------------------------------------------------- 0.42s
nova : Checking free port for Nova API ---------------------------------------------- 0.42s
openvswitch : Checking free port for OVSDB ------------------------------------------ 0.37s
service-precheck : neutron | Validate inventory groups ------------------------------ 0.36s
keystone : Get container facts ------------------------------------------------------ 0.36s
[root@controller ~]#
```

图 3-3 执行结果和用时（单节点部署模式下进行部署前检查）

在 all-in-one 节点上使用以下命令下载 OpenStack 集群 all-in-one 节点所需要的全部镜像，执行结果和用时如图 3-4 所示。

```
[root@controller ~]# docker pull 99cloud/skyline:latest
[root@controller ~]# kolla-ansible pull
```

在 all-in-one 节点上使用以下命令部署 OpenStack 集群，执行结果和用时如图 3-5 所示。

```
[root@controller ~]# kolla-ansible deploy
```

```
PLAY RECAP *********************************************************************
localhost          : ok=31   changed=9    unreachable=0    failed=0    skipped=10   rescued=0    ignored=0

Monday 17 April 2023  13:50:02 +0800 (0:00:24.137)       0:06:25.987 **********
===============================================================================
service-images-pull : nova_cell | Pull images ---------------------------- 106.01s
service-images-pull : keystone | Pull images ------------------------------ 66.37s
service-images-pull : neutron | Pull images ------------------------------- 53.44s
service-images-pull : heat | Pull images ---------------------------------- 29.12s
service-images-pull : nova | Pull images ---------------------------------- 27.80s
service-images-pull : horizon | Pull images ------------------------------- 24.14s
service-images-pull : glance | Pull images -------------------------------- 21.39s
service-images-pull : openvswitch | Pull images --------------------------- 20.82s
service-images-pull : placement | Pull images ----------------------------- 11.91s
service-images-pull : common | Pull images -------------------------------- 10.08s
service-images-pull : mariadb | Pull images -------------------------------- 3.67s
service-images-pull : memcached | Pull images ------------------------------ 3.51s
service-images-pull : rabbitmq | Pull images ------------------------------- 3.37s
Gather facts --------------------------------------------------------------- 1.15s
Group hosts based on enabled services -------------------------------------- 0.47s
nova : Reload nova API services to remove RPC version pin ------------------ 0.24s
nova : Run Nova API online database migrations ----------------------------- 0.22s
nova-cell : Run Nova cell online database migrations ----------------------- 0.19s
nova-cell : Reload nova cell services to remove RPC version cap ------------ 0.18s
Bootstrap upgrade ---------------------------------------------------------- 0.15s
[root@controller ~]#
```

图 3-4　执行结果和用时（单节点部署模式下下载镜像）

```
PLAY RECAP *********************************************************************
localhost          : ok=273  changed=192  unreachable=0    failed=0    skipped=124  rescued=0 ignored=1

Monday 17 April 2023  13:57:55 +0800 (0:00:01.164)       0:06:29.157 **********
===============================================================================
service-ks-register : keystone | Creating services ------------------------ 12.63s
heat : Running Heat bootstrap container ----------------------------------- 10.95s
rabbitmq : Waiting for rabbitmq to start ---------------------------------- 10.63s
mariadb : Check MariaDB service port liveness ----------------------------- 10.42s
nova-cell : Waiting for nova-compute services to register themselves ------- 9.60s
service-ks-register : heat | Creating endpoints ---------------------------- 9.02s
nova : Running Nova API bootstrap container -------------------------------- 6.74s
neutron : Running Neutron bootstrap container ------------------------------ 6.65s
service-ks-register : heat | Granting user roles --------------------------- 6.33s
service-ks-register : heat | Creating roles -------------------------------- 5.41s
mariadb : Running MariaDB bootstrap container ------------------------------ 4.99s
service-ks-register : heat | Creating services ----------------------------- 4.84s
nova : Running Nova API bootstrap container -------------------------------- 4.71s
glance : Running Glance bootstrap container -------------------------------- 4.62s
service-ks-register : placement | Creating endpoints ----------------------- 4.48s
service-ks-register : neutron | Creating endpoints ------------------------- 4.41s
service-ks-register : glance | Creating endpoints -------------------------- 4.29s
nova-cell : Running Nova cell bootstrap container -------------------------- 4.22s
service-ks-register : nova | Creating endpoints ---------------------------- 4.20s
service-rabbitmq : nova | Ensure RabbitMQ users exist ---------------------- 3.85s
[root@controller ~]#
```

图 3-5　执行结果和用时（单节点部署模式下部署 OpenStack 集群）

上述步骤完成后，OpenStack 集群部署结束，所有的服务已经启动并正常运行。在浏览器地址栏中输入 globals.yml 文件中定义的 kolla_internal_vip_address 地址，进入 Horizon 登录界面，此时，用户名为 admin，密码为 passwords.yml 文件中 keystone_admin_password 的值 wxic@2024，单节点部署模式下的 Horizon 登录和概览界面分别如图 3-6 和图 3-7 所示。

图 3-6　Horizon 登录界面（单节点部署模式下）

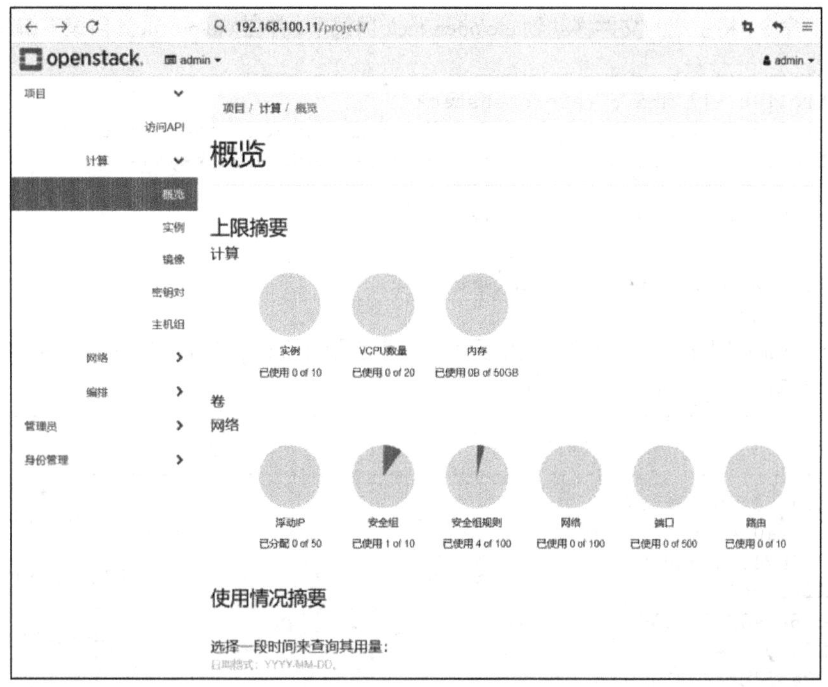

图 3-7　Horizon 概览界面（单节点部署模式下）

任务 3.1.8　OpenStack CLI 客户端设置

OpenStack 集群部署完成后，客户端执行命令之前需要生成 clouds.yaml 和 admin-openrc.sh 文件，这些是管理员（admin）用户的凭据，执行结果和用时如图 3-8 所示。

```
[root@controller ~]# kolla-ansible post-deploy
```

```
[root@controller ~]# kolla-ansible post-deploy
Post-Deploying Playbooks : ansible-playbook -e @/etc/kolla/globals.yml  -e @/etc/kolla/passwords.yml
-e CONFIG_DIR=/etc/kolla  /usr/local/share/kolla-ansible/ansible/post-deploy.yml
[WARNING]: Invalid characters were found in group names but not replaced, use -vvvv to see details

PLAY [Creating clouds.yaml file on the deploy node] ****************************************************

TASK [Gathering Facts] *********************************************************************************
Monday 17 April 2023  15:46:30 +0800 (0:00:00.042)       0:00:00.042 **********
ok: [localhost]

TASK [Create /etc/openstack directory] *****************************************************************
Monday 17 April 2023  15:46:32 +0800 (0:00:01.275)       0:00:01.318 **********
changed: [localhost]

TASK [Template out clouds.yaml] ************************************************************************
Monday 17 April 2023  15:46:32 +0800 (0:00:00.310)       0:00:01.628 **********
changed: [localhost]

PLAY [Creating admin openrc file on the deploy node] ***************************************************

TASK [Gathering Facts] *********************************************************************************
Monday 17 April 2023  15:46:33 +0800 (0:00:00.673)       0:00:02.302 **********
ok: [localhost]

TASK [Template out admin-openrc.sh] ********************************************************************
Monday 17 April 2023  15:46:34 +0800 (0:00:00.896)       0:00:03.198 **********
changed: [localhost]

TASK [octavia : Template out octavia-openrc.sh] ********************************************************
Monday 17 April 2023  15:46:34 +0800 (0:00:00.455)       0:00:03.654 **********
skipping: [localhost]

PLAY RECAP *********************************************************************************************
localhost      : ok=5    changed=3    unreachable=0    failed=0    skipped=1    rescued=0    ignored=0

Monday 17 April 2023  15:46:34 +0800 (0:00:00.101)       0:00:03.755 **********
===============================================================================
Gathering Facts ------------------------------------------------------------------------------- 1.28s
Gathering Facts ------------------------------------------------------------------------------- 0.90s
Template out clouds.yaml ---------------------------------------------------------------------- 0.67s
Template out admin-openrc.sh ------------------------------------------------------------------ 0.46s
Create /etc/openstack directory --------------------------------------------------------------- 0.31s
octavia : Template out octavia-openrc.sh ------------------------------------------------------ 0.10s
[root@controller ~]#
```

图 3-8　执行结果和用时（单节点部署模式下生成 admin 用户的凭据）

使用以下命令，将生成的文件移动到/etc/openstack目录下，并在/etc/profile.d目录下编写openstack-yoga.sh脚本。

```
[root@controller ~]# mkdir /etc/openstack
[root@controller ~]# mv /etc/kolla/admin-openrc.sh /etc/openstack/
[root@controller ~]# cat << MXD > /etc/profile.d/openstack-yoga.sh
#!/usr/bin/env bash
source /etc/openstack/admin-openrc.sh
MXD
# 重启终端后便可以正常使用 OpenStack 相关命令
[root@controller ~]# logout
```

使用OpenStack相关命令验证客户端是否可以正常使用及服务能否正常启动，命令和结果如下。

```
[root@controller ~]# openstack service list
+------------------------+----------------------------+------------------------+
| ID                     | Name                       | Type                   |
+------------------------+----------------------------+------------------------+
| 14dfe01d14ef44a5a736257|                            |                        |
|        1ce840a4f       |           heat             |     orchestration      |
| 323028e9c7bd4c6d8369891|                            |                        |
|        2f4129e30       |         placement          |       placement        |
| 719d391e36664029923c30e|                            |                        |
|        229fd94c4       |          keystone          |       identity         |
| 9bd2b6503aef4cfa8031309|                            |                        |
|        ce981fbd3       |          neutron           |        network         |
| a4301fd01ee24ec8bb70aca|                            |                        |
|        c18eb43bc       |           glance           |         image          |
| ab532616b11941f8996eebb|                            |                        |
|        30dadbd33       |            nova            |        compute         |
| cf49592d78ee4198b19f85b|                            |                        |
|        d7f7b4ef4       |         heat-cfn           |     cloudformation     |
+------------------------+----------------------------+------------------------+
[root@controller ~]# openstack compute service list
+----------------+----------+------------+---------+---------+-----+------------+
| ID             | Binary   | Host       | Zone    | Status  |State| Updated At |
+----------------+----------+------------+---------+---------+-----+------------+
| fb6b76c1-7d67- |  nova-   |            |         |         |     | 2023-04-17 |
|   4414-89d9-   | scheduler| controller | internal| enabled | up  | T07:48:46. |
|  fee50fb72ff3  |          |            |         |         |     |   000000   |
| 1e8b67c1-1092- |  nova-   |            |         |         |     | 2023-04-17 |
|   4274-ad2b-   | conductor| controller | internal| enabled | up  | T07:48:45. |
|  43e8cea9cce1  |          |            |         |         |     |   000000   |
| 5106b3e5-e73d- |  nova-   |            |         |         |     | 2023-04-17 |
|   4f2f-aa66-   | compute  | controller |   nova  | enabled | up  |  T07:48:45 |
|  47bbec61b268  |          |            |         |         |     |   000000   |
+----------------+----------+------------+---------+---------+-----+------------+
[root@controller ~]# openstack network agent list
+----------------+----------+------------+---------+---------+-----+------------+
|                |  Agent   |            | Availab |         |     |            |
|       ID       |  Type    |    Host    |  ility  |  Alive  |State|   Binary   |
|                |          |            |  Zone   |         |     |            |
+----------------+----------+------------+---------+---------+-----+------------+
| 4c274495-d2b2- |  Open    |            |         |         |     |  neutron-  |
|   42e3-80af-   | vSwitch  | controller |  None   |   :-)   | UP  |openvswitch |
|  e9e61d94ca8e  |  agent   |            |         |         |     |   -agent   |
```

```
   766ec7e7-7897-                                                         neutron-l3
|    4698-b7b2-    | L3 agent | controller |   none   |   :-)   |  UP  |  -agent   |
   4e945261703a
   d65643+0-4d80-      DHCP                                                 neutron-
|    4e35-a95e-    |   agent  | controller |   none   |   :-)   |  UP  | dhcp-agent |
   31d55829876f
   e15437e0-e2a8-    Metadata                                               neutron-
|    4693-97bc-    |   agent  | controller |   None   |   :-)   |  UP  | metadata-  |
   06109dc00523                                                              agent
+------------------+----------+------------+----------+---------+------+-----------+
```

可以发现，所有命令都可以正常使用，且服务状态正常，下面查看此次部署 OpenStack 版本的详细信息，包括各个组件的版本号，命令和结果如下。

```
[root@controller ~]# openstack version show
+----------+---------------+---------+----------+--------------------+-------+-------+
|  Region  | Service Type  | Version |  Status  |      Endpoint      |  Min  |  Max  |
|   Name   |               |         |          |                    | Microv| Microv|
|          |               |         |          |                    | ersion| ersion|
+----------+---------------+---------+----------+--------------------+-------+-------+
|RegionWxic|orchestration  |   1.0   | CURRENT  | http://192.168.    | None  | None  |
|          |               |         |          | 100.11:8004/v1/    |       |       |
|RegionWxic|  placement    |   1.0   | CURRENT  | http://192.168.    |  1.0  | 1.39  |
|          |               |         |          | 100.11:8780/       |       |       |
|RegionWxic|  identity     |  3.14   | CURRENT  | http://192.        | None  | None  |
|          |               |         |          | 168.100.11:5000    |       |       |
|          |               |         |          | /v3/               |       |       |
|RegionWxic|  network      |   2.0   | CURRENT  | http://192.168.    | None  | None  |
|          |               |         |          | 100.11:9696/v2.    |       |       |
|          |               |         |          | 0/                 |       |       |
|RegionWxic|   image       |   2.0   |SUPPORTED | http://192.168.    | None  | None  |
|          |               |         |          | 100.11:9292/v2/    |       |       |
|RegionWxic|   image       |   2.1   |SUPPORTED | http://192.168.    | None  | None  |
|          |               |         |          | 100.11:9292/v2/    |       |       |
|RegionWxic|   image       |   2.2   |SUPPORTED | http://192.168.    | None  | None  |
|          |               |         |          | 100.11:9292/v2/    |       |       |
|RegionWxic|   image       |   2.3   |SUPPORTED | http://192.168.    | None  | None  |
|          |               |         |          | 100.11:9292/v2/    |       |       |
|RegionWxic|   image       |   2.4   |SUPPORTED | http://192.168.    | None  | None  |
|          |               |         |          | 100.11:9292/v2/    |       |       |
|RegionWxic|   image       |   2.5   |SUPPORTED | http://192.168.    | None  | None  |
|          |               |         |          | 100.11:9292/v2/    |       |       |
|RegionWxic|   image       |   2.6   |SUPPORTED | http://192.168.    | None  | None  |
|          |               |         |          | 100.11:9292/v2/    |       |       |
|RegionWxic|   image       |   2.7   |SUPPORTED | http://192.168.    | None  | None  |
|          |               |         |          | 100.11:9292/v2/    |       |       |
|RegionWxic|   image       |   2.8   |SUPPORTED | http://192.168.    | None  | None  |
|          |               |         |          | 100.11:9292/v2/    |       |       |
|RegionWxic|   image       |   2.9   |SUPPORTED | http://192.168.    | None  | None  |
|          |               |         |          | 100.11:9292/v2/    |       |       |
|RegionWxic|   image       |  2.10   |SUPPORTED | http://192.168.    | None  | None  |
|          |               |         |          | 100.11:9292/v2/    |       |       |
|RegionWxic|   image       |  2.11   |SUPPORTED | http://192.168.    | None  | None  |
|          |               |         |          | 100.11:9292/v2/    |       |       |
|RegionWxic|   image       |  2.12   |SUPPORTED | http://192.168.    | None  | None  |
|          |               |         |          | 100.11:9292/v2/    |       |       |
```

```
|RegionWxic |     image     |  2.13  |SUPPORTED |http://192.168.  | None | None |
                                                 100.11:9292/v2/
|RegionWxic |     image     |  2.15  | CURRENT  |http://192.168.  | None | None |
                                                 100.11:9292/v2/
|RegionWxic |    compute    |  2.0   |SUPPORTED |http://192.168.  | None | None |
                                                 100.11:8774/v2/
                                                 http://192.168.
|RegionWxic |    compute    |  2.1   | CURRENT  | 100.11:8774/    | 2.1  | 2.93 |
                                                     v2.1/
|RegionWxic |cloudformation |  1.0   | CURRENT  |http://192.168.  | None | None |
                                                 100.11:8000/v1/
+-----------+---------------+--------+----------+-----------------+------+------+
```

任务 3.1.9 安装 Skyline 服务

微课 3.4 云基础环境构建 4

Skyline 是新一代的 OpenStack 管理界面，由九州云公司于 2021 年 9 月捐献给 OpenStack 社区。同年 12 月末，Skyline 孵化完成，成为 OpenStack 正式项目。2022 年，Skyline 开发团队完成了 OpenStack 的代码重构，并增加了对 Octavia、Manila、Swift、Barbican、Zun、Trove 等社区模块的支持。Skyline 开发团队也通过企业微信群与社区开发人员、社区用户进行了多轮互动讨论。2022 年 10 月 5 日，Skyline 的第一个正式版本随 OpenStack Yoga 正式发布。

Skyline 不仅提供了 OpenStack 基础服务（如计算、存储、网络的操作界面等），还支持许多增值服务（如文件存储、对象存储、负载均衡、数据库等服务）。一旦完成部署，Skyline 不依赖任何插件就能迅速调用各种云服务接口，满足企业级的生产需求。云上的虚拟机、容器、Kubernetes 集群、关系数据库服务（Relational Database Service，RDS）数据库等各种资源，都能在 Skyline 的平台上完成全生命周期管理。

Skyline 1.0.0 已完成表 3-2 中组件的对接，并支持完整的图形化操作界面。

表 3-2 Skyline 1.0.0 组件说明

组件名称	服务类型	是否实现
Keystone	用户认证	实现
Glance	镜像服务	实现
Nova	计算服务	实现
Placement	调度服务	实现
Neutron	网络服务	实现
Cinder	块存储服务	实现
Swift	对象服务	实现
Zun	容器服务	实现
Manila	文件存储	实现
Heat	编排服务	实现
Trove	数据库服务	实现
Octavia	负载均衡	实现
Magnum	Kubernetes 集群服务	实现
Prometheus	监控服务	实现
Barbican	证书服务	实现
VPNaaS	虚拟专用网络（Virtual Private Network，VPN）服务	实现

1. 创建 Skyline 服务的数据库

在 MariaDB 中创建 Skyline 服务的数据库并赋予其远程访问权限，命令及结果如下。

```
# 查询数据库登录密码
[root@controller ~]# grep ^database /etc/kolla/passwords.yml
database_password: BraVkrGCC4hj59EXRYp9viZj8X8YM5CBC3v6l6Bn
# 查询运行数据库服务的容器 ID
[root@controller ~]# docker container ls -f name=mariadb --format='{{.ID}}'
58bd2b1faf08
# 进入数据库容器，创建 Skyline 服务的数据库并赋予其远程访问权限
[root@controller ~]# docker exec -it 58bd2b1faf08 sh
(mariadb)[mysql@controller /]$ mysql -uroot
-pBraVkrGCC4hj59EXRYp9viZj8X8YM5CBC3v6l6Bn
Welcome to the MariaDB monitor.  Commands end with ; or \g.
Your MariaDB connection id is 30361
Server version: 10.6.11-MariaDB-1:10.6.11+maria~deb11-log mariadb.org binary
distribution
Copyright (c) 2000, 2018, Oracle, MariaDB Corporation Ab and others.
Type 'help;' or '\h' for help. Type '\c' to clear the current input statement.
MariaDB [(none)]> CREATE DATABASE skyline DEFAULT CHARACTER SET utf8 DEFAULT
COLLATE utf8_general_ci;
Query OK, 1 row affected (0.004 sec)
MariaDB [(none)]> GRANT ALL PRIVILEGES ON skyline.* TO 'skyline'@'localhost'
IDENTIFIED BY 'mariadb_yoga';
Query OK, 0 rows affected (0.006 sec)
MariaDB [(none)]> GRANT ALL PRIVILEGES ON skyline.* TO 'skyline'@'%' IDENTIFIED
BY 'mariadb_yoga';
Query OK, 0 rows affected (0.005 sec)
MariaDB [(none)]> flush privileges;
Query OK, 0 rows affected (0.005 sec)
```

2. 创建 Skyline 服务的新用户并向其分配 admin 角色

在默认域 default 中创建 Skyline 服务的新用户 skyline，设置其密码为 wxic@yoga，并向该用户分配 admin 角色。

```
[root@controller ~]# openstack user create --domain default --password wxic@yoga skyline
+---------------+----------------------------------------------------+
| Field         | Value                                              |
+---------------+----------------------------------------------------+
| domain_id     | default                                            |
| enabled       | True                                               |
| id            | 3e575726d1634a6497b12892479ec63e                   |
| name          | skyline                                            |
| options       | {}                                                 |
| password_expires_at | None                                         |
+-------------------+------------------------------------------------+
[root@controller ~]# openstack role add --project service --user skyline admin
```

3. 修改 Skyline 配置文件

创建 Skyline 服务需要的配置文件目录和日志文件目录。

```
[root@controller ~]# mkdir -p /etc/skyline /var/log/skyline /var/lib/skyline \
/var/log/nginx
```

在/etc/skyline 目录下创建并修改 skyline.yaml 配置文件，该配置文件的最新版示例文件可在 OpenStack 官方帮助文档的 settings 界面中找到，修改 skyline.yaml 配置文件中的以下几项内容。

① database_url：定义数据库名称和远程连接地址。

② debug：是否启用 debug 功能。建议不启用该功能，因为启用该功能之后会产生大量的日志文件，非开发人员没有必要启用该功能。

③ log_dir：定义日志存放目录（前面的步骤中创建的/var/log/skyline 目录）。

④ keystone_url：Keystone 服务的认证地址。

⑤ default_region：集群区域名。

⑥ system_user_password：已创建的新用户的密码。

命令及结果如下。

```
# 查询 Keystone 内部服务端点地址
[root@controller ~]# openstack endpoint list --interface internal \
--service keystone -f value -c URL
http://192.168.100.11:5000
# 创建配置文件 skyline.yaml
[root@controller ~]# cat << MXD > /etc/skyline/skyline.yaml
default:
 access_token_expire: 3600
 access_token_renew: 1800
 cors_allow_origins: []
  # 设置 MySQL 连接地址及密码
 database_url: mysql://skyline:mariadb_yoga@192.168.100.11:3306/skyline
 debug: false
 log_dir: /var/log/skyline
 log_file: skyline_wxic.log
 prometheus_basic_auth_password: 'wxic@yoga'
 prometheus_basic_auth_user: ''
 prometheus_enable_basic_auth: false
 prometheus_endpoint: http://192.168.100.11:9091
 secret_key: aCtmgbcUqYUy_HNVg5BDXCaeJgJQzHJXwqbXr0Nmb2o
 session_name: session
 ssl_enabled: true
openstack:
 base_domains:
 - heat_user_domain
 # 修改默认区域
 default_region: RegionWxic
 enforce_new_defaults: true
 extension_mapping:
  floating-ip-port-forwarding: neutron_port_forwarding
  fwaas_v2: neutron_firewall
  qos: neutron_qos
  vpnaas: neutron_vpn
 interface_type: public
 # Keystone 认证地址
 keystone_url: http://192.168.100.11:5000/v3/
 nginx_prefix: /api/openstack
 reclaim_instance_interval: 604800
 service_mapping:
  baremetal: ironic
  compute: nova
  container: zun
  container-infra: magnum
  database: trove
  identity: keystone
  image: glance
  key-manager: barbican
  load-balancer: octavia
```

```
  network: neutron
  object-store: swift
  orchestration: heat
  placement: placement
  sharev2: manilav2
  volumev3: cinder
 sso_enabled: false
 sso_protocols:
 - openid
 # 修改区域名
 sso_region: RegionWxic
 system_admin_roles:
 - admin
 - system_admin
 system_project: service
 system_project_domain: Default
 system_reader_roles:
 - system_reader
 system_user_domain: Default
 system_user_name: skyline
 # 用户密码
 system_user_password: 'wxic@yoga'
setting:
 base_settings:
 - flavor_families
 - gpu_models
 - usb_models
 flavor_families:
 - architecture: x86_architecture
   categories:
   - name: general_purpose
     properties: []
   - name: compute_optimized
     properties: []
   - name: memory_optimized
     properties: []
   - name: high_clock_speed
     properties: []
   - architecture: heterogeneous_computing
     categories:
     - name: compute_optimized_type_with_gpu
       properties: []
     - name: visualization_compute_optimized_type_with_gpu
       properties: []
   gpu_models:
   - nvidia_t4
   usb_models:
   - usb_c
MXD
```

4. 运行 Skyline 服务

运行初始化引导容器生成数据库 Skyline 的表结构，并查看日志以验证数据库能否正常连接和表结构能否正常生成，命令和结果如下。

```
[root@controller ~]# docker run -d --name skyline_bootstrap \
-e KOLLA_BOOTSTRAP="" \
-v /etc/skyline/skyline.yaml:/etc/skyline/skyline.yaml \
```

```
-v /var/log:/var/log \
--net=host 99cloud/skyline:latest
6d78f3a1c491f199000d7a4cc03d785486ac62b58e29c14e6b924ef1efa28f58
# 查看skyline_bootstrap容器的日志
[root@controller ~]# docker logs -f skyline_bootstrap
+ echo '/usr/local/bin/gunicorn -c /etc/skyline/gunicorn.py skyline_apiserver.
main:app'
+ mapfile -t CMD
++ tail /run_command
++ xargs -n 1
+ [[ -n 0 ]]
+ cd /skyline-apiserver/
+ make db_sync
alembic -c skyline_apiserver/db/alembic/alembic.ini upgrade head
2023-04-17 08:02:51.271 | INFO     | alembic.runtime.migration:__init__:205 -
Context impl MySQLImpl.
2023-04-17 08:02:51.272 | INFO     | alembic.runtime.migration:__init__:208 - Will
assume non-transactional DDL.
2023-04-17 08:02:51.281 | INFO     | alembic.runtime.migration:run_migrations:619
- Running upgrade  -> 000, init
+ exit 0
```

该容器的作用是生成Skyline服务的数据库的表结构，经查询可知生成了alembic_version、revoked_token、settings这3张表的表结构，结果如下。

```
MariaDB [skyline]> show tables;
+---------------------+
| Tables_in_skyline |
+---------------------+
| alembic_version  |
| revoked_token   |
| settings        |
+---------------------+
3 rows in set (0.000 sec)

MariaDB [skyline]> describe alembic_version;
+----------+------------+----------+----------------+-----------------+--------+
|  Field   |   Type     |  Null    |      Key       |    Default      | Extra  |
+----------+------------+----------+----------------+-----------------+--------+
| version_ |  Varchar   |   NO     |      PRI       |     NULL        |        |
|   num    |   (32)     |          |                |                 |        |
+----------+------------+----------+----------------+-----------------+--------+
1 row in set (0.001 sec)

MariaDB [skyline]> describe revoked_token;
+----------+------------+----------+----------------+-----------------+--------+
|  Field   |   Type     |  Null    |      Key       |    Default      | Extra  |
+----------+------------+----------+----------------+-----------------+--------+
|   uuid   |  Varchar   |   NO     |      MUL       |     NULL        |        |
|          |   (128)    |          |                |                 |        |
|  expire  |  int(11)   |   NO     |                |     NULL        |        |
+----------+------------+----------+----------------+-----------------+--------+
2 rows in set (0.001 sec)

MariaDB [skyline]> describe settings;
+----------+------------+----------+----------------+-----------------+--------+
|  Field   |   Type     |  Null    |      Key       |    Default      | Extra  |
```

```
+----------+------------+----------+----------------+-----------------+--------+
|   key    |  Varchar   |   NO     |      MUL       |      NULL       |        |
|          |   (128)    |          |                |                 |        |
|  value   | longtext   |   YES    |                |      NULL       |        |
+----------+------------+----------+----------------+-----------------+--------+
2 rows in set (0.001 sec)
```

表结构生成成功后，删除数据库初始化引导容器 skyline_bootstrap，命令和结果如下。

```
[root@controller ~]# docker rm -f skyline_bootstrap
skyline_bootstrap
```

运行 skyline-apiserver 服务容器 skyline，设置重启策略为 always，并挂载配置文件和日志目录的数据卷，将容器连接的网络设置为主机的网络，命令和结果如下。

```
[root@controller ~]# docker run -d --name skyline --restart=always \
-v /etc/skyline/skyline.yaml:/etc/skyline/skyline.yaml \
-v /var/log:/var/log \
-e LISTEN_ADDRESS=192.168.100.11:9942 \
--net=host 99cloud/skyline:latest
bffe9d5a70c144716e8cabf1940e092f0560ad6e62bfa5218c49eca62c7416bd
# 查询容器运行日志
[root@controller ~]# docker logs -f skyline
+ echo '/usr/local/bin/gunicorn -c /etc/skyline/gunicorn.py skyline_apiserver.main:app'
+ mapfile -t CMD
++ tail /run_command
++ xargs -n 1
+ [[ -n '' ]]
+ GENERATOR_ARGS='--output-file /etc/nginx/nginx.conf'
+ [[ -n 192.168.100.11:9942 ]]
+ GENERATOR_ARGS+=' --listen-address 192.168.100.11:9942'
+ [[ -n '' ]]
+  skyline-nginx-generator  --output-file  /etc/nginx/nginx.conf  --listen-address
192.168.100.11:9942
+ nginx
+ echo 'Running command: /usr/local/bin/gunicorn -c /etc/skyline/gunicorn.py
skyline_apiserver.main:app'
Running command: /usr/local/bin/gunicorn -c /etc/skyline/gunicorn.py skyline_
apiserver.main:app
+ exec /usr/local/bin/gunicorn -c /etc/skyline/gunicorn.py skyline_apiserver.main:app
[2023-04-17 08:05:07 +0000].710 23 DEBUG [-] Using selector: EpollSelector
[2023-04-17 08:05:07 +0000].743 24 DEBUG [-] Using selector: EpollSelector
```

Skyline 服务组件安装结束后，在浏览器中访问 http://192.168.100.11:9942 便可以进入图 3-9 所示的 Skyline 登录界面，输入用户名和密码进行登录后，可看到图 3-10 所示的 Skyline 首页。

图 3-9　Skyline 登录界面

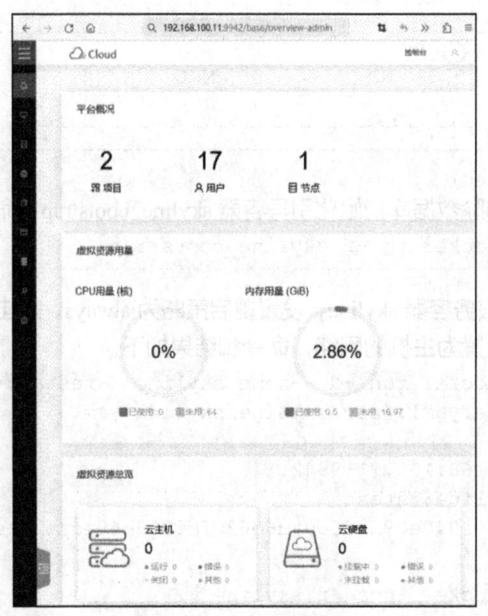

图 3-10　Skyline 首页

任务 3.2　典型云平台部署

本任务实施聚焦于利用先进的 Kolla-ansible 部署工具，在双节点部署模式下部署 OpenStack 集群，通过详尽的步骤展示高可用私有云平台的构建过程，为企业级私有云建设提供切实可行的技术参考与实践指导。

微课 3.5　典型云平台部署 1

微课 3.6　典型云平台部署 2

任务 3.2.1　规划节点

双节点部署 OpenStack 云平台时，各节点主机名和 IP 地址规划如表 3-3 所示。

表 3-3　各节点主机名和 IP 地址规划（双节点部署模式下）

节点类型	主机名	IP 地址规划	
		内部管理	实例通信
控制节点	controller	192.168.100.10	192.168.200.10
计算节点	compute01	192.168.100.20	192.168.200.20

任务 3.2.2　环境准备

在物理机上申请两台安装了 openEuler 22.09 操作系统的虚拟机分别作为 OpenStack 控制节点和计算节点，在设置处理器处勾选“虚拟化 IntelVT-x/EPT 或 AMD-V/RVI(V)”复选框。控制节点类型为 4 个 vCPU/16GB 内存/120GB 硬盘；计算节点类型为 2 个 vCPU/8GB 内存/120GB 硬盘及 4 个 20GB 额外硬盘；需要给虚拟机设置两个网络接口，网络接口 1 设置为内部网络，其网卡使用仅主机模式，用于控制节点的通信和管理，网络接口 2 设置为外部网络，其网卡使用 NAT 模式，主要用于控制节点的数据网络，在集群部署完成后，创建的云主机使用网络接口 2 的网卡。创建的控制节点和计算节点分别如图 3-11 和图 3-12 所示。

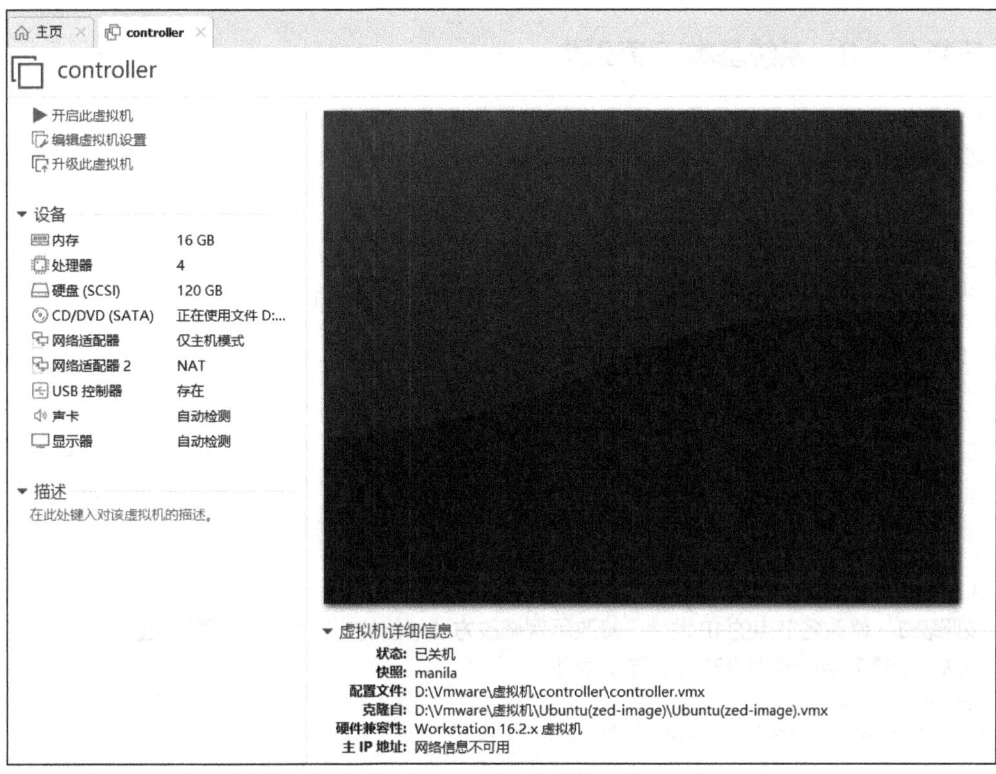

图 3-11 创建的控制节点

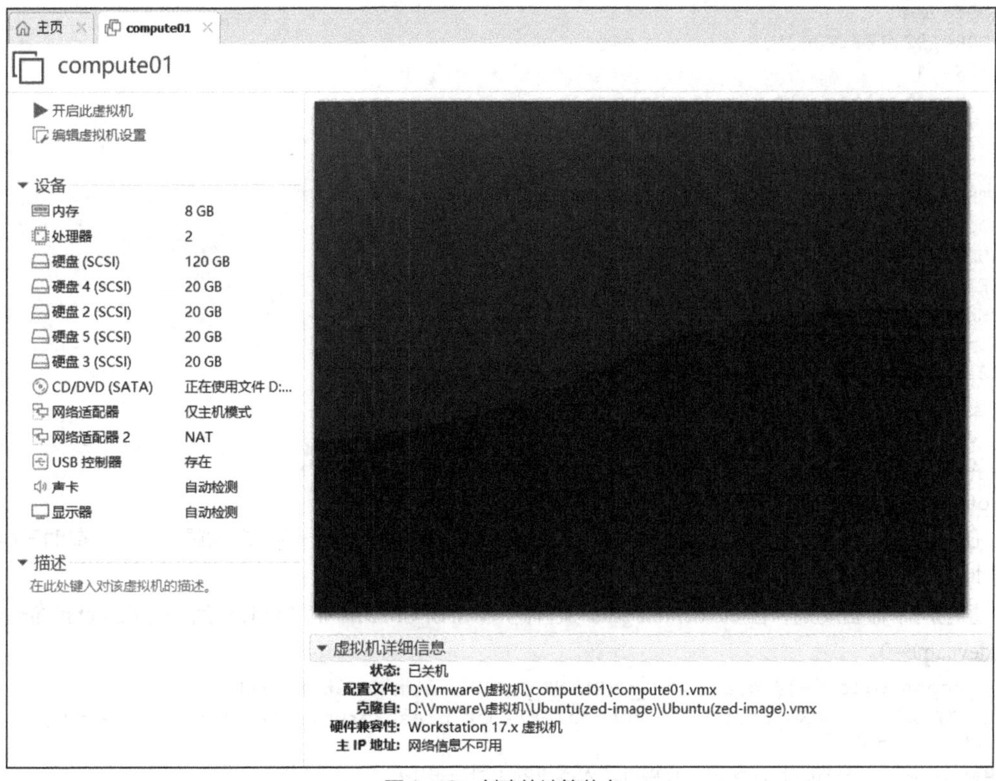

图 3-12 创建的计算节点

任务 3.2.3 系统基本环境设置

1. 修改主机名

修改主机名，命令如下。

```
# 控制节点
[root@localhost ~]# hostnamectl set-hostname controller
[root@localhost ~]# exec bash
[root@controller ~]#
# 计算节点
[root@localhost ~]# hostnamectl set-hostname compute01
[root@localhost ~]# exec bash
[root@compute01 ~]#
```

2. 更新系统软件包

所有节点更新系统软件包以获得最新的功能支持和漏洞修复。

```
[root@controller ~]# dnf -y update && dnf -y upgrade
[root@compute01 ~]# dnf -y update && dnf -y upgrade
```

3. 修改 IP 地址和网卡名

按照表 3-3 修改各节点的 IP 地址，修改示例修改方法给出的内容即可，下面给出控制节点的示例修改方法，计算节点的示例修改方法与此类似。

```
[root@controller ~]# cat \
/etc/sysconfig/network-scripts/ifcfg-ens160
……
BOOTPROTO=none
ONBOOT=yes
IPADDR=192.168.100.10
PREFIX=24
[root@controller ~]# cat \
/etc/sysconfig/network-scripts/ifcfg-ens224
……
BOOTPROTO=none
ONBOOT=yes
IPADDR=192.168.200.10
PREFIX=24
GATEWAY=192.168.200.2
DNS1=223.6.6.6
DNS2=119.29.29.29
```

载入网卡配置并启动相应的网卡。

```
[root@controller ~]# nmcli c reload
[root@controller ~]# nmcli c up ens160
[root@controller ~]# nmcli c up ens224
```

在使用 Kolla-ansible 部署 OpenStack 时，确保所有节点的网卡名一致是非常重要的，这有助于更加方便地进行脚本的维护和管理。

禁用网卡命名规则，在/etc/default/grub 文件的 GRUB_CMDLINE_LINUX 后面追加 net.ifnames=0 biosdevname=0。

```
[root@controller ~]# grep GRUB_CMDLINE_LINUX /etc/default/grub
GRUB_CMDLINE_LINUX="rd.lvm.lv=openeuler/root crashkernel=512M net.ifnames=0
biosdevname=0"
# 更新 GRUB 2 引导加载程序配置，并将其写入系统的引导记录中
```

```
[root@controller network-scripts]# grub2-mkconfig -o /boot/efi/EFI/openEuler/
grub.cfg
Generating grub configuration file ...
Found linux image: /boot/vmlinuz-5.10.0-106.18.0.68.oe2209.x86_64
Found initrd image: /boot/initramfs-5.10.0-106.18.0.68.oe2209.x86_64.img
Found linux image: /boot/vmlinuz-0-rescue-dc74f2bed8c44bf3adcbbfbe4e0d7039
Found initrd image: /boot/initramfs-0-rescue-dc74f2bed8c44bf3adcbbfbe4e0d7039.img
Adding boot menu entry for UEFI Firmware Settings ...
done
# 修改网卡名为 eth0、eth1
[root@controller ~]# cd /etc/sysconfig/network-scripts/
[root@controller network-scripts]# mv ifcfg-ens160 ifcfg-eth0
[root@controller network-scripts]# mv ifcfg-ens224 ifcfg-eth1
[root@controller network-scripts]# sed -i "s#ens160#eth0#" ifcfg-eth0
[root@controller network-scripts]# sed -i "s#ens224#eth1#" ifcfg-eth1
# 重启系统
[root@controller ~] reboot
# 验证网卡名是否修改成功
[root@controller ~]# nmcli connection show
NAME  UUID                                  TYPE      DEVICE
eth1  3e626e76-d8f6-3ea8-8c02-f1d22acf2b6f ethernet  eth1
eth0  64b87afc-b476-3601-a1bb-a6aa5fd15997 ethernet  eth0
```

可以发现，网卡名已经修改完成，重复这个操作，将所有节点的网卡名修改一致。

任务 3.2.4 安装 Ansible 和 Kolla-ansible

使用以下命令下载并安装 pip3。

```
[root@controller ~]# dnf -y install python3-pip
```

使用镜像源来加快 pip3 安装包的下载速度。

```
[root@controller ~]# mkdir .pip
[root@controller ~]# cat << WXIC > .pip/pip.conf
[global]
index-url = https://pypi.tuna.tsinghua.edu.cn/simple
[install]
trusted-host=pypi.tuna.tsinghua.edu.cn
WXIC
```

更新 Python 3 中的 pip3 工具到最新版本。

```
[root@controller ~]# pip3 install --ignore-installed --upgrade pip
```

使用以下命令安装 Ansible，并查看所安装的 Ansible 的版本号。

```
[root@controller ~]# pip3 install -U 'ansible>=4,<6'
[root@controller ~]# ansible --version
ansible [core 2.12.10]
```

安装 Kolla-ansible 和 Kolla-ansible 环境必需的依赖项。

```
#控制节点
[root@controller ~]# dnf -y install \
git python3-devel libffi-devel gcc openssl-devel python3-libselinux
[root@controller ~]# dnf -y install openstack-kolla-ansible
[root@controller ~]# kolla-ansible --version
14.2.0
#计算节点
[root@compute01 ~]# dnf -y install python3-libselinux
```

创建Kolla-ansible配置文件目录。

```
[root@controller ~]# mkdir -p /etc/kolla/{globals.d,config}
[root@controller ~]# chown $USER:$USER /etc/kolla
```

将inventory文件复制到/etc/ansible目录下。

```
[root@controller ~]# mkdir /etc/ansible
[root@controller ~]# cp \
/usr/share/kolla-ansible/ansible/inventory/* /etc/ansible
```

使用以下命令安装Ansible Galaxy的依赖项。

```
[root@controller ~]# pip3 install cryptography==38.0.4
[root@controller ~]# tar -xvf kolla-ansible-deps.tar.gz -C /root/.ansible
```

任务 3.2.5 Ansible 运行配置优化

对Ansible运行配置进行调优以加快执行速度，具体优化如下。

```
[root@controller ~]# cat << MXD > /etc/ansible/ansible.cfg
[defaults]
#SSH服务关闭密钥检测
host_key_checking=False
#如果不使用 sudo，则建议将其启用
pipelining=True
#执行任务的并发数
forks=100
timeout=800
#禁用警告
devel_warning = False
deprecation_warnings=False
#显示每个任务花费的时间
callback_whitelist = profile_tasks
#记录Ansible的输出，通过相对路径表示
log_path= wxic_cloud.log
#主机清单文件，通过相对路径表示
inventory = openstack_cluster
#命令执行环境，也可更改为/bin/bash
executable = /bin/sh
remote_port = 22
remote_user = root
#默认输出的详细程度
#可选值为 0、1、2、3、4 等
#值越大表示输出越详细
verbosity = 0
show_custom_stats = True
interpreter_python = auto_legacy_silent
[colors]
#成功的任务使用绿色显示
ok = green
#跳过的任务使用亮灰色显示
skip = bright gray
#警告使用亮紫色显示
warn = bright purple
```

```
[privilege_escalation]
become_user = root
[galaxy]
display_progress = True
MXD
```

使用 ansible-config view 命令查看修改后的配置。

```
[root@controller ~]# ansible-config view
[defaults]
#SSH 服务关闭密钥检测
host_key_checking=False
……
```

任务 3.2.6 Kolla-ansible 环境初始配置

1. 修改主机清单文件

进入/etc/ansible 目录，编辑 openstack_cluster 主机清单文件来指定集群节点的主机及其所属组。在这个主机清单文件中还可以指定控制节点连接集群各个节点的用户名、密码（注意，ansible_password 的值为 root 用户的密码，所有节点的 root 用户的密码不可以是纯数字）等。

```
[root@controller ~]# cd /etc/ansible/
[root@controller ansible]# awk '!/^#/ && !/^$/' multinode > openstack_cluster
[root@controller ansible]# cat -n openstack_cluster
  1  [all:vars]
  2  ansible_password=wxic@2024
  3  ansible_become=false
  4  [control]
  5  192.168.100.10
  6  [network]
  7  192.168.100.10
  8  [compute]
  9  192.168.100.20
 10  [monitoring]
 11  192.168.100.10
 12  [storage]
 13  192.168.100.20
……
……
```

在上面的主机清单文件中定义了 control、network、compute、monitoring 和 storage 这 5 个组，指定了集群各个节点需要承担的角色，在组 all:vars 中定义了全局变量，各组中会有一些变量配置信息，这些变量主要用来连接服务器。其中，ansible_password 用来指定登录服务器的用户的密码，ansible_become 用来指定是否使用 sudo 来执行命令，其他组的内容保持默认即可，不用修改。

2. 检查主机清单文件是否配置正确

使用以下命令测试各主机之间能否连通。

```
[root@controller ~]# dnf -y install sshpass
[root@controller ~]# ansible all -m ping
localhost | SUCCESS => {
  "ansible_facts": {
    "discovered_interpreter_python": "/usr/bin/python3.10"
  },
  "changed": false,
  "ping": "pong"
```

```
}
192.168.100.20 | SUCCESS => {
  "ansible_facts": {
    "discovered_interpreter_python": "/usr/bin/python3.10"
  },
  "changed": false,
  "ping": "pong"
}
192.168.100.10 | SUCCESS => {
  "ansible_facts": {
    "discovered_interpreter_python": "/usr/bin/python3.10"
  },
  "changed": false,
  "ping": "pong"
}
```

3. 配置 OpenStack 各服务组件的密码

在使用 Kolla-ansible 部署 OpenStack Yoga 平台时，建议使用随机密码生成器来生成各个服务组件的密码，命令如下。

```
[root@controller ~]# kolla-genpwd
```

修改 Horizon 界面登录密码为 wxic@2024，命令如下。

```
[root@controller ~]# sed -i \
's/keystone_admin_password: .*/keystone_admin_password: \
wxic@2024/g' /etc/kolla/passwords.yml
```

验证修改结果，命令如下。

```
[root@controller ~]# grep keystone_admin /etc/kolla/passwords.yml
keystone_admin_password: wxic@2024
```

4. 编辑 globals.yml 文件

在使用 Kolla-ansible 方式部署 OpenStack 平台时，最重要的操作之一便是 globals.yml 文件的修改，通过阅读 OpenStack 官方文档的服务指南，用户可以按照自己的需求选择安装相关的组件。

此次部署时安装了较多组件，具体的组件列表可查看修改后的 globals.yml 文件，其中需要注意 kolla_internal_vip_address 的地址，该 IP 地址为 192.168.100.0/24 网段中的任何一个未被使用的 IP 地址（此次部署时使用 IP 地址 192.168.100.100），部署完成后使用该 IP 地址登录 Horizon。

```
[root@controller ~]# cd /etc/kolla/
[root@controller kolla]# cp globals.yml{,_bak}
root@kolla-ansible:/etc/kolla# cat << MXD > globals.yml
---
kolla_base_distro: "ubuntu"
kolla_install_type: "source"
openstack_release: "yoga"
kolla_internal_vip_address: "192.168.100.100"
docker_registry: "quay.nju.edu.cn"
network_interface: "eth0"
neutron_external_interface: "eth1"
neutron_plugin_agent: "openvswitch"
openstack_region_name: "RegionWxic"
enable_aodh: "yes"
enable_barbican: "yes"
enable_ceilometer: "yes"
enable_ceilometer_ipmi: "yes"
enable_cinder: "yes"
enable_cinder_backup: "yes"
```

```
enable_cinder_backend_lvm: "yes"
enable_cloudkitty: "yes"
enable_gnocchi: "yes"
enable_gnocchi_statsd: "yes"
enable_manila: "yes"
enable_manila_backend_generic: "yes"
enable_neutron_vpnaas: "yes"
enable_neutron_qos: "yes"
enable_neutron_bgp_dragent: "yes"
enable_neutron_provider_networks: "yes"
enable_redis: "yes"
enable_swift: "yes"
glance_backend_file: "yes"
glance_file_datadir_volume: "/var/lib/glance/wxic/"
barbican_crypto_plugin: "simple_crypto"
barbican_library_path: "/usr/lib/libCryptoki2_64.so"
cinder_volume_group: "cinder-wxic"
cloudkitty_collector_backend: "gnocchi"
cloudkitty_storage_backend: "influxdb"
nova_compute_virt_type: "kvm"
swift_devices_name: "KOLLA_SWIFT_DATA"
MXD
```

在/etc/kolla/config 目录下自定义 Neutron 和 Manila 服务的一些配置，在部署集群时使用自定义的配置覆盖默认的配置。

```
[root@controller ~]# cd /etc/kolla/config/
[root@controller config]# mkdir neutron
[root@controller config]# cat << MXD > neutron/dhcp_agent.ini
[DEFAULT]
dnsmasq_dns_servers = 8.8.8.8,8.8.4.4,223.6.6.6,119.29.29.29
MXD
[root@controller config]# cat << MXD > neutron/ml2_conf.ini
[ml2]
tenant_network_types = flat,vxlan,vlan
[ml2_type_vlan]
network_vlan_ranges = provider:10:1000
[ml2_type_flat]
flat_networks = provider
MXD
[root@controller config]# cat << MXD > neutron/openvswitch_agent.ini
[securitygroup]
firewall_driver = openvswitch
[ovs]
bridge_mappings = provider:br-ex
MXD
[root@controller config]# cat << MXD > manila-share.conf
[generic]
service_instance_flavor_id = 100
MXD
```

任务 3.2.7 存储节点磁盘初始化

1. 初始化 Cinder 服务磁盘

在 compute01 节点上使用一块存储容量为 20GB 的磁盘创建 cinder-volumes 卷组，该卷组名和

globals.yml 中“cinder_volume_group”指定的参数一致。

```
[root@compute01 ~]# pvcreate /dev/nvme0n2
  Physical volume "/dev/nvme0n2" successfully created.
[root@compute01 ~]# vgcreate cinder-wxic /dev/nvme0n2
  Volume group "cinder-wxic" successfully created
[root@compute01 ~]# vgs cinder-wxic
  VG          #PV #LV #SN Attr   VSize   VFree
  cinder-wxic   1   0   0 wz--n- <20.00g <20.00g
```

2. 初始化 Swift 服务磁盘

在 compute01 节点上使用 3 块存储容量 20GB 的磁盘作为 Swift 存储设备的磁盘，并添加特殊的分区名称和文件系统标签，编写 Swift_disk_init.sh 脚本初始化磁盘。其中，设备名 KOLLA_SWIFT_DATA 和 globals.yml 文件中“swift_devices_name”指定的参数一致。

```
[root@compute01 ~]# cat << MXD > Swift_disk_init.sh
index=0
for d in nvme0n3 nvme0n4 nvme0n5; do
   parted /dev/${d} -s -- mklabel gpt mkpart KOLLA_SWIFT_DATA 1 -1
   sudo mkfs.xfs -f -L d${index} /dev/${d}p1
   (( index++ ))
done
MXD
[root@compute01 ~]# chmod +x Swift_disk_init.sh
[root@compute01 ~]# ./Swift_disk_init.sh
[root@compute01 ~]# lsblk
NAME             MAJ:MIN RM  SIZE RO TYPE MOUNTPOINTS
nvme0n3           259:5    0  20G  0 disk
└─nvme0n3p1       259:6    0  20G  0 part
nvme0n4           259:7    0  20G  0 disk
└─nvme0n4p1       259:8    0  20G  0 part
nvme0n5           259:9    0  20G  0 disk
└─nvme0n5p1       259:10   0  20G  0 part
[root@compute01 ~]# parted /dev/nvme0n3 print
Model: VMware Virtual NVMe Disk (nvme)
Disk /dev/nvme0n3: 21.5GB
Sector size (logical/physical): 512B/512B
Partition Table: gpt
Disk Flags:

Number  Start   End     Size    File system  Name              Flags
 1      1049kB  21.5GB  21.5GB  xfs          KOLLA_SWIFT_DATA
```

任务 3.2.8　部署集群环境

在 controller 节点上安装 OpenStack CLI 客户端。

```
[root@controller ~]# dnf -y install python3-openstackclient
```

为了使部署的 controller 节点网络路由正常工作，需要在 openEuler 操作系统中启用 IP 地址转发功能，修改 controller 和 compute01 节点的/etc/sysctl.conf 文件，并配置在系统启动时自动加载 br_netfilter 模块，具体操作如下。

```
#控制节点
[root@controller ~]# cat << MXD >> /etc/sysctl.conf
net.ipv4.ip_forward=1
```

```
net.bridge.bridge-nf-call-ip6tables=1
net.bridge.bridge-nf-call-iptables=1
MXD
# 临时加载模块，重启后模块失效
[root@controller ~]# modprobe br_netfilter
# 重新加载配置
[root@controller ~]# sysctl -p /etc/sysctl.conf
# 创建 yoga.service 文件，设置系统启动时自动加载 br_netfilter 模块
[root@controller ~]# cat << MXD > /usr/lib/systemd/system/yoga.service
[Unit]
Description=Load br_netfilter and sysctl settings for OpenStack

[Service]
Type=oneshot
RemainAfterExit=yes
ExecStart=/sbin/modprobe br_netfilter
ExecStart=/usr/sbin/sysctl -p

[Install]
WantedBy=multi-user.target
MXD
[root@controller ~]# systemctl enable --now yoga.service
Created symlink /etc/systemd/system/multi-user.target.wants/yoga.service →
/lib/systemd/system/yoga.service.
# 计算节点
[root@compute01 ~]# cat << MXD >> /etc/sysctl.conf
net.ipv4.ip_forward=1
net.bridge.bridge-nf-call-ip6tables=1
net.bridge.bridge-nf-call-iptables=1
MXD
# 临时加载模块，重启后模块失效
[root@compute01 ~]# modprobe br_netfilter
# 重新加载配置
[root@compute01 ~]# sysctl -p /etc/sysctl.conf
# 创建 yoga.service 文件，设置系统启动时自动加载 br_netfilter 模块
[root@compute01~]# cat << MXD > /usr/lib/systemd/system/yoga.service
[Unit]
Description=Load br_netfilter and sysctl settings for OpenStack

[Service]
Type=oneshot
RemainAfterExit=yes
ExecStart=/sbin/modprobe br_netfilter
ExecStart=/usr/sbin/sysctl -p

[Install]
WantedBy=multi-user.target
MXD
[root@compute01 ~]# systemctl enable --now yoga.service
Created symlink /etc/systemd/system/multi-user.target.wants/yoga.service →
/lib/systemd/system/yoga.service.
```

在控制节点上使用以下命令安装 OpenStack 集群所需要的基础依赖项并修改配置文件（如安装

Docker 和修改 Hosts 文件等），执行结果和用时如图 3-13 所示。

```
[root@controller ~]# kolla-ansible bootstrap-servers
```

```
PLAY RECAP ********************************************************************************************
192.168.100.10      : ok=26  changed=12  unreachable=0  failed=0  skipped=22  rescued=0  ignored=0
192.168.100.20      : ok=26  changed=15  unreachable=0  failed=0  skipped=22  rescued=0  ignored=0
localhost           : ok=2   changed=0   unreachable=0  failed=0  skipped=0   rescued=0  ignored=0

Sunday 16 April 2023  17:41:58 +0800 (0:00:00.131)       0:08:40.122 **********
===============================================================================
openstack.kolla.docker : Install packages ------------------------------------------------- 404.13s
openstack.kolla.docker_sdk : Install docker SDK for python --------------------------------- 86.40s
openstack.kolla.packages : Install packages ------------------------------------------------ 12.13s
openstack.kolla.docker_sdk : Install packages ---------------------------------------------- 2.93s
Gather facts ------------------------------------------------------------------------------- 1.99s
openstack.kolla.docker : Install docker rpm gpg key ---------------------------------------- 1.27s
openstack.kolla.packages : Remove packages ------------------------------------------------- 0.95s
openstack.kolla.baremetal : Disable firewalld ---------------------------------------------- 0.89s
openstack.kolla.baremetal : Change state of selinux ---------------------------------------- 0.82s
openstack.kolla.docker : Restart docker ---------------------------------------------------- 0.78s
openstack.kolla.docker : Write docker config ----------------------------------------------- 0.75s
openstack.kolla.docker : Ensure docker config directory exists ----------------------------- 0.64s
openstack.kolla.docker : Start and enable docker ------------------------------------------- 0.62s
openstack.kolla.baremetal : Ensure node_config_directory directory exists ------------------ 0.56s
openstack.kolla.baremetal : Generate /etc/hosts for all of the nodes ----------------------- 0.45s
openstack.kolla.baremetal : Ensure /etc/cloud/cloud.cfg exists ----------------------------- 0.39s
openstack.kolla.baremetal : Check if firewalld is installed -------------------------------- 0.39s
openstack.kolla.docker : Ensure yum repos directory exists --------------------------------- 0.37s
openstack.kolla.baremetal : Ensure localhost in /etc/hosts --------------------------------- 0.34s
openstack.kolla.baremetal : Ensure unprivileged users can use ping ------------------------- 0.33s
[root@controller ~]#
```

图 3-13　执行结果和用时（双节点部署模式下使用命令安装 OpenStack 集群所需要的基础依赖项并修改配置文件）

将国外镜像拉取地址修改为国内镜像拉取地址，以加速 Docker 镜像的拉取，编辑控制节点和计算节点的/etc/docker/daemon.json 文件，添加 registry-mirrors 部分的内容。下面给出控制节点的/etc/docker/daemon.json 文件的配置示例，计算节点进行同样的修改即可。

```
[root@controller ~]# cat /etc/docker/daemon.json
{
   "bridge": "none",
   "default-ulimits": {
     "nofile": {
       "hard": 1048576,
       "name": "nofile",
       "soft": 1048576
     }
   },
   "ip-forward": false,
   "iptables": false,
   "registry-mirrors": [
        "https://registry.docker-cn.com",
        "http://hub-mirror.c.163.com",
        "https://docker.mirrors.ustc.edu.cn",
        "https://mirror.ccs.tencentyun.com"
   ]
 }
[root@controller ~]# systemctl daemon-reload
[root@controller ~]# systemctl restart docker
```

在控制节点上生成 Swift 服务所需要的环，编写 Swift-init.sh 脚本，其中，STORAGE_NODES 的 IP 地址为 Swift 磁盘所在节点的 IP 地址。

```
[root@controller ~]# cat << MXD > Swift-init.sh
#!/usr/bin/env bash
STORAGE_NODES=(192.168.100.20)
KOLLA_SWIFT_BASE_IMAGE="kolla/swift-base:master-ubuntu-jammy"
mkdir -p /etc/kolla/config/swift
docker run \
 --rm \
```

```
 -v /etc/kolla/config/swift/:/etc/kolla/config/swift/ \
 $KOLLA_SWIFT_BASE_IMAGE \
 swift-ring-builder \
  /etc/kolla/config/swift/object.builder create 10 3 1

for node in ${STORAGE_NODES[@]}; do
  for i in {0..2}; do
   docker run \
    --rm \
    -v /etc/kolla/config/swift/:/etc/kolla/config/swift/ \
    $KOLLA_SWIFT_BASE_IMAGE \
    swift-ring-builder \
     /etc/kolla/config/swift/object.builder add r1z1-${node}:6000/d${i} 1;
    done
done
docker run \
 --rm \
 -v /etc/kolla/config/swift/:/etc/kolla/config/swift/ \
 $KOLLA_SWIFT_BASE_IMAGE \
 swift-ring-builder \
  /etc/kolla/config/swift/account.builder create 10 3 1

for node in ${STORAGE_NODES[@]}; do
  for i in {0..2}; do
   docker run \
    --rm \
    -v /etc/kolla/config/swift/:/etc/kolla/config/swift/ \
    $KOLLA_SWIFT_BASE_IMAGE \
    swift-ring-builder \
     /etc/kolla/config/swift/account.builder add r1z1-${node}:6001/d${i} 1;
  done
done
docker run \
 --rm \
 -v /etc/kolla/config/swift/:/etc/kolla/config/swift/ \
 $KOLLA_SWIFT_BASE_IMAGE \
 swift-ring-builder \
  /etc/kolla/config/swift/container.builder create 10 3 1

for node in ${STORAGE_NODES[@]}; do
  for i in {0..2}; do
   docker run \
    --rm \
    -v /etc/kolla/config/swift/:/etc/kolla/config/swift/ \
    $KOLLA_SWIFT_BASE_IMAGE \
    swift-ring-builder \
     /etc/kolla/config/swift/container.builder add r1z1-${node}:6002/d${i} 1;
  done
done
for ring in object account container; do
  docker run \
   --rm \
   -v /etc/kolla/config/swift/:/etc/kolla/config/swift/ \
   $KOLLA_SWIFT_BASE_IMAGE \
   swift-ring-builder \
```

```
    /etc/kolla/config/swift/${ring}.builder rebalance;
done
MXD
root@kolla-ansible:/etc/ansible# ./Swift-init.sh
```

在控制节点上进行部署前检查，openEuler 22.09 操作系统执行过程中如果出现“openEuler release NA version 22.09 is not supported. Supported releases are: 20.03”提示，则不必理会，继续进行接下来的操作即可，它不会影响后面的部署，执行结果和用时如图 3-14 所示。

```
[root@controller ~]# kolla-ansible prechecks
```

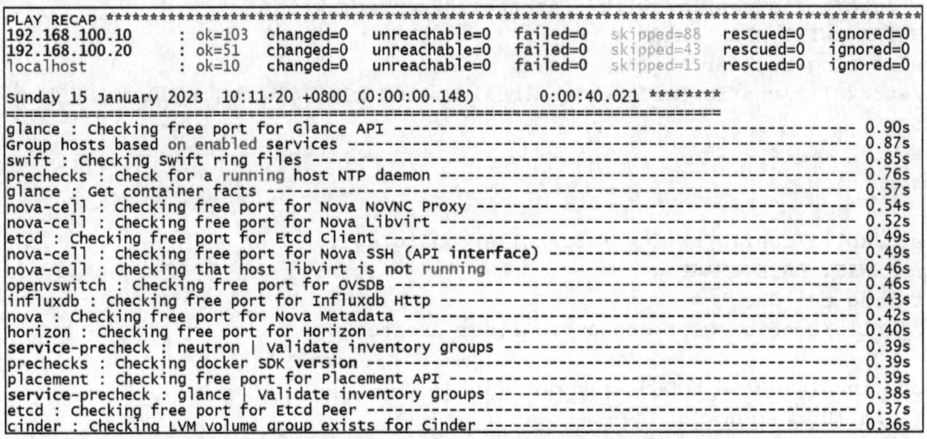

```
PLAY RECAP ***********************************************************************************************
192.168.100.10      : ok=103  changed=0   unreachable=0   failed=0   skipped=88   rescued=0   ignored=0
192.168.100.20      : ok=51   changed=0   unreachable=0   failed=0   skipped=43   rescued=0   ignored=0
localhost           : ok=10   changed=0   unreachable=0   failed=0   skipped=15   rescued=0   ignored=0

Sunday 15 January 2023  10:11:20 +0800 (0:00:00.148)       0:00:40.021 ********
===============================================================================
glance : Checking free port for Glance API ------------------------------------------------------ 0.90s
Group hosts based on enabled services ---------------------------------------------------------- 0.87s
swift : Checking Swift ring files -------------------------------------------------------------- 0.85s
prechecks : Check for a running host NTP daemon ------------------------------------------------ 0.76s
glance : Get container facts ------------------------------------------------------------------- 0.57s
nova-cell : Checking free port for Nova NoVNC Proxy -------------------------------------------- 0.54s
nova-cell : Checking free port for Nova Libvirt ------------------------------------------------ 0.52s
etcd : Checking free port for Etcd Client ------------------------------------------------------ 0.49s
nova-cell : Checking free port for Nova SSH (API interface) ------------------------------------ 0.49s
nova-cell : Checking that host libvirt is not running ------------------------------------------ 0.46s
openvswitch : Checking free port for OVSDB ----------------------------------------------------- 0.46s
influxdb : Checking free port for Influxdb Http ------------------------------------------------ 0.43s
nova : Checking free port for Nova Metadata ---------------------------------------------------- 0.42s
horizon : Checking free port for Horizon ------------------------------------------------------- 0.40s
service-precheck : neutron | Validate inventory groups ----------------------------------------- 0.39s
prechecks : Checking docker SDK version -------------------------------------------------------- 0.39s
placement : Checking free port for Placement API ----------------------------------------------- 0.39s
service-precheck : glance | Validate inventory groups ------------------------------------------ 0.38s
etcd : Checking free port for Etcd Peer -------------------------------------------------------- 0.37s
cinder : Checking LVM volume group exists for Cinder ------------------------------------------- 0.36s
```

图 3-14　执行结果和用时（双节点部署模式下进行部署前检查）

在控制节点上使用以下命令下载OpenStack集群各个节点所需要的全部镜像,执行结果和用时如图3-15所示。

```
[root@controller ~]# docker pull 99cloud/skyline:latest
[root@controller ~]# kolla-ansible pull
```

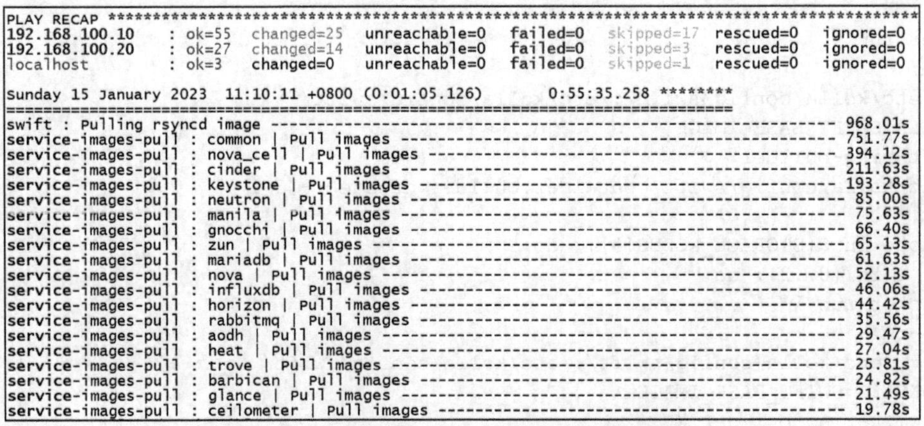

```
PLAY RECAP ***********************************************************************************************
192.168.100.10      : ok=55   changed=25   unreachable=0   failed=0   skipped=17   rescued=0   ignored=0
192.168.100.20      : ok=27   changed=14   unreachable=0   failed=0   skipped=3    rescued=0   ignored=0
localhost           : ok=3    changed=0    unreachable=0   failed=0   skipped=1    rescued=0   ignored=0

Sunday 15 January 2023  11:10:11 +0800 (0:01:05.126)       0:55:35.258 ********
===============================================================================
swift : Pulling rsyncd image ------------------------------------------------------------------ 968.01s
service-images-pull : common | Pull images ---------------------------------------------------- 751.77s
service-images-pull : nova_cell | Pull images ------------------------------------------------- 394.12s
service-images-pull : cinder | Pull images ---------------------------------------------------- 211.63s
service-images-pull : keystone | Pull images -------------------------------------------------- 193.28s
service-images-pull : neutron | Pull images ---------------------------------------------------- 85.00s
service-images-pull : manila | Pull images ----------------------------------------------------- 75.63s
service-images-pull : gnocchi | Pull images ---------------------------------------------------- 66.40s
service-images-pull : zun | Pull images -------------------------------------------------------- 65.13s
service-images-pull : mariadb | Pull images ---------------------------------------------------- 61.63s
service-images-pull : nova | Pull images ------------------------------------------------------- 52.13s
service-images-pull : influxdb | Pull images --------------------------------------------------- 46.06s
service-images-pull : horizon | Pull images ---------------------------------------------------- 44.42s
service-images-pull : rabbitmq | Pull images --------------------------------------------------- 35.56s
service-images-pull : aodh | Pull images ------------------------------------------------------- 29.47s
service-images-pull : heat | Pull images ------------------------------------------------------- 27.04s
service-images-pull : trove | Pull images ------------------------------------------------------ 25.81s
service-images-pull : barbican | Pull images --------------------------------------------------- 24.82s
service-images-pull : glance | Pull images ----------------------------------------------------- 21.49s
service-images-pull : ceilometer | Pull images ------------------------------------------------- 19.78s
```

图 3-15　执行结果和用时（双节点部署模式下下载镜像）

在控制节点上执行以下命令部署 OpenStack 集群，执行结果和用时如图 3-16 所示。

```
[root@controller ~]# kolla-ansible deploy
```

当上述步骤完成后，OpenStack 集群部署完成，所有的服务已经启动并可正常运行。在浏览器地址栏中输入 globals.yml 文件中定义的 kolla_internal_vip_address 地址，进入 Horizon 登录界面，此时，用户名为 admin，密码为 passwords.yml 文件中 keystone_admin_password 的值 wxic@2024，双节点部署模式下的 Horizon 登录界面和概览界面分别如图 3-17 和图 3-18 所示。

```
PLAY RECAP *************************************************************************************
192.168.100.10      : ok=457  changed=197  unreachable=0  failed=0  skipped=235  rescued=0  ignored=0
192.168.100.20      : ok=150  changed=94   unreachable=0  failed=0  skipped=92   rescued=0  ignored=0
localhost           : ok=3    changed=0    unreachable=0  failed=0  skipped=1    rescued=0  ignored=0

Sunday 15 January 2023  11:45:58 +0800 (0:00:00.498)       0:15:24.725 ********
===============================================================================
glance : Running Glance bootstrap container ------------------------------------------- 35.64s
gnocchi : Running gnocchi bootstrap container ----------------------------------------- 21.88s
trove : Running trove bootstrap container --------------------------------------------- 19.06s
keystone : Running Keystone bootstrap container --------------------------------------- 18.55s
aodh : Running aodh bootstrap container ----------------------------------------------- 17.66s
barbican : Running barbican bootstrap container --------------------------------------- 17.54s
neutron : Running Neutron bootstrap container ----------------------------------------- 16.55s
neutron : Restart neutron-dhcp-agent container ---------------------------------------- 16.33s
etcd : Restart etcd container --------------------------------------------------------- 15.91s
ceilometer : Restart ceilometer-central container ------------------------------------- 15.74s
openvswitch : Restart openvswitch-db-server container --------------------------------- 15.61s
mariadb : Create MariaDB volume ------------------------------------------------------- 15.34s
rabbitmq : Creating rabbitmq volume --------------------------------------------------- 15.30s
common : Creating log volume ---------------------------------------------------------- 15.30s
ceilometer : Running Ceilometer bootstrap container ----------------------------------- 11.96s
heat : Running Heat bootstrap container ----------------------------------------------- 11.40s
service-ks-register : keystone | Creating services ------------------------------------ 10.47s
nova : Running Nova API bootstrap container ------------------------------------------- 9.00s
service-ks-register : barbican | Creating roles --------------------------------------- 8.45s
manila : Running Manila bootstrap container ------------------------------------------- 7.95s
```

图 3-16　执行结果和用时（双节点部署模式下部署 OpenStack 集群）

图 3-17　Horizon 登录界面（双节点部署模式下）

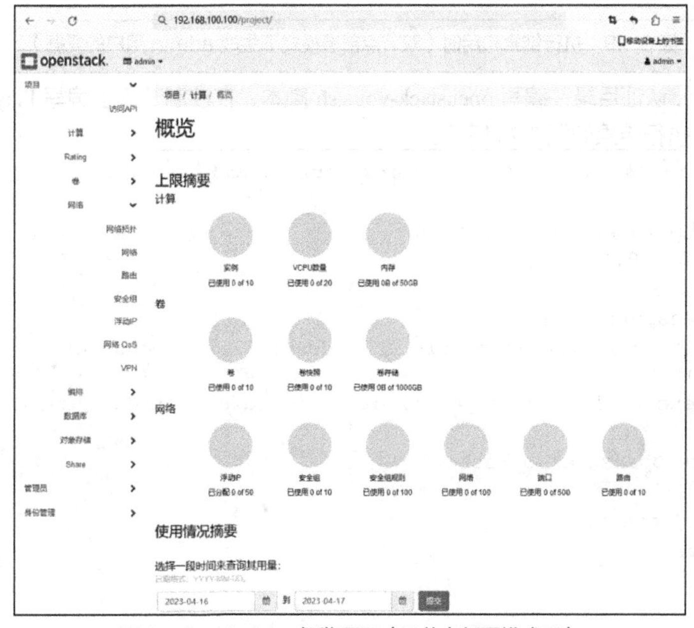

图 3-18　Horizon 概览界面（双节点部署模式下）

任务 3.2.9　OpenStack CLI 客户端设置

OpenStack 集群部署完成后，客户端执行命令之前需要生成 clouds.yaml 和 admin-openrc.sh 文件，这些是 admin 用户的凭据，执行结果和用时如图 3-19 所示。

```
[root@controller ~]# kolla-ansible post-deploy
```

```
[root@controller ~]# kolla-ansible post-deploy
Post-Deploying Playbooks : ansible-playbook -e @/etc/kolla/globals.yml  -e @/etc/kolla/passwords.yml
 -e CONFIG_DIR=/etc/kolla  /usr/local/share/kolla-ansible/ansible/post-deploy.yml
[WARNING]: Invalid characters were found in group names but not replaced, use -vvvv to see details

PLAY [Creating clouds.yaml file on the deploy node] ******************************************************

TASK [Gathering Facts] ***********************************************************************************
Monday 17 April 2023  12:14:24 +0800 (0:00:00.023)       0:00:00.023 **********
ok: [localhost]

TASK [Create /etc/openstack directory] *******************************************************************
Monday 17 April 2023  12:14:25 +0800 (0:00:01.231)       0:00:01.254 **********
changed: [localhost]

TASK [Template out clouds.yaml] **************************************************************************
Monday 17 April 2023  12:14:26 +0800 (0:00:00.305)       0:00:01.560 **********
changed: [localhost]

PLAY [Creating admin openrc file on the deploy node] *****************************************************

TASK [Gathering Facts] ***********************************************************************************
Monday 17 April 2023  12:14:26 +0800 (0:00:00.695)       0:00:02.256 **********
ok: [localhost]

TASK [Template out admin-openrc.sh] **********************************************************************
Monday 17 April 2023  12:14:27 +0800 (0:00:00.938)       0:00:03.194 **********
changed: [localhost]

TASK [octavia : Template out octavia-openrc.sh] **********************************************************
Monday 17 April 2023  12:14:28 +0800 (0:00:00.470)       0:00:03.665 **********
skipping: [localhost]

PLAY RECAP ***********************************************************************************************
localhost              : ok=5    changed=3    unreachable=0    failed=0    skipped=1    rescued=0    ignored=0

Monday 17 April 2023  12:14:28 +0800 (0:00:00.076)       0:00:03.742 **********
===============================================================================
Gathering Facts ------------------------------------------------------------------------------- 1.23s
Gathering Facts ------------------------------------------------------------------------------- 0.94s
Template out clouds.yaml ---------------------------------------------------------------------- 0.70s
Template out admin-openrc.sh ------------------------------------------------------------------ 0.47s
Create /etc/openstack directory --------------------------------------------------------------- 0.31s
octavia : Template out octavia-openrc.sh ------------------------------------------------------ 0.08s
[root@controller ~]#
```

图 3-19　执行结果和用时（双节点部署模式下生成 admin 用户的凭据）

创建 OpenStack 认证目录并编写 openstack-yoga.sh 脚本，在控制节点上编写 Playbook 脚本完成以上操作，Playbook 执行结果如图 3-20 所示。

```
[root@controller ~]# cat << MXD > openrc_script.yaml
---
- name: copy files and create script
  hosts: 192.168.100.10
  vars:
    files_to_transfer:
    - { src: '/etc/kolla/admin-openrc.sh', dest: '/etc/openstack/
admin-openrc.sh' }
    - { src: '/etc/kolla/clouds.yaml', dest: '/etc/openstack/clouds.yaml' }
  tasks:
  - name: Create openstack directory
    file:
      path: /etc/openstack
      state: directory

  - name: copy file
    copy:
```

```
      src: "{{ item.src }}"
      dest: "{{ item.dest }}"
    with_items: "{{ files_to_transfer }}"

  - name: Create script
    copy:
      content: |
        #!/usr/bin/env bash
        export OS_CLOUD=kolla-admin-internal
        source /etc/openstack/admin-openrc.sh
      dest: /etc/profile.d/openstack-yoga.sh
      mode: '0755'
MXD
[root@controller ~]# ansible-playbook openrc_script.yaml
```

```
[root@controller ~]# ansible-playbook openrc_script.yaml
[WARNING]: Invalid characters were found in group names but not replaced, use -vvvv to see details

PLAY [copy files and create script] ****************************************************************

TASK [Gathering Facts] *****************************************************************************
Monday 17 April 2023  12:42:09 +0800 (0:00:00.023)       0:00:00.023 **********
ok: [192.168.100.10]

TASK [Create openstack directory] ******************************************************************
Monday 17 April 2023  12:42:10 +0800 (0:00:01.663)       0:00:01.686 **********
ok: [192.168.100.10]

TASK [copy file] ***********************************************************************************
Monday 17 April 2023  12:42:11 +0800 (0:00:00.351)       0:00:02.038 **********
changed: [192.168.100.10] => (item={'src': '/etc/kolla/admin-openrc.sh', 'dest': '/etc/openstack/admin-openrc.sh'})
changed: [192.168.100.10] => (item={'src': '/etc/kolla/clouds.yaml', 'dest': '/etc/openstack/clouds.yaml'})

TASK [Create script] *******************************************************************************
Monday 17 April 2023  12:42:12 +0800 (0:00:01.481)       0:00:03.520 **********
changed: [192.168.100.10]

PLAY RECAP *****************************************************************************************
192.168.100.10             : ok=4    changed=2    unreachable=0    failed=0    skipped=0    rescued=0    ignored=0

Monday 17 April 2023  12:42:13 +0800 (0:00:00.712)       0:00:04.232 **********
===============================================================================
Gathering Facts ------------------------------------------------------------------------------- 1.66s
copy file ------------------------------------------------------------------------------------- 1.48s
Create script --------------------------------------------------------------------------------- 0.71s
Create openstack directory -------------------------------------------------------------------- 0.35s
[root@controller ~]#
```

图 3-20　Playbook 执行结果

使用 OpenStack 相关命令验证客户端是否可以正常使用。

```
# 重启终端后即可正常使用 OpenStack 相关命令
[root@controller ~]# logout
# 或者
# 按 Ctrl+D 快捷键
[root@controller ~]# openstack region list
+------------+---------------+-------------+
| Region     | Parent Region | Description |
+------------+---------------+-------------+
| RegionWxic | None          |             |
+------------+---------------+-------------+
[root@controller ~]# openstack container create wxic-cloud
+---------------------------------------+------------+
-----------------------------------+----
| account                               | container  |
x-trans-id                         |
+---------------------------------------+------------+
-----------------------------------+
| AUTH_0c79e73967184c52b495f145544d9995 | wxic-cloud |
txe8a166999bb141bba5483-00643d00be |
```

```
+-------------------------------------------+-------------
-----------------------------------+
[root@controller ~]# openstack container list
+------------+
| Name       |
+------------+
| wxic-cloud |
+------------+
```

可以发现，所有命令都可以正常使用。

使用 OpenStack 相关命令验证服务能否正常启动，命令和结果如下。

```
[root@controller ~]# openstack service list
+-----------------------+-----------------------------+-------------------------+
|          ID           |            Name             |           Type          |
+-----------------------+-----------------------------+-------------------------+
| 0662c7595fc243b9a34f6e4 |                           |                         |
|        870624ea0      |          octavia            |       load-balancer     |
| 3f61C46311e443b6a448abf |                           |                         |
|        36229ac38      |          keystone           |         identity        |
| 3f9939302b1f43edb8fb00c |                           |                         |
|        ae664b077      |            nova             |          compute        |
| 4e4595f549d8420d930b6b8 |                           |                         |
|        8611c65e2      |           glance            |           image         |
| 4fa87ac3448f44ed8a0519e |                           |                         |
|        5230efd78      |          gnocchi            |           metric        |
| 7a258eaa8d95463cbbf290c |                           |                         |
|        28ddb9241      |          cinderv3           |          volumev3       |
| 7dd49be322c34704a8eda4a |                           |                         |
|        8ab11d094      |          manilav2           |           sharev2       |
| 9445d80d06004826ba139bd |                           |                         |
|        db4c6063b      |          placement          |          placement      |
| a09b179f1a2a42e6876172d |                           |                         |
|        f0940bad2      |            heat             |        orchestration    |
| aa350fee7bb3402c89e4b33 |                           |                         |
|        1e1a507a1      |          neutron            |          network        |
| aaf95660f5ed4f38a096408 |                           |                         |
|        1579aebel      |           swift             |        object-store     |
| ac38d7162112422aa0e46b9 |                           |                         |
|        2c9ea8e7f      |         cloudkitty          |           rating        |
| c944cf8fde784174b51815b |                           |                         |
|        3f4eb39d1      |          heat-cfn           |       cloudformation    |
| df027b50206b4ee4bc69cc3 |                           |                         |
|        d7f91b3aa      |           trove             |          database       |
| e1229b3200fe4b90b448273 |                           |                         |
|        a66ae3dde      |            aodh             |          alarming       |
| f09d2cf1b3cb499fb7ba932 |                           |                         |
|        477b64655      |           mani7a            |           share         |
+-----------------------+-----------------------------+-------------------------+
[root@controller ~]# openstack compute service list
+----------------+----------+-----------+---------+---------+------+-----------+
|       ID       |  Binary  |   Host    |  Zone   | Status  |State | Updated At|
+----------------+----------+-----------+---------+---------+------+-----------+
| 327c7b0c-440f- |  nova-   |           |         |         |      | 2023-04-17|
| 4262-8799-     |          |controller |internal | enabled |  up  |T04:48:25. |
| f30be415cf41   |scheduler |           |         |         |      |   000000  |
```

```
   c6db64de-b36c-                                                          2023-04-17
|    4f33-b7ce-    |    nova-     | controller | internal | enabled |  up  | T04:48:21.|
   4805af551abd       conductor                                             000000
   379904ab-2925-                                                          2023-04-17
|    451b-b9a2-    |    nova-     | compute01  |   nova   | enabled |  up  | T04:48:18.|
    cb338a9b2d47      compute                                               000000
+------------------+-------------+------------+----------+---------+------+-----------+
[root@controller ~]# openstack network agent list
+------------------+-------------+------------+----------+---------+------+-----------+
                                               Availab
|        ID        |    Agent    |    Host    |  ility   |  Alive  |State |  Binary   |
                        Type                      Zone
+------------------+-------------+------------+----------+---------+------+-----------+
   1aa2e1af-f3b1-                                                          neutron-
|    48c3-b7be-    |   Metadata  | controller |   None   |   :-)   |  UP  | metadata- |
   53319fae2b5b        agent                                                agent
   1c27cb46-0ade-                                                          neutron-
|    4352-add3-    |  L3 agent   | controller |   none   |   :-)   |  UP  | l3-agent  |
    0302e65122f2
                        BGP
   81802e2e-028a-      dynamic                                             neutron-
|    4ab8-a8e0-    |   routing   | controller |   None   |   :-)   |  UP  |   bgp-    |
    d7eb6a3e027c        agent                                              dragent
   8b6bac83-c5b0-       Open                                               neutron-
|    4079-a546-    |   vSwitch   | controller |   None   |   :-)   |  UP  |openvswitch|
    4f0de95d8f47        agent                                               -agent
   8dc54066-aa13-                                                          neutron-
|    4cef-8a6a-    |    DHCP     | controller |   none   |   :-)   |  UP  | ancp-agent|
    134305d1ea65        agent
   bf82f5fe-6738-       Open                                               neutron-
|    4fb5-926c-    |   vSwitch   | compute01  |   None   |   :-)   |  UP  |openvswitch|
    93e5585866e6        agent                                               -agent
+------------------+-------------+------------+----------+---------+------+-----------+
```

可以发现，所有命令都可以正常使用，且服务状态正常。

查看此次部署 OpenStack 版本的详细信息，包括各个组件的版本号，结果如下。

```
[root@controller ~]# openstack network agent list
+------------+----------+--------+----------+-------------------+--------+-------+
                Service                                              Min     Max
| Region Name |  Type   |Version |  Status  |     Endpoint      | Microv | Microv |
                                                                   ersion   ersion
+------------+----------+--------+----------+-------------------+--------+-------+
                 load-                          http://192.168.
| RegionWxic | balancer |  None  | CURRENT  |   100.100:9876/   |  None  |  None  |
                                                http://192.168.
| RegionWxic | identity |  3.14  | CURRENT  | 100.100:5000/v3/  |  None  |  None  |
                                                http://192.168.
| RegionWxic | compute  |  2.0   |SUPPORTED | 100.100:8774/v2/  |  None  |  None  |
                                                http://192.168.
| RegionWxic | compute  |  2.1   | CURRENT  |100.100:8774/v2.1/ |  2.1   |  2.93  |
                                                http://192.168.
| RegionWxic |  image   |  2.0   |SUPPORTED |  100.100:9292/v2  |  None  |  None  |
                                                http://192.168.
| RegionWxic |  image   |  2.1   |SUPPORTED |  100.100:9292/v2  |  None  |  None  |
```

```
| RegionWxic | image | 2.2 | SUPPORTED | http://192.168.100.100:9292/v2 | None | None |
| RegionWxic | image | 2.3 | SUPPORTED | http://192.168.100.100:9292/v2 | None | None |
| RegionWxic | image | 2.4 | SUPPORTED | http://192.168.100.100:9292/v2 | None | None |
| RegionWxic | image | 2.5 | SUPPORTED | http://192.168.100.100:9292/v2 | None | None |
| RegionWxic | image | 2.6 | SUPPORTED | http://192.168.100.100:9292/v2 | None | None |
| RegionWxic | image | 2.7 | SUPPORTED | http://192.168.100.100:9292/v2 | None | None |
| RegionWxic | image | 2.8 | SUPPORTED | http://192.168.100.100:9292/v2 | None | None |
| RegionWxic | image | 2.9 | SUPPORTED | http://192.168.100.100:9292/v2 | None | None |
| RegionWxic | image | 2.10 | SUPPORTED | http://192.168.100.100:9292/v2 | None | None |
| RegionWxic | image | 2.11 | SUPPORTED | http://192.168.100.100:9292/v2 | None | None |
| RegionWxic | image | 2.12 | SUPPORTED | http://192.168.100.100:9292/v2 | None | None |
| RegionWxic | image | 2.13 | SUPPORTED | http://192.168.100.100:9292/v2 | None | None |
| RegionWxic | image | 2.15 | CURRENT | http://192.168.100.100:9292/v2 | None | None |
| RegionWxic | metric | 1.0 | CURRENT | http://192.168.100.100:8041/v1/ | None | None |
| RegionWxic | block-storage | 3.0 | CURRENT | http://192.168.100.100:8776/v3/ | 3.0 | 3.70 |
| RegionWxic | shared-file-system | 1.0 | DEPRECATED | http://192.168.100.100:8786/v1/ | None | None |
| RegionWxic | shared-file-system | 2.0 | CURRENT | http://192.168.100.100:8786/v2/ | 2.0 | 2.73 |
| RegionWxic | placement | 1.0 | CURRENT | http://192.168.100.100:8780/ | 1.0 | 1.39 |
| RegionWxic | orchestration | 1.0 | CURRENT | http://192.168.100.100:8004/v1/ | None | None |
| RegionWxic | network | 2.0 | CURRENT | http://192.168.100.100:9696/v2.0/ | None | None |
| RegionWxic | object-store | 1.0 | CURRENT | http://192.168.100.100:8080/v1/ | None | None |
| RegionWxic | rating | 1.0 | CURRENT | http://192.168.100.100:8889/v1/ | None | None |
| RegionWxic | cloudFormation | 1.0 | CURRENT | http://192.168.100.100:8000/v1/ | None | None |
| RegionWxic | database | 1.0 | CURRENT | http://192.168.100.100:8779/v1.0/ | None | None |
| RegionWxic | alarm | 2.0 | CURRENT | http://192.168.100.100:8042/v2/ | None | None |
+------------+----------+--------+---------+------------------+--------+-------+
```

接下来安装 Skyline 组件，可以参考任务 3.1 中的内容进行安装，因为此次安装比单节点部署模式下安装时多了很多组件，所以在安装完成登录 Skyline 首页查看服务列表时，可以发现其功能更加丰富。Skyline 服务列表如图 3-21 所示。

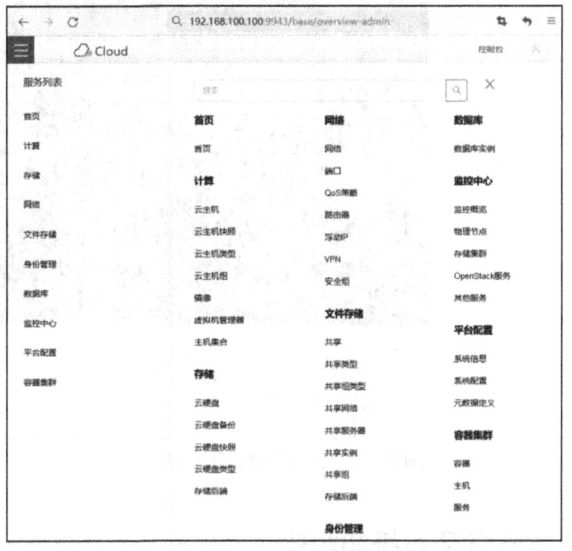

图 3-21　Skyline 服务列表

任务 3.3　OpenStack 云基础架构平台扩容

本任务为 OpenStack 基础平台进行扩容，核心在于在 OpenStack 已有环境中新增一个计算节点，通过详细的操作流程展示如何将新硬件资源融入 OpenStack 环境，以增强 OpenStack 平台整体的计算效率，提高其资源调度的灵活性，从而确保平台能够灵活应对日益攀升的应用处理压力，始终保持高效稳定的运行状态。

微课 3.7
OpenStack 云基础架构平台扩容

任务 3.3.1　规划节点

多节点部署 OpenStack 云平台时，各节点主机名和 IP 地址规划如表 3-4 所示。

表 3-4　各节点主机名和 IP 地址规划（多节点部署模式下）

节点类型	主机名	IP 地址规划	
		内部管理	实例通信
控制节点	controller	192.168.100.10	192.168.200.10
计算节点 1	compute01	192.168.100.20	192.168.200.20
计算节点 2	compute02	192.168.100.30	192.168.200.30

任务 3.3.2　环境准备

在任务 3.2 的基础上增加一个计算节点，其类型为 2 个 vCPU/8GB 内存/120GB 硬盘，在设置处理器处勾选“虚拟化 IntelVT-x/EPT 或 AMD-V/RVI(V)”复选框；同样，需要给虚拟机设置两个网络接口，网络接口 1 设置为内部网络，其网卡使用仅主机模式，用于控制节点通信和管理，网络接口 2 设置为外部网络，其网卡使用 NAT 模式，主要用于控制节点的数据网络。在集群部署完成后，创建的云主机使用网络接口 2 的网卡。创建新增计算节点 2，如图 3-22 所示。

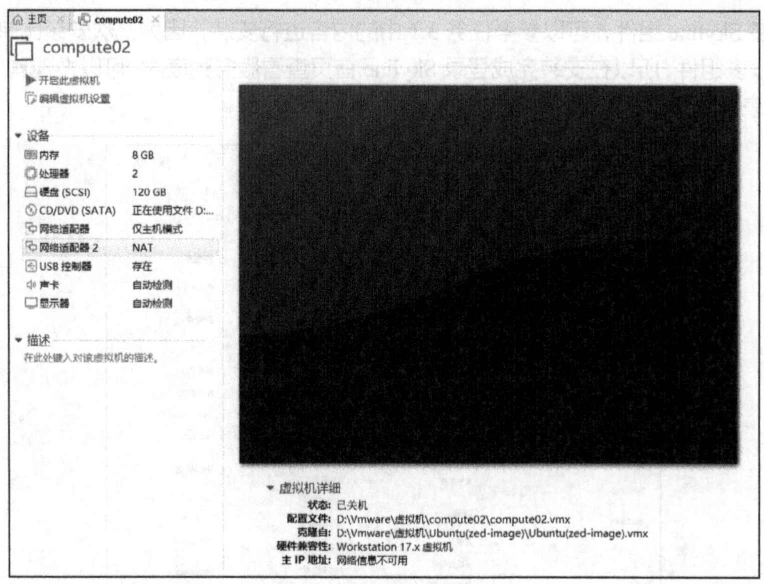

图 3-22 创建新增计算节点 2

任务 3.3.3 计算节点 2 系统初始化

1. 修改主机名

修改主机名，命令如下。

```
# 计算节点 2
[root@localhost ~]# hostnamectl set-hostname compute02
[root@localhost ~]# exec bash
[root@compute02 ~]#
```

2. 更新系统软件包

命令如下。

```
[root@compute02 ~]# dnf -y upgrade && dnf -y update
```

3. 修改 IP 地址和网卡名

按照表 3-4 修改计算节点 2 的 IP 地址。下面给出示例修改方法。

```
[root@compute02 ~]# cat \
/etc/sysconfig/network-scripts/ifcfg-ens160
......
BOOTPROTO=none
ONBOOT=yes
IPADDR=192.168.100.30
PREFIX=24
[root@compute02 ~]# cat \
/etc/sysconfig/network-scripts/ifcfg-ens224
......
BOOTPROTO=none
ONBOOT=yes
IPADDR=192.168.200.30
PREFIX=24
GATEWAY=192.168.200.2
DNS1=223.6.6.6
DNS2=119.29.29.29
```

载入网卡配置并启动相应的网卡。

```
[root@compute02 ~]# nmcli c reload
[root@compute02 ~]# nmcli c up ens160
[root@compute02 ~]# nmcli c up ens224
```

在任务 3.2 的 3.2.3 小节系统基本环境设置中已经说明了集群环境网卡名统一的必要性，故需要禁用计算节点 2 的网卡命名规则，使之与现有集群环境保持统一的网卡名。下面给出计算节点 2 的修改示例。

```
# 在 GRUB_CMDLINE_LINUX 的后面追加 net.ifnames=0 biosdevname=0
[root@compute02 ~]# grep GRUB_CMDLINE_LINUX /etc/default/grub
GRUB_CMDLINE_LINUX="rd.lvm.lv=openeuler/root crashkernel=512M net.ifnames=0
biosdevname=0"
# 更新 GRUB 2 引导加载程序配置，并将其写入系统的引导记录中
[root@compute02 ~]# grub2-mkconfig -o /boot/efi/EFI/openEuler/grub.cfg
Generating grub configuration file ...
Adding boot menu entry for UEFI Firmware Settings ...
done
# 修改网卡名为 eth0、eth1
[root@compute02 ~]# cd /etc/sysconfig/network-scripts/
[root@compute02 network-scripts]# mv ifcfg-ens160 ifcfg-eth0
[root@compute02 network-scripts]# mv ifcfg-ens224 ifcfg-eth1
[root@compute02 network-scripts]# sed -i "s#ens160#eth0#" ifcfg-eth0
[root@compute02 network-scripts]# sed -i "s#ens224#eth1#" ifcfg-eth1
# 重启系统
[root@compute02 ~] reboot
# 验证网卡名是否修改成功
[root@compute02 ~]# nmcli c show
NAME   UUID                                  TYPE      DEVICE
eth1  07400800-2242-32b9-88bb-5fcd323c9550 ethernet  eth1
eth0  139eb82e-8cee-3395-81e7-ce6f3cc5cc40 ethernet  eth0
```

可以发现，网卡名已经修改完成。

任务 3.3.4　集群增加计算节点 2

1. 修改主机清单文件

在控制节点上修改 openstack_cluster 主机清单文件，在组 compute 中加入计算节点 2 的 IP 地址和 root 用户的密码，修改后的文件如下。

```
[root@controller ~]# cat /etc/ansible/openstack_cluster
......

[compute]
192.168.100.20
192.168.100.30
......

# 验证 compute02 节点能否与 controller 节点连通
[root@controller ~]# ansible -m ping all
localhost | SUCCESS => {
  "ansible_facts": {
    "discovered_interpreter_python": "/usr/bin/python3"
  },
  "changed": false,
  "ping": "pong"
}
```

```
192.168.100.20 | SUCCESS => {
  "ansible_facts": {
    "discovered_interpreter_python": "/usr/bin/python3"
  },
  "changed": false,
  "ping": "pong"
}
192.168.100.30 | SUCCESS => {
  "ansible_facts": {
    "discovered_interpreter_python": "/usr/bin/python3"
  },
  "changed": false,
  "ping": "pong"
}
192.168.100.10 | SUCCESS => {
  "ansible_facts": {
    "discovered_interpreter_python": "/usr/bin/python3"
  },
  "changed": false,
  "ping": "pong"
}
```

2. 为现有集群加入 compute02 节点

为了使部署的计算节点 2 网络路由正常工作，需要在 openEuler 操作系统中启用 IP 地址转发功能，修改计算节点 2 的/etc/sysctl.conf 文件，并配置在系统启动时自动加载 br_netfilter 模块，具体操作如下。

```
[root@compute02 ~]# cat << MXD >> /etc/sysctl.conf
net.ipv4.ip_forward=1
net.bridge.bridge-nf-call-ip6tables=1
net.bridge.bridge-nf-call-iptables=1
MXD
# 临时加载模块，重启后模块失效
[root@compute02 ~]# modprobe br_netfilter
# 重新加载配置
[root@compute02 ~]# sysctl -p /etc/sysctl.conf
# 创建 yoga.service 文件，设置系统启动时自动加载 br_netfilter 模块
[root@compute02 ~]# cat /usr/lib/systemd/system/yoga.service
[Unit]
Description=Load br_netfilter and sysctl settings for OpenStack

[Service]
Type=oneshot
RemainAfterExit=yes
ExecStart=/sbin/modprobe br_netfilter
ExecStart=/usr/sbin/sysctl -p

[Install]
WantedBy=multi-user.target
[root@compute02 ~]# systemctl enable --now yoga.service
Created symlink /etc/systemd/system/multi-user.target.wants/yoga.service →
/lib/systemd/system/yoga.service.
```

在 controller 节点上使用以下命令安装计算节点 2 所需要的基础依赖项并修改配置文件，如在计算节点 2 上安装 Docker 和修改 Hosts 文件等，执行结果和用时如图 3-23 所示。

```
[root@controller ~]#  kolla-ansible --limit 192.168.100.30 bootstrap-servers
```

```
PLAY RECAP ***********************************************************************************************
192.168.100.10        : ok=29  changed=3   unreachable=0   failed=0   skipped=22  rescued=0   ignored=0
192.168.100.20        : ok=36  changed=3   unreachable=0   failed=0   skipped=20  rescued=0   ignored=0
192.168.100.30        : ok=39  changed=23  unreachable=0   failed=0   skipped=20  rescued=0   ignored=0
localhost             : ok=2   changed=0   unreachable=0   failed=0   skipped=0   rescued=0   ignored=0

Sunday 15 January 2023  17:04:23 +0800 (0:00:00.388)       0:01:30.907 ********
===============================================================================
openstack.kolla.packages : Remove packages ---------------------------------------------------- 25.70s
openstack.kolla.docker : Restart docker ------------------------------------------------------- 16.80s
openstack.kolla.docker : Install packages ----------------------------------------------------- 15.94s
openstack.kolla.docker_sdk : Install docker SDK for python ------------------------------------- 9.82s
openstack.kolla.docker : Enable docker apt repository ------------------------------------------ 4.85s
Gather facts ----------------------------------------------------------------------------------- 1.87s
openstack.kolla.docker : Start and enable docker ----------------------------------------------- 1.09s
openstack.kolla.docker : Install docker apt gpg key -------------------------------------------- 1.06s
openstack.kolla.packages : Install packages ---------------------------------------------------- 1.00s
openstack.kolla.docker : Install CA certificates and gnupg packages ---------------------------- 0.89s
openstack.kolla.docker_sdk : Install packages -------------------------------------------------- 0.83s
openstack.kolla.docker : Reload docker service file -------------------------------------------- 0.82s
openstack.kolla.baremetal : Set firewall default policy ---------------------------------------- 0.64s
openstack.kolla.docker : Write docker config --------------------------------------------------- 0.56s
openstack.kolla.baremetal : Disable cloud-init manage_etc_hosts -------------------------------- 0.56s
openstack.kolla.docker : Copying CNI config file ----------------------------------------------- 0.47s
openstack.kolla.docker : Copying over containerd config ---------------------------------------- 0.46s
openstack.kolla.baremetal : Generate /etc/hosts for all of the nodes --------------------------- 0.46s
openstack.kolla.docker : Configure docker service ---------------------------------------------- 0.46s
openstack.kolla.docker : Copy zun-cni script --------------------------------------------------- 0.42s
```

图 3-23 执行结果和用时（多节点部署模式下使用命令安装 OpenStack 所需要的基础依赖项和修改配置文件）

在计算节点 2 上验证 Docker 是否安装完毕，且 Hosts 文件是否已经修改完成。

```
[root@compute02 ~] docker -v
Docker version 18.09.0, build d1134d1
[root@compute02 ~] cat /etc/hosts
127.0.0.1 localhost
# BEGIN ANSIBLE GENERATED HOSTS
192.168.100.10 controller
192.168.100.20 compute01
192.168.100.30 compute02
# END ANSIBLE GENERATED HOSTS
```

配置 Docker 镜像拉取加速，编辑计算节点 2 的/etc/docker/daemon.json 文件，添加 registry-mirrors 部分的内容，计算节点 2 的修改示例如下。

```
[root@compute02 ~] cat /etc/docker/daemon.json
{
  "bridge": "none",
  "default-ulimits": {
    "nofile": {
      "hard": 1048576,
      "name": "nofile",
      "soft": 1048576
    }
  },
  "ip-forward": false,
  "iptables": false,
  "registry-mirrors": [
      "https://registry.docker-cn.com",
      "http://hub-mirror.c.163.com",
      "https://docker.mirrors.ustc.edu.cn",
      "https://mirror.ccs.tencentyun.com"
  ]
}
[root@compute02 ~] systemctl daemon-reload
[root@compute02 ~] systemctl restart docker
```

在 controller 节点上使用以下命令进行部署前检查，执行结果和用时如图 3-24 所示。

```
[root@controller ~]# kolla-ansible prechecks
```

在 controller 节点上使用以下命令下载计算节点 2 所需要的全部镜像，执行结果和用时如图 3-25 所示。

```
[root@controller ~]# kolla-ansible --limit 192.168.100.30 pull
```

```
PLAY RECAP ******************************************************************************************
192.168.100.10        : ok=70  changed=0  unreachable=0  failed=0  skipped=121  rescued=0  ignored=0
192.168.100.20        : ok=38  changed=0  unreachable=0  failed=0  skipped=56   rescued=0  ignored=0
192.168.100.30        : ok=38  changed=0  unreachable=0  failed=0  skipped=34   rescued=0  ignored=0
localhost             : ok=11  changed=0  unreachable=0  failed=0  skipped=14   rescued=0  ignored=0

Sunday 15 January 2023  17:10:55 +0800 (0:00:00.129)       0:00:37.192 ********
===============================================================================
Gather facts ----------------------------------------------------------- 1.72s
Group hosts based on enabled services ---------------------------------- 0.97s
swift : Checking Swift ring files -------------------------------------- 0.86s
prechecks : Checking timedatectl status -------------------------------- 0.85s
glance : Checking free port for Glance API ----------------------------- 0.74s
glance : Get container facts ------------------------------------------- 0.63s
nova-cell : Checking free port for Nova Libvirt ------------------------ 0.50s
nova-cell : Checking free port for Nova SSH (API interface) ------------ 0.50s
nova-cell : Checking that host libvirt is not running ------------------ 0.49s
openvswitch : Checking free port for OVSDB ----------------------------- 0.46s
cinder : Checking LVM volume group exists for Cinder ------------------- 0.44s
service-precheck : neutron | Validate inventory groups ----------------- 0.44s
iscsi : Checking free port for iscsi ----------------------------------- 0.43s
influxdb : Get container facts ----------------------------------------- 0.40s
zun : Get container facts ---------------------------------------------- 0.39s
service-precheck : glance | Validate inventory groups ------------------ 0.38s
prechecks : Checking docker SDK version -------------------------------- 0.37s
heat : Get container facts --------------------------------------------- 0.36s
service-precheck : nova | Validate inventory groups -------------------- 0.36s
etcd : Get container facts --------------------------------------------- 0.36s
```

图3-24 执行结果和用时（多节点部署模式下进行部署前检查）

```
PLAY RECAP ******************************************************************************************
192.168.100.30        : ok=21  changed=8  unreachable=0  failed=0  skipped=1  rescued=0  ignored=0

Sunday 15 January 2023  17:48:16 +0800 (0:00:25.834)       0:04:16.601 ********
===============================================================================
service-images-pull : nova_cell | Pull images -------------------------- 72.08s
service-images-pull : common | Pull images ----------------------------- 68.59s
service-images-pull : neutron | Pull images ---------------------------- 28.19s
service-images-pull : zun | Pull images -------------------------------- 25.83s
service-images-pull : ceilometer | Pull images ------------------------- 18.29s
service-images-pull : openvswitch | Pull images ------------------------ 16.39s
service-images-pull : kuryr | Pull images ------------------------------ 11.87s
service-images-pull : iscsi | Pull images ------------------------------- 8.26s
Gather facts ------------------------------------------------------------ 4.08s
Gather facts ------------------------------------------------------------ 1.62s
Group hosts based on enabled services ----------------------------------- 0.75s
Group hosts to determine when using --limit ----------------------------- 0.12s
nova-cell : Reload nova cell services to remove RPC version cap --------- 0.11s
neutron : include_tasks ------------------------------------------------- 0.05s
openvswitch : include_tasks --------------------------------------------- 0.05s
nova-cell : include_tasks ----------------------------------------------- 0.05s
zun : include_tasks ----------------------------------------------------- 0.04s
ceilometer : include_tasks ---------------------------------------------- 0.04s
common : include_tasks -------------------------------------------------- 0.04s
kuryr : include_tasks --------------------------------------------------- 0.04s
```

图3-25 执行结果和用时（多节点部署模式下下载镜像）

在controller节点上使用以下命令，将计算节点2加入现有OpenStack集群，执行结果和用时如图3-26所示。

```
[root@controller ~]# kolla-ansible deploy
```

```
PLAY RECAP ******************************************************************************************
192.168.100.10        : ok=429  changed=65  unreachable=0  failed=0  skipped=238  rescued=0  ignored=0
192.168.100.20        : ok=143  changed=29  unreachable=0  failed=0  skipped=90   rescued=0  ignored=0
192.168.100.30        : ok=104  changed=78  unreachable=0  failed=0  skipped=73   rescued=0  ignored=0
localhost             : ok=4    changed=0   unreachable=0  failed=0  skipped=0    rescued=0  ignored=0

Sunday 15 January 2023  18:03:35 +0800 (0:00:00.687)       0:12:28.808 ********
===============================================================================
common : Creating log volume ------------------------------------------- 15.32s
mariadb : Create MariaDB volume ---------------------------------------- 15.29s
rabbitmq : Creating rabbitmq volume ------------------------------------ 15.29s
iscsi : Restart tgtd container ----------------------------------------- 11.01s
neutron : Running Neutron bootstrap container --------------------------- 8.92s
service-ks-register : keystone | Creating services ---------------------- 8.59s
service-ks-register : barbican | Creating roles ------------------------- 8.32s
service-ks-register : keystone | Creating endpoints --------------------- 8.11s
service-ks-register : heat | Creating endpoints ------------------------- 7.10s
rabbitmq : Restart rabbitmq container ----------------------------------- 7.09s
nova-cell : Waiting for nova-compute services to register themselves ---- 6.97s
service-ks-register : manila | Creating endpoints ----------------------- 6.95s
neutron : Restart neutron-openvswitch-agent container ------------------- 6.73s
neutron : Restart neutron-server container ------------------------------ 6.71s
swift : Copying over config.json files for services --------------------- 6.27s
swift : Copying over swift.conf ----------------------------------------- 6.01s
nova : Running Nova API bootstrap container ----------------------------- 5.13s
heat : Running Heat bootstrap container --------------------------------- 5.02s
service-ks-register : heat | Creating roles ----------------------------- 5.00s
rabbitmq : Waiting for rabbitmq to start -------------------------------- 4.86s
```

图3-26 执行结果和用时（将计算节点2加入现有OpenStack集群）

任务3.3.5 验证计算节点2是否加入集群

1. 使用命令进行验证

使用以下命令查看OpenStack环境中计算服务（Nova）的状态，运行中的计算服务列表如图3-27所示，可以看到计算节点2处于运行状态。

```
[root@controller ~]# openstack compute service list -c Binary -c Host -c Status
+------------------+-------------+---------+
| Binary           | Host        | Status  |
+------------------+-------------+---------+
| nova-scheduler   | controller  | enabled |
| nova-conductor   | controller  | enabled |
| nova-compute     | compute01   | enabled |
| nova-compute     | compute02   | enabled |
+------------------+-------------+---------+
```

```
root@controller:~# openstack compute service list
+--------------------------------------+----------------+------------+----------+---------+-------+----------------------------+
| ID                                   | Binary         | Host       | Zone     | Status  | State | Updated At                 |
+--------------------------------------+----------------+------------+----------+---------+-------+----------------------------+
| 69fae9ee-ecc5-4911-a7ba-183a23516e0e | nova-scheduler | controller | internal | enabled | up    | 2023-01-15T10:16:35.000000 |
| d16b4432-275d-4f6b-8983-944ad88acf36 | nova-conductor | controller | internal | enabled | up    | 2023-01-15T10:16:32.000000 |
| e74de89a-4374-461c-90e5-2454a9bf2de9 | nova-compute   | compute01  | nova     | enabled | up    | 2023-01-15T10:16:32.000000 |
| 8a586214-2177-4acf-a61b-4a3239ce8cfb | nova-compute   | compute02  | nova     | enabled | up    | 2023-01-15T10:16:31.000000 |
+--------------------------------------+----------------+------------+----------+---------+-------+----------------------------+
```

图 3-27　运行中的计算服务列表

2. 在 Horizon 界面中查看计算节点的状态

在 Horizon 界面的左侧导航栏中选择“管理员→计算→虚拟机管理器”选项，在右侧可以查看计算节点的状态，如图 3-28 所示，此时发现计算节点 2 为激活状态，表示 OpenStack 云基础架构平台扩容成功。

图 3-28　在 Horizon 界面中查看计算节点的状态

项目小结

本项目首先介绍了 OpenStack 云平台的多种部署模式，其次针对目前较为流行的容器技术讲解了如何使用 Kolla-ansible 模式部署 OpenStack 云平台，最后深入介绍了 Ansible 的相关知识，以加深读者对 Kolla-ansible 的理解。

本项目在讲解了这些基础知识之后，首先通过 Kolla-ansible 模式，以单节点部署模式完成 OpenStack 的快速构建，使读者能快速体验云平台的相关服务；其次以双节点部署模式完成典型 OpenStack 的平台部署，使读者对云平台构建的工作流程有了全面的了解；最后在上述任务的基础上，讲解了如何添加计算节点，以实现云平台的扩容。

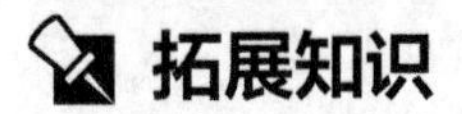

拓展知识

Ansible Playbook 基本编写方法

Playbook 是以 YAML（YAML Ain't a Markup Language）格式编写的文本文件，其通常使用的扩展名为.yml。Playbook 使用空格字符缩进表示其数据结构，YAML 对于空格数量没有要求。

Playbook 语法有如下特性。

（1）以---（3 个-）开始，必须顶行编写。

（2）次行开始编写 Playbook 的内容，但是一般要求写明该 Playbook 的功能。

（3）严格缩进，并且不能使用 Tab 键缩进。

（4）缩进级别必须一致，同样的缩进代表同样的级别，程序判别配置的级别是通过缩进结合换行来实现的。

（5）键值对的值可同行写，也可换行写。同行写以“:”分隔，换行写以“-”分隔。

Playbook 文本文件示例如下。

```
---
- name: Configure important user consistently
  hosts: servera.lab.example.com
  tasks:
   - name: newbie exists with UID 4000
     user:
       name:newbie
       uid:4000
       state: present
       - name:******  （可以忽略，但是一定要在前面加一个-，以表示这是一个新的命令）
       模块：***
```

Playbook 的---后的第一行以-加空格开头（表示该 Playbook 是列表的第一项）。

```
- name: Configure important user consistently
```

name 属性是可选的，它用于记录编写的 Playbook 的用途。

```
hosts: servera.lab.example.com
```

hosts 属性用于指定对其运行的 Playbook 中的任务的主机。

```
tasks:
  - name: newbie exists with UID 4000
    user:
      name:newbie
      uid:4000
      state: present
    - name:******  （可以忽略，但是一定要在前面加一个-，以表示这是一个新的命令）
      模块：***
```

tasks 属性用于指定要为 Playbook 运行的任务列表，user 是需要运行的模块，其参数是一组键值对，它们是模块的子项。

知识巩固

1. 单选

（1）OpenStack 手动部署支持两种部署架构，即（　　）和 multi-node。

A. 手工部署　　B. 自动化部署　　C. all-in-one　　D. Fuel

（2）OpenStack 部署工具有很多，不属于这个范畴的是（　　）。

A. Fuel　　B. host　　C. RDO　　D. PackStack

2. 填空

（1）在 Kolla-ansible 中部署 OpenStack Yoga 平台时，各个服务组件的密码存储在______文件中，此文件默认所有的密码是_______的，必须手动或者通过运行随机密码生成器来填写，在部署时建议使用_______生成器来生成各个服务组件的密码。

（2）Skyline 不仅提供了 OpenStack 基础服务，如计算、存储、网络的操作界面，还支持许多增值服务，如_______、对象存储、_______和数据库等服务。

3. 简答

（1）在云平台部署过程中，当安装 Ansible 和 Kolla-ansible 时为何要安装 pip3?

（2）简述 Ansible 运行配置优化过程。

（3）由于 Docker 默认的镜像拉取地址在国外，在国内拉取镜像的速度比较慢，如何加快镜像拉取?

拓展任务

云平台主机聚合

在任务 3.2 的基础上继续完成云平台主机聚合的实验。将控制节点的资源合并到计算节点中，以扩大云平台的资源池。

进入云平台，在界面左侧导航栏中选择“管理员→虚拟机管理器”选项，界面右侧显示的是未进行主机聚合的云平台资源概况，如图 3-29 所示。

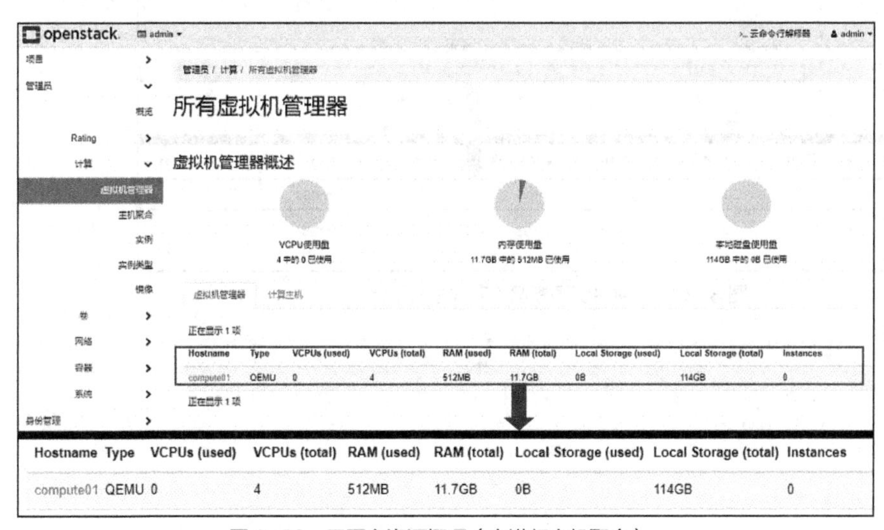

图 3-29　云平台资源概况（未进行主机聚合）

微课 3.8　云平台主机聚合

任务步骤

1. 在控制节点上修改 inventory 文件

命令如下。

```
[root@controller ~]# vi /etc/ansible/openstcak_cluster
# 在 compute 主机组下添加控制节点的 IP 地址
[compute]
# 原有的计算节点的 IP 地址
192.168.100.20
# 新添加的控制节点的 IP 地址
192.168.100.10
```

2. 将控制节点的资源加入集群中

命令如下。

```
# 在控制节点上下载 Nova 服务所需要的镜像
[root@controller ~]# kolla-ansible --limit 192.168.100.10 pull
# 在控制节点上部署 nova-compute 服务
[root@controller ~]# kolla-ansible deploy --tags nova
```

3. 云平台验证主机聚合

进入云平台，在界面左侧导航栏中选择“管理员→虚拟机管理器”选项，界面右侧显示的是完成主机聚合后的云平台资源概况，如图 3-30 所示。

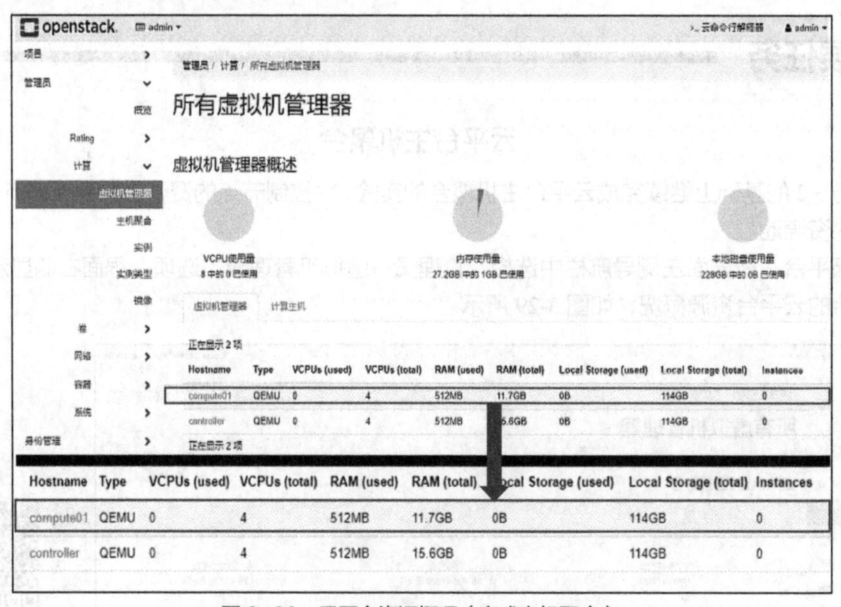

图 3-30 云平台资源概况（完成主机聚合）

项目4 云基础架构平台运维

04

学习目标

【知识目标】

① 学习云服务组件的定义。
② 学习云服务组件的基本概念。
③ 学习云服务组件的运维方法。

【技能目标】

① 掌握云服务组件基础运维技能。
② 具备云服务组件的运维和管理的综合能力。
③ 掌握更多高级服务组件的运维方法。

【素养目标】

① 培养创新创意的科学精神。
② 培养诚实守信的优良品质。
③ 培养认真严谨的工作作风。

项目概述

在前期的工作中，小张通过 Kolla-ansible 模式完整部署了 OpenStack 云平台并结合公司实际工作业务需求对平台进行了扩容。接下来小张要配合技术部门的员工进一步熟悉平台中的各种云服务组件，如云基础服务组件、存储服务组件和高级服务组件。小张需确保技术部门的员工在熟悉组件基本概念的基础上，掌握云服务组件的运维管理方法，熟练应用和管理 OpenStack 云平台。

知识准备

OpenStack 发布的一份 2022 年用户调查报告指出，OpenStack 部署在 2022 年达到了一个新的里程碑：现在拥有超过 4000 万个生产核心，与 2021 年相比增长了 60%，自 2020 年以来增长了 166%，且全球共有 300 多个公有云数据中心。由于 OpenStack 对混合云环境和 Kubernetes 集成支持的依赖增强，OpenStack 核心数量呈指数增长，目前 Kubernetes 部署在超过 85%的 OpenStack 部署中。OpenStack 核心服务（Nova、Neutron、Keystone、Glance 和 Ironic）的采用率仍然很高，但随着运营商发展其架构以适应新的工作负载，其逐渐转向 OpenStack 的支持服务，如 Octavia 和 Magnum。

鉴于 OpenStack 版本的快速迭代和更新，为了使该技术能更好地服务于未来的工作和生产，读者需要更深入地学习和实践 OpenStack-Yoga 的各类组件。接下来本项目将详细介绍 OpenStack 的云基础服务组件、存储服务组件和高级服务组件等基本概念。

4.1 云基础服务组件

OpenStack 是一个开源的云计算管理平台，由多个主要的组件组合起来完成具体工作。OpenStack 的云基础服务组件包括 Keystone 认证服务组件、Glance 镜像服务组件、Neutron 网络服务组件和 Nova 计算服务组件。

4.1.1 Keystone 认证服务

Keystone 是 OpenStack 平台框架中的一个重要组成部分，负责身份认证（通过 Identity API 实现）、服务管理、服务规则设定和服务令牌分发。Keystone 类似于服务总线，或者可将其看作整个 OpenStack 平台框架的注册表，其他服务通过 Keystone 来进行注册，任何服务之间的相互调用都需要经过 Keystone 的身份认证来完成。

Keystone 涉及用户、证书、认证、令牌、角色、租户和端点等主要概念。

4.1.2 Glance 镜像服务

Glance 依赖于存储服务和数据库服务，存储服务用于存储镜像（Image）本身，数据库服务主要用于存储与镜像相关的各种元数据。此外，Glance 主要为 Nova 计算服务组件提供服务，通过 Nova 创建虚拟机时，必须使用 Glance 获取相应的镜像，并根据镜像创建虚拟机实例。

Glance 支持的镜像格式较多，如 raw、qcow2、VHD、VMDK 和开放虚拟化格式（Open Virtualization Format，OVF）等。由于镜像文件都比较大，镜像从创建到成功上传到 Glance 中需要通过异步任务的方式完成。镜像状态包括 Queued（排队）、Saving（保存中）、Uploading（上传中）、Importing（导入中）、Active（有效）、Deactivated（无效）、Killed（错误）、Deleted（被删除）和 Pending_delete（等待删除），任务状态包括 Pending（等待中）、Processing（处理中）、Success（成功）和 Failure（失败）。

4.1.3 Neutron 网络服务

Neutron 是 OpenStack 核心项目之一，它提供云计算环境下的虚拟网络功能。它将网络（Network）、子网（Subnet）、端口（Port）和路由器（Router）抽象化为虚拟网络，并将启动的虚拟机实例连接到这个虚拟网络上。

Neutron 网络服务由网络和子网组成，其中网络是一个隔离的二层广播域。Neutron 支持多种类型的网络，包括本地网络（Local Network）、扁平网络（Flat Network）、虚拟局域网、虚拟拓展局域网（Virtual eXtensible Local Area Network，VXLAN）和通用路由封装（Generic Routing Encapsulation，GRE）。子网是一个 IPv4 或者 IPv6 地址段，虚拟机实例的 IP 地址从子网中分配。每个子网需要定义 IP 地址的范围和掩码。网络与子网是一对多关系，一个子网只能属于某个网络，而一个网络可以有多个子网。这些子网可以属于不同的 IP 地址段，但不能重叠。

4.1.4 Nova 计算服务

Nova 是 OpenStack 计算的弹性控制器，是整个云平台最重要的组件之一，其功能包括运行虚拟机实例、管理网络及通过用户和项目控制其对云的访问。

创建 Nova 实例一般先由用户通过 Web 界面或 CLI 界面发送一个启动实例的请求，由 Keystone 对其进行身份认证。通过认证后，用户将获得一个令牌，用户使用所获得的令牌向 Nova API 发送请求，验证镜像和实例类型（Flavor）是否存在。通过上述认证后该启动实例的请求被发送给计算服务的调度节点，调度节点随机将此请求发送给一个计算节点让其启动实例，计算节点接收请求之后，开始下载镜像并启动实例。计算节点启动虚拟机实例时会通过 Neutron 的动态主机配置协议（Dynamic Host Configuration Protocol，DHCP）获取一个 IP 网络资源，在 OVS（Open vSwitch）网桥上获取相应的端口绑定虚拟机实例的虚拟网卡接口，启动虚拟机实例。

4.2 存储服务组件

OpenStack 的存储服务组件包括 Cinder 块存储服务组件、Swift 对象存储服务组件和 Manila 共享文件系统服务组件。

4.2.1 Cinder 块存储服务

Cinder 是 OpenStack 块存储（OpenStack Block Storage）的项目名称，Cinder 的核心功能是对卷（Volume）的管理，它允许对卷、卷的类型、快照进行处理。但是，Cinder 并没有实现对块设备的管理和实际服务（提供逻辑卷），而是通过后端的统一存储接口来支持不同块设备厂商的块存储服务，实现其驱动（Driver）支持并与 OpenStack 进行整合。Cinder 可为运行实例提供稳定的数据块存储服务，并提供对卷从创建到删除的整个生命周期的管理。

Cinder 采用的是松散的架构理念，由 cinder-api 统一管理外部对 Cinder 的调用，由 cinder-scheduler 负责调度合适的节点来构建卷存储。volume-provider 通过驱动负责具体的存储空间，Cinder 内部依旧通过消息队列（Queue）沟通，解耦各子服务以支持异步调用。

Cinder 得到请求后会自动访问块存储服务，它有两个显著的特点，第一，必须由用户提出请求，服务才会进行响应；第二，用户可以使用自定义的方式实现半自动化服务。简而言之，Cinder 虚拟化块存储设备池，提供给用户自助服务的 API，使用户得以请求和使用块存储设备池中的资源，而 Cinder 本身并不能获取具体的存储形式或物理设备信息。

4.2.2 Swift 对象存储服务

Swift 最初是由 Rackspace 公司开发的高可用分布式对象存储服务（Object Storage Service，OSS），于 2010 年被贡献给 OpenStack 开源社区，作为其最初的核心子项目之一。Swift 为 OpenStack 的 Nova 子项目提供虚拟机镜像存储服务，它支持多租户模式、容器（Container）和对象读写操作，适合解决互联网应用场景下的非结构化数据存储问题。

Swift 包括两个组成部分，一个是代理（Proxy）服务，另一个是存储（Storage）服务。

代理服务是 Swift 内部存储的拓扑逻辑，即一个具体文件位于哪个存储节点的哪个区上。它也是一台 Web 服务器，可以通过超文本传输协议（Hypertext Transfer Protocol，HTTP）或超文本传输安全协议（Hypertext Transfer Protocol Secure，HTTPS）对外提供 RESTful API 服务。存储服务是负责文件存储的服务，由 3 个组件组成，分别是 object-server、container-server 和 account-server，其中，object-server 负责具体的文件存储，container-server 包含每个对象的索引，而 account-server 包含每个容器的索引。

4.2.3 Manila 共享文件系统服务

Manila 提供对共享或分布式文件系统的协调访问服务，它提供的是带有完整文件系统的存储服务，服务类型有网络文件系统（Network File System，NFS）、通用互联网文件系统（Common Internet File System，CIFS）、Gluster 文件系统（Gluster File System，GlusterFS）和 HDFS 等。云主机可以直接在系统中挂载 Manila 启动的实例。

Manila 的核心组件包括 manila-api、manila-data、manila-scheduler、manila-share 和 messaging queue 这 5 个组件。manila-api 是一个 Web 服务器网关接口（Web Server Gateway Interface，WSGI）应用程序，用于对整个共享文件系统服务进行身份认证和路由请求，它支持 OpenStack API。manila-data 是一个独立的服务，其目的是接收请求，处理潜在长时间运行的数据操作，如复制、共享迁移或备份。manila-scheduler 处理来自用户的请求，并将这些请求路由（分发）到适当的共享服务。manila-share 负责管理提供共享文件系统的后端设备。messaging queue 使共享文件系统的进程之间能够异步传递路由信息。

4.3 高级服务组件

OpenStack 的高级服务组件包括 Heat 编排服务组件、Ceilometer 监控服务组件和 Cloudkitty 计费服务组件。

4.3.1 Heat 编排服务

Heat 提供基于模板（Template）来编排复合云应用的服务。Heat 模板的使用简化了复杂的基础设施、服务和应用的定义及部署。Heat 支持丰富的资源类型，不仅覆盖了常用的基础架构资源（如计算、网络、存储、镜像），还覆盖了 Ceilometer 的警报、Sahara 的集群、Trove 的实例等高级资源。它可以基于模板来实现云环境中资源的初始化和依赖关系的处理、部署等基本操作，也具备解决自动收缩问题、维持负载均衡等高级特性。

OpenStack 以命令行和 Horizon 的方式为用户提供资源管理渠道，然而，这两种方式的工作效率并不高。即使把命令行保存为脚本，再输入/输出，依赖关系之间仍需要编写额外的脚本以进行维护，且不易于扩展。如果用户直接通过 RESTful API 编写程序，则同样会引发额外的复杂性问题。因此，这两种方式都不利于用户通过 OpenStack 管理批量资源和编排各种资源。

Heat 在这种情况下应运而生，它采用了业界流行的模板方式对资源进行设计和定义编排。用户只需要打开文本编辑器，编写一段基于键值对的模板，就能够方便地得到想要的编排方式。为了方便用户的使用，Heat 提供了大量的模板示例，通常用户选择想要的模板示例，并通过复制、粘贴的方式即可完成模板的编写。

Heat 的编排方式如下。

首先，用户可以通过编排 OpenStack 自身提供的基础架构资源来得到最基本的虚拟机。在编排虚拟机的过程中，用户可以编写简单脚本，以便对虚拟机进行简单的配置。

其次，用户可以通过 Heat 提供的 Software Configuration 和 Software Deployment 等资源类型对虚拟机进行复杂的配置，如安装软件和配置软件等。

最后，当用户有一些高级的功能需求（如需要一组能够根据负载自动伸缩的虚拟机，或者一组负载均衡的虚拟机）时，Heat 提供了 Autoscaling 和 Load Balance 等模板对其进行支持。在 Heat 中，只需要一定长度的模板，就可以满足这些高级的功能需求。

4.3.2 Ceilometer 监控服务

Ceilometer 是 OpenStack 的子项目，它能把 OpenStack 内部发生的大部分事件收集起来，为计费和监控及其他服务提供数据支撑。

Ceilometer 监控通过在计算节点上部署 Compute 服务，轮询其计算节点上的虚拟实例，获取各实例的 CPU、网络、磁盘等监控信息，并将这些监控信息发送到 RabbitMQ（开源的消息队列软件）中，由 Collector（Ceilometer 的核心组件）服务负责接收信息并对其进行持久化存储。Ceilometer 项目创建的最初目的是实现一个能为计费系统采集数据的框架。目前，OpenStack 社区已经更新了 Ceilometer 的最初目标，其新目标是希望 Ceilometer 成为 OpenStack 中数据采集（监控数据、计费数据）的唯一基础设施，采集到的数据提供给监控、计费、面板等项目使用。

4.3.3 Cloudkitty 计费服务

Cloudkitty 计费服务可完成对虚拟机实例、云硬盘、镜像、网络进出流量、浮动 IP 地址的计费，这得益于 Cloudkitty 巧妙而优秀的设计。Cloudkitty 将软件插件化，使得添加新的计费源十分容易，版本升级也十分方便。Cloudkitty 主要依赖于与遥测相关的项目，包括 Ceilometer 和 Gnocchi。计费策略和散列表（hashmap）计费模型是其核心，模块插件化是其设计灵魂。

当前 Cloudkitty 的整体架构包括计费服务的对象获取（Tenant Fetcher）、计费数据源的收集（Collector）、计费引擎（Rating）的实现和计费费用数据的存储（Storage）。

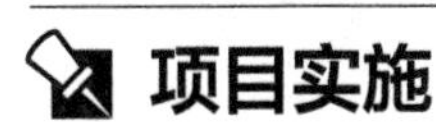

项目实施

任务 4.1 云基础服务组件运维管理

云基础服务组件运维管理涉及 Keystone 认证服务运维管理、Glance 镜像服务运维管理、Neutron 网络服务运维管理和 Nova 计算服务运维管理。

任务 4.1.1 Keystone 认证服务运维管理

微课 4.1 Keystone 认证服务运维管理

以 Kolla-ansible 部署的单节点 OpenStack 云平台为实验平台，配置并启用 Keystone 认证服务。假设 OpenStack 云平台上有两个租户，即租户 A 和租户 B，其分别属于两个部门，该公司对镜像的管理比较严格，只有管理员有权限对镜像进行管理。但公司有一个镜像需要共享给租户 A，而对租户 B 不可见。要实现这样的资源隔离最简单的方法之一就是由租户 A 中的用户自行上传镜像，这样租户 A 中的用户就能看到这个镜像，而租户 B 中的用户是看不到这个镜像的。但由于租户 A 中的用户是普通用户，无权管理镜像，只能由管理员通过命令将相应权限开放给租户 A，其中的用户才可管理镜像。接下来通过以下步骤实现上述目标。

1. 创建租户

创建租户 A，命令如下。

```
[root@controller ~]# openstack project create --domain default A
```

在 OpenStack 界面中查看租户 A，如图 4-1 所示。

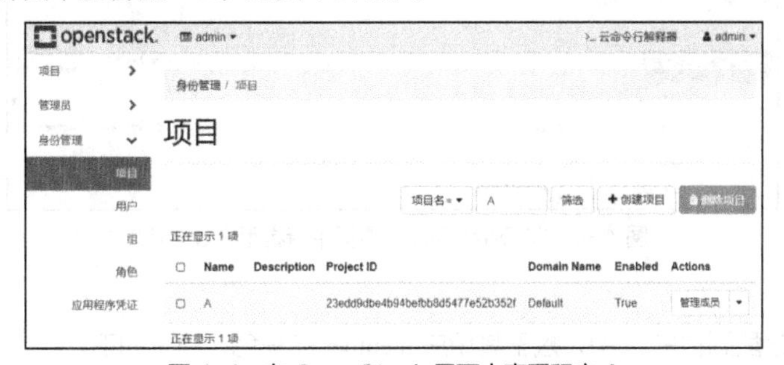

图 4-1 在 OpenStack 界面中查看租户 A

创建租户 B，命令如下。

```
[root@controller ~]# openstack project create --domain default B
```

在 OpenStack 界面中查看租户 B，如图 4-2 所示。

图 4-2 在 OpenStack 界面中查看租户 B

2. 创建用户

创建普通用户 userA，密码为 123456，命令如下。

```
[root@controller ~]# openstack user create --domain default \
--password 123456 userA
```

在 OpenStack 界面中查看用户 userA，如图 4-3 所示。

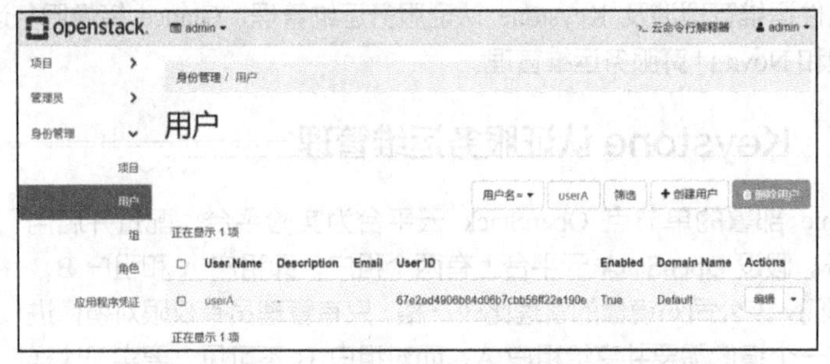

图 4-3 在 OpenStack 界面中查看用户 userA

创建普通用户 userB，密码为 123456，命令如下。

```
[root@controller ~]# openstack user create --domain default \
--password 123456 userB
```

在 OpenStack 界面中查看用户 userB，如图 4-4 所示。

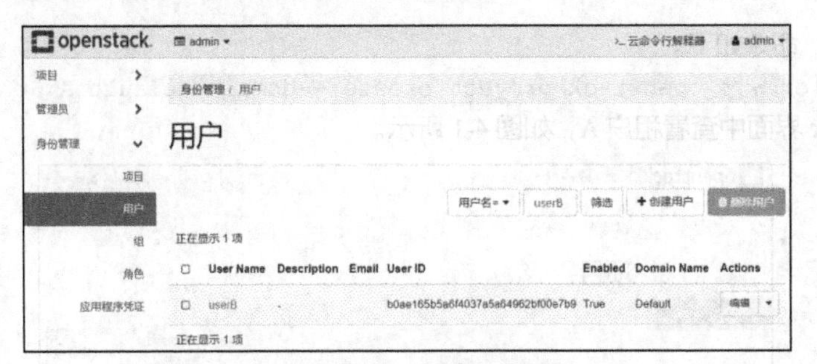

图 4-4 在 OpenStack 界面中查看用户 userB

3. 赋予角色

将用户 userA 分配到租户 A 中，赋予其用户 member 的角色，命令如下。

```
[root@controller ~]# openstack role add --project A --user userA member
```

将用户 userB 分配到租户 B 中，赋予其用户 member 的角色，命令如下。

```
[root@controller ~]# openstack role add --project B --user userB member
```

使用命令查询租户列表信息和用户列表信息，命令和结果如下。

```
[root@controller ~]# openstack project list
+----------------------------------+-----------+
| ID                               | Name    |
+----------------------------------+-----------+
| 06267a21b9494e31b2e30b1ee5f0fdc8 | admin   |
| 23edd9dbe4b94befbb8d5477e52b352f | A       |
| 92dec96a7ff14ff5a73ff70943df7fc3 | B       |
| c9f4c50ba58045929a90572f5701d7e9 | service  |
+----------------------------------+------------+
[root@controller ~]# openstack user list
+----------------------------------+----------+
| ID                               | Name   |
+----------------------------------+----------+
| 002279a051e7467a9abfcd6807effd71 | admin  |
……
| 67e2ed4906b84d06b7cbb56ff22a190e | userA  |
| b0ae165b5a6f4037a5a64962bf00e7b9 | userB  |
+----------------------------------+----------+
```

通过前面的操作可以看到两个租户和两个用户都已存在。

4. 上传镜像

将 cirros-0.6.1-x86_64-disk.img 下载至控制节点的/root 目录下，命令如下。

```
[root@controller ~]# wget \
https://github.com/cirros-dev/cirros/releases/download/0.6.1/cirros-0.6.1-x86_64-disk.img
```

创建一个 cirros-0.6.1 镜像并查看镜像列表，命令和结果如下。

```
[root@controller ~]# openstack image create --disk-format qcow2 \
--container-format bare \
--progress --file cirros-0.6.1-x86_64-disk.img "cirros-0.6.1"
[root@controller ~]# openstack image list
+--------------------------------------+--------------+--------+
| ID                                   | Name         | Status |
+--------------------------------------+--------------+--------+
| 24d60cc1-9396-47d3-bc54-5c31bf092c4d | cirros-0.6.1 | active |
+--------------------------------------+--------------+--------+
```

在 OpenStack 界面中查看 userA、userB 镜像内容，结果分别如图 4-5 和图 4-6 所示。

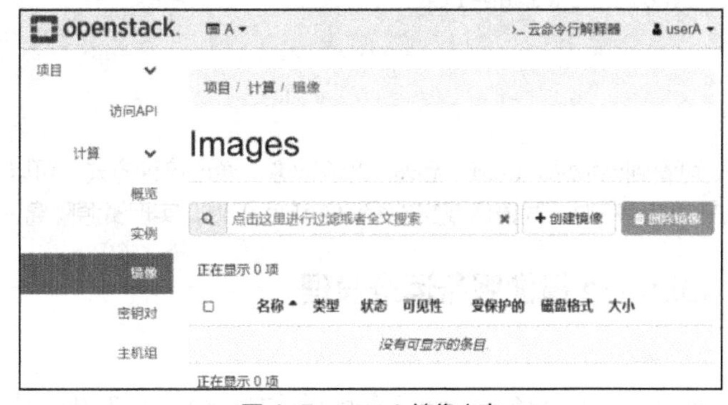

图 4-5　userA 镜像内容

图 4-6 userB 镜像内容

上传镜像后，userA 和 userB 都无法看到该镜像。接下来做相关配置，使得租户 A 中的用户可以看到该镜像。

5. 权限配置

将镜像共享给租户 A，命令格式为 openstack image add project < image name or ID > < project name or ID >，命令如下。

```
[root@controller ~]# openstack image add project cirros-0.6.1 A
```

在共享镜像之后，镜像的状态是 Pending，此时需要激活镜像，命令如下。

```
[root@controller ~]# openstack image set cirros-0.6.1 --project A --accept
[root@controller ~]# openstack image member list cirros-0.6.1
```

此时镜像的状态变为 Accepted，切换至用户 userA 和用户 userB 中，分别查看镜像列表信息，命令和结果如下。

```
[root@controller ~]# export OS_PROJECT_NAME=A
[root@controller ~]# export OS_USERNAME=userA
[root@controller ~]# export OS_PASSWORD=123456
[root@controller ~]# glance image-list
+--------------------------------------+--------------+
|                  ID                  |   Name   |
+--------------------------------------+--------------+
| 24d60cc1-9396-47d3-bc54-5c31bf092c4d | cirros-0.6.1 |
+--------------------------------------+--------------+
[root@controller ~]# export OS_PROJECT_NAME=B
[root@controller ~]# export OS_USERNAME=userB
[root@controller ~]# export OS_PASSWORD=123456
[root@controller ~]# glance image-list
+----+------+
| ID | Name |
+----+------+
+----+------+
```

可以发现，userA 可看到该镜像，userB 无法看到该镜像。通过这种方式，可以使管理员设置不同租户中的用户对不同镜像或文件资源的访问权限，实现资源隔离。

任务 4.1.2 Glance 镜像服务运维管理

微课 4.2 Glance 镜像服务运维管理

1. 上传镜像

下载镜像文件 openEuler-22.09-x86_64.qcow2 并创建名为 openEuler-22.09、格

式为 qcow2 的镜像，命令和结果如下。

```
[root@controller ~]# wget \
 https://mirrors.nju.edu.cn/openeuler/openEuler-22.09/virtual_machine_img/
x86_64/openEuler-22.09-x86_64.qcow2.xz
[root@controller ~]# xz -d openEuler-22.09-x86_64.qcow2.xz
[root@controller ~]# openstack image create --disk-format qcow2 \
--container-format bare \
--progress --file openEuler-22.09-x86_64.qcow2 \
"openEuler-22.09"
```

2. 查看镜像列表

使用相关命令查看镜像列表，并查看 openEuler-22.09 镜像的详细信息，命令和结果如下。

```
[root@controller ~]# openstack image list
+--------------------------------------+-----------------+--------+
| ID                                   | Name            | Status |
+--------------------------------------+-----------------+--------+
| 7f200158-bd9b-4e87-88fd-8042b21f198b | openEuler-22.09 | active |
+--------------------------------------+-----------------+--------+
```

3. 使用 Skyline 界面创建镜像

通过 Skyline 界面，使用镜像统一资源定位符（Uniform Resource Locator，URL）链接创建镜像，如图 4-7 所示。

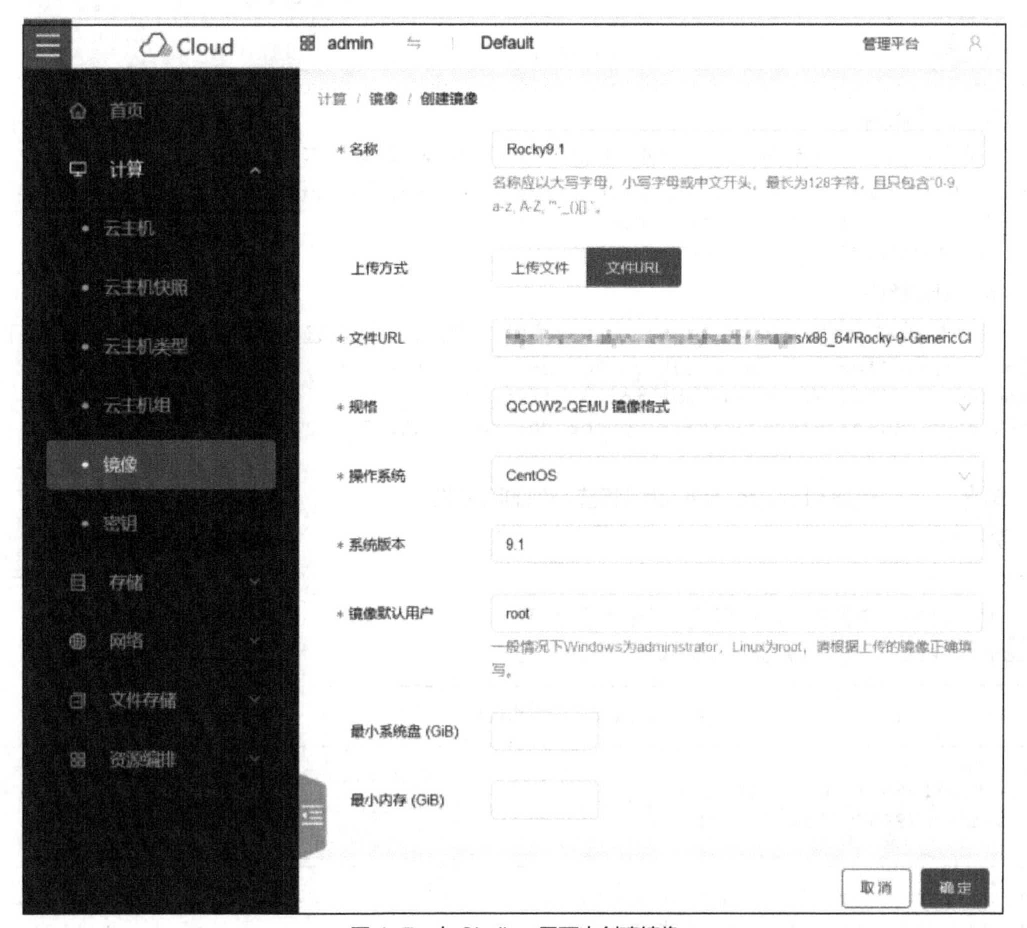

图 4-7 在 Skyline 界面中创建镜像

4. 删除镜像

使用相关命令删除刚刚创建的镜像，命令如下。

```
[root@controller ~]# openstack image \
delete 7f200158-bd9b-4e87-88fd-8042b21f198b
```

注意　删除的镜像 ID 为前面查询到的实际镜像 ID。

任务 4.1.3 Neutron 网络服务运维管理

微课 4.3 Neutron 网络服务运维管理

1. 创建网络

使用 OpenStack 相关命令创建名为 neutron-test 的网络，命令如下。

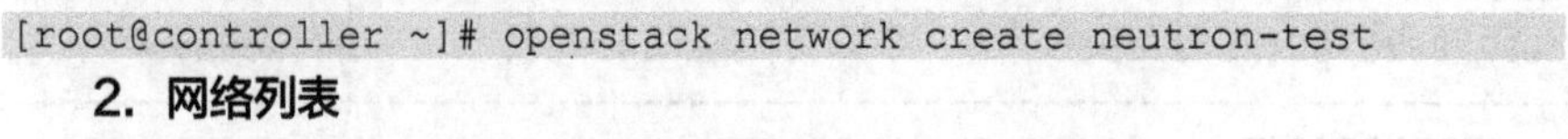

```
[root@controller ~]# openstack network create neutron-test
```

2. 网络列表

使用相关命令查询所创建网络的列表信息，命令和结果如下。

```
[root@controller ~]# openstack network list
+--------------------------------------+--------------+-----------+
| ID                                   | Name         | Subnets   |
+--------------------------------------+--------------+-----------+
| 5340f2e8-1227-4d53-94db-abd3de11d7f8 | neutron-test |           |
+--------------------------------------+--------------+-----------+
```

3. 删除网络

使用相关命令删除前面创建的网络，并进行验证，命令和结果如下。

```
[root@controller ~]# openstack network delete neutron-test
[root@controller ~]# openstack network list
```

此处并没有返回值，表示网络已经被删除。

4. 创建路由

使用相关命令创建一个名为 route-test 的路由，查看路由列表之后便删除该路由，命令如下。

```
[root@controller ~]# openstack router create route-test
[root@controller ~]# openstack router list
[root@controller ~]# openstack router delete [routeID]
```

注意　此处删除的[routeID]为实际查询到的路由 ID。

再次查看路由列表，具体示例命令和结果如下。

```
[root@controller ~]# openstack router list
+--------------------+-------+--------+-------+--------------+-------------+-------+
|        ID          | Name  | Status | State |   Project    | Distributed |  HA   |
+--------------------+-------+--------+-------+--------------+-------------+-------+
|                    |       |        |       | 06267a21b94  |             |       |
| 2e16a07f-6627-409a-| route-| ACTIVE |  UP   | 94e31b2e30b  |    True     | False |
|  aedf-d8794589f4b3 | test  |        |       | 1ee5f0fdc8   |             |       |
+--------------------+-------+--------+-------+--------------+-------------+-------+
[root@controller ~]# openstack router delete 2e16a07f-6627-409a-aedf-d8794589f4b3
[root@controller ~]# openstack router  list
[root@controller ~]#
```

任务 4.1.4　Nova 计算服务运维管理

1. 创建实例

微课 4.4　Nova 计算服务运维管理

创建实例类型，命令如下。

```
[root@controller ~]# openstack flavor create 2v_4G_20G \
--vcpus 2 --ram 4096 --disk 20
```

查看可用实例类型列表，命令如下。

```
[root@controller ~]# openstack flavor list
```

查看可用镜像列表，命令和结果如下。

```
[root@controller ~]# openstack image list
+--------------------------------------+--------------+--------+
| ID                                   | Name         | Status |
+--------------------------------------+--------------+--------+
| 55758bd0-031b-455d-aea8-ca5c9e9e19e0 | cirros-0.6.1 | active |
+--------------------------------------+--------------+--------+
```

查看可用网络列表，命令和结果如下。

```
[root@controller ~]# openstack network list
+--------------------------------------+---------+--------------------------------------+
| ID                                   | Name    | Subnets                              |
+--------------------------------------+---------+--------------------------------------+
| 68be109f-9f38-48c8-b65d-0f42f1b14f32 | int-net | 268cf0e6-6af2-45d2-98cf-9fc75bb45b3d |
| ccde2b6e-96ab-48a5-af7a-9dc26bf6294c | ext-net | e54e217c-6d99-4063-969e-3c9a5137570b |
+--------------------------------------+---------+--------------------------------------+
```

创建实例使用的安全组（Security Group），命令如下。

```
[root@controller ~]# openstack security group create secgroup01
[root@controller ~]# openstack security group list
```

创建用于连接实例的安全外壳（Secure Shell，SSH）密钥对并查看密钥对列表，命令和结果如下。

```
[root@controller ~]# openstack keypair create \
--public-key .ssh/id_rsa.pub controller_key
[root@controller ~]# openstack keypair list
+----------------+-------------------------------------------------+------+
| Name           | Fingerprint                                     | Type |
+----------------+-------------------------------------------------+------+
| controller_key | 61:99:00:64:39:c6:1e:c3:9b:e7:6a:42:58:b2:14:3e | ssh  |
+----------------+-------------------------------------------------+------+
```

创建实例 cirros-0.6.1，命令如下。

```
[root@controller ~]# netID=$(openstack network list | grep int-net | awk '{ print $2 }')
[root@controller ~]# openstack server create --flavor 2V_4G_20G --image
cirros-0.6.1 \
--security-group secgroup01 --nic net-id=$netID \
--key-name controller_key  "cirros-0.6.1"
```

查看外部网络列表，申请浮动 IP 地址并将其绑定至实例上。首先，查看外部网络列表，命令和结果如下。

```
[root@controller ~]# openstack network list --external
+--------------------------------------+---------+--------------------------------------+
| ID                                   |Name     |Subnets                               |
+--------------------------------------+---------+--------------------------------------+
| ccde2b6e-96ab-48a5-af7a-9dc26bf6294c | ext-net |e5e217c-6d99-4063-969e-3c9a5137570b |
+--------------------------------------+---------+--------------------------------------+
```

其次，申请浮动IP地址192.168.200.23，命令和结果如下。

```
[root@controller ~]# openstack floating ip create ext-net \
--floating-ip-address 192.168.200.23
+---------------------+--------------------------------------+
|        Field        |                 Value                |
+---------------------+--------------------------------------+
|     created_at      | 2023-01-28T09:21:50Z                 |
|     description     |                                      |
|     dns_domain      | None                                 |
|      dns_name       | None                                 |
|  fixed_ip_address   | None                                 |
| floating_ip_address | 192.168.200.23                       |
| floating_network_id | ccde2b6e-96ab-48a5-af7a-9dc26bf6294c |
|         id          | 53798a67-4501-454d-8d37-cf073ba24a42 |
|        name         | 192.168.200.23                       |
|    port_details     | None                                 |
|       port_id       | None                                 |
|     project_id      | d502ed6ca8304eb9a7eee463f5e8a924     |
|   qos_policy_id     | None                                 |
|  revision_number    | 0                                    |
|     router_id       | None                                 |
|       status        | DOWN                                 |
|     subnet_id       | None                                 |
|        tags         | []                                   |
|     updated_at      | 2023-01-28T09:21:50Z                 |
+---------------------+--------------------------------------+
```

最后，绑定浮动IP地址到实例上，命令和结果如下。

```
[root@controller ~]# openstack server list
+------------------+---------+--------+-------------+---------------+-----------+
|        ID        |  Name   | Status |  Networks   |     Image     |  Flavor   |
+------------------+---------+--------+-------------+---------------+-----------+
| 46ec44a0-ad70-42 | cirros- |        | int-net=    |               |           |
| b3-9d29-d7050c85 | 0.6.1   | ACTIVE | 10.0_0.185  | cirros-0.6.1  | 2V_4G_20G |
| 00fc             |         |        |             |               |           |
+------------------+---------+--------+-------------+---------------+-----------+
[root@controller ~]# openstack port list --device-id \
46ec44a0-ad70-42b3-9d29-d7050c8500fc
+-------------------+------+-------------+--------------------------------+--------+
|        ID         | Name | MAC Address |       Fixed IP Addresses       | Image  |
+-------------------+------+-------------+--------------------------------+--------+
| 1c7f3208-b375-495f- |    | fa:16:3e:   | ip_address='10.0.0.185,        |        |
| 985f-e67c596652bd |      | bb:2e:0c    | subnet_id='268cf0e6-6af2-45    | ACTIVE |
|                   |      |             | d2-98cf-9fc75bb45b3d'          |        |
+-------------------+------+-------------+--------------------------------+--------+
[root@controller ~]# openstack floating ip set --port \
1c7f3208-b375-495f-985f-e67c596652bd 192.168.200.23
[root@controller ~]# openstack server list
+------------------+---------+--------+-----------------+---------+-----------+
|        ID        |  Name   | Status |    Networks     |  Image  |  Flavor   |
+------------------+---------+--------+-----------------+---------+-----------+
| 46ec44a0-ad70-42 | cirros- |        | int-net=10.0.0  | cirros- |           |
| b3-9d29-d7050c85 | 0.6.1   | ACTIVE | .185, 192.168.  | 0.6.1   | 2V_4G_20G |
| 00fc             |         |        | 200.23          |         |           |
+------------------+---------+--------+-----------------+---------+-----------+
```

为安全组 secgroup01 添加互联网控制报文协议（Internet Control Message Protocol，ICMP）入口访问规则，命令和结果如下。

```
[root@controller ~]# openstack security group rule create --protocol icmp \
--ingress secgroup01
+---------------------------+--------------------------------------------------+
|          Field            |                    Value                         |
+---------------------------+--------------------------------------------------+
|        created_at         | 2023-01-28T09:33:24Z                             |
|        description        |                                                  |
|         direction         | ingress                                          |
|        ether_type         | IPv4                                             |
|            id             | 09ebf8ac-0bdd-4760-a48b-362d6c60e61b             |
|           name            | None                                             |
|      port_range_max       | None                                             |
|      port_range_min       | None                                             |
|        project_id         | d502ed6ca8304eb9a7eee463f5e8a924                 |
|         protocol          | icmp                                             |
|  remote_address_group_id  | None                                             |
|      remote_group_id      | None                                             |
|      remote_ip_prefix     | 0.0.0.0/0                                        |
|      revision_number      | 0                                                |
|     security_group_id     | 48624140-b3ff-4fab-a093-6725452d3e5a             |
|           tags            | []                                               |
|        updated_at         | 2023-01-28T09:33:24Z                             |
+---------------------------+--------------------------------------------------+
```

为安全组 secgroup01 添加传输控制协议（Transmission Control Protocol，TCP）入口访问规则，命令和结果如下。

```
[root@controller ~]# openstack security group rule create --protocol tcp \
--dst-port 22:22 secgroup01
+---------------------------+------------------------------------------------+
|          Field            |                    Value                       |
+---------------------------+------------------------------------------------+
|        created_at         |   2023-01-28T09:33:40Z                         |
|        description        |                                                |
|         direction         |   ingress                                      |
|        ether_type         |   IPv4                                         |
|            id             |   06eebd8b-60f3-4412-b66f-72e292583C49         |
|           name            |   None                                         |
|      port_range_max       |   22                                           |
|      port_range_min       |   22                                           |
|        project_id         |   d502ed6ca8304eb9a7eee463f5e8a924             |
|         protocol          |   tcp                                          |
|  remote_address_group_id  |   None                                         |
|      remote_group_id      |   None                                         |
|      remote_ip_prefix     |   0.0.0.0/0                                    |
|      revision_number      |   0                                            |
|     security_group_id     |   48624140-b3ff-4fab-a093-6725452d3e5a         |
|           tags            |   []                                           |
|        updated_at         |   2023-01-28T09:33:40Z                         |
+---------------------------+------------------------------------------------+
[root@controller ~]# openstack security group rule list secgroup01
+-----------+---------+----------+---------+-------+---------+---------+---------+
|           |   IP    |          |   IP    | Port  |         | Remote  | Remote  |
|    ID     |Protocol |Ethertype |  Range  | Range |Direction| Security| Address |
|           |         |          |         |       |         |  Group  |  Group  |
```

```
+-----------+--------+---------+--------+------+---------+---------+--------+
|  06eebd8b-|        |         |        |      |         |         |        |
|  60f3-4412-|  tcp  |  IPv4   |  0.0.  |22: 22| ingress |  None   |  None  |
|  b66f-72e29|       |         | 0.0/0  |      |         |         |        |
|  2583c49  |        |         |        |      |         |         |        |
|  09ebf8ac-|        |         |        |      |         |         |        |
|  0bdd-4760-|  icmp |  IPv4   |  0.0.  |      | ingress |  None   |  None  |
|  a48b-362d6|       |         | 0.0/0  |      |         |         |        |
|  c60e61b  |        |         |        |      |         |         |        |
|  45a6d6c9-|        |         |        |      |         |         |        |
|  dd43-4612-|  None |  IPv6   |  ::/0  |      | egress  |  None   |  None  |
|  b97d-e9abc|       |         |        |      |         |         |        |
|  99c00f4  |        |         |        |      |         |         |        |
|  94234529-|        |         |        |      |         |         |        |
|  e73b-44f4-|  None |  IPv4   |  0.0.  |      | egress  |  None   |  None  |
|  8358-6cd42|       |         | 0.0/0  |      |         |         |        |
|  4ad5076  |        |         |        |      |         |         |        |
+-----------+--------+---------+--------+------+---------+---------+--------+
```

使用SSH登录实例进行测试，命令和结果如下。

```
[root@controller ~]# openstack server list
+--------------------+--------+--------+--------------------+---------+----------+
|         ID         |  Name  | Status |      Networks      |  Image  |  Flavor  |
+--------------------+--------+--------+--------------------+---------+----------+
|                    |        |        |  int-net=10.0.0.   |         |          |
| 46ec44a0-ad70- 42b3-| cirros-| ACTIVE | 185,192.168.200.   | cirros- | 2v_4G_20G|
|  9d29-d7050c8500fc |  0.6.1 |        |        23          |  0.6.1  |          |
+--------------------+--------+--------+--------------------+---------+----------+
[root@controller ~]# ssh cirros@192.168.200.23
The authenticity of host '192.168.200.23 (192.168.200.23)' can't be established.
ED25519 key fingerprint is SHA256:Kv3rZ9TlzPokBJ+RSPGukzvv0kMs p6e46eiAeqx/e7I.
This key is not known by any other names
Are you sure you want to continue connecting. (yes/no/ [fingerprint])? yes
warning: Permanently added '192.168.200.23' (ED25519) to the list of known hosts.
$ ping mi.com -c 3
PING 58.83.160.156 (58.83.160.156) 56(84) bytes of data.
64 bytes from 58.83.160.156(58.83.160.156): icmp_seq=1 ttl=127 time=65.1 ms
64 bytes from 58.83.160.156(58.83.160.156): icmp_seq=2 ttl=127 time=78.9 ms
64 bytes from 58.83.160.156(58.83.160.156): icmp_seq=3 ttl=127 time=112 ms
--- 58.83.160.156 ping statistics ---
3 packets transmitted, 3 received, 0% packet loss, time 2003ms
rtt min/avg/max/mdev = 65.089/85.274/111.796/19.587 ms
$ cat /etc/os-release
PRETTY_NAME= "cirros 0.6.1"
NAME= "cirros"
VERSION_ID= "0.6.1"
ID=cirros
HOME_URL="https://cirros-cloud.net"
BUG_REPORT_URL="https://github. com/cirros-dev/cirros/issues"
$ route -n
Kernel IP routing table
Destination     Gateway   Genmask        Flags  Metric Ref  Use Iface
0.0.0.0         10.0.0.1  0.0.0.0         UG    1002  0      0 eth0
10.0.0.0        0.0.0.0   255.255.255.0   U     1002  0      0 eth0
169.254.169.254 10.0.0.1  255.255.255.255 UGH  1002  0      0 eth0
```

2. 查看实例详细信息

命令如下。

```
[root@controller ~]# openstack console url show cirros-0.6.1
[root@controller ~]# openstack server show cirros-0.6.1
```

3. 使用 OpenStack API 关闭、启动和重建实例

curl 命令是一种在命令行中使用的网络工具，用于向服务器发送请求并获取响应，它可以模拟各种 HTTP 请求，常用于调试 Web 服务器和 API。curl 命令的基本参数如下。

① -s：在请求的过程中不显示进度条。

② -H：添加请求头部。

③ -X：指定请求方法（如 GET、POST、PUT、DELETE）。

使用 curl 命令调用 OpenStack API 后返回的 JS 对象简谱（JavaScript Object Notation，JSON）数据不易读，此时可以使用 python3 -m json.tool 命令将响应数据格式化输出。下面使用 curl 命令进行部分 OpenStack API 的调用。

首先，查询实例列表，命令和结果如下。

```
[root@controller ~]# curl -s \
-H "X-Auth-Token:`openstack token issue -f json | jq -r '.id'`" \
http://controller:8774/v2.1/servers | python3 -m json.tool
http://controller:8774/v2.1/serversI python3 -m json. tool
 {
      "servers": [
           {
                 "id" : "46ec44a0-ad70-42b3-9d29-d7050c8500fc" ,
                 "name" : "cirros-0.6.1",
                 "links" : [
                      {
                           "rel" : "self",
                           "href" : "http://controller: 8774/v2.1/servers/
46ec44a0-ad70-42b3-9d29-d7050c8500fc"
                      }
                      {
                           "rel" : "bookmark",
                           "href" : "http://controller:8774/servers/46ec44a0-
ad70-42b3-9d29-d7050c8500fc"
                      }
                 ]
           }
      ]
 }
```

其次，查询实例 cirros-0.6.1 的 ID、状态等，命令和结果如下。

```
[root@controller ~]# SERVER_ID=`openstack server \
show cirros-0.6.1 -f json | jq -r ".id"`
[root@controller ~]# curl -s -X GET \
-H "X-Auth-Token: `openstack token issue -f json | jq -r '.id'`" \
http://controller:8774/v2.1/servers/$SERVER_ID |\
python3 -m json.tool | \
jq -r '.server.name,.server.id,.server.status,.server.addresses'
cirros-0.6.1
46ec44a0-ad70-42b3-9d29-d7050c8500fc
ACTIVE
{
   "int-net": [
```

```
        {
            "version": 4,
            "addr": "10.0.0.185",
            "OS-EXT-IPS:type": "fixed",
            "OS-EXT-IPS-MAC:mac_addr": "fa:16: 3e:bb:2e:0c"
        },
        {
            "version": 4,
            "addr": "192.168.200.23" ,
            "OS-EXT-IPS:type": "floating",
            "OS-EXT-IPS-MAC:mac_ addr": "fa:16: 3e:bb:2e:0c"
        }
    ]
}
```

再次，关闭和启动实例 cirros-0.6.1，命令和结果如下。

```
[root@controller ~]# curl -s -X POST -H "Content-Type: application/json" \
-H "X-Auth-Token: `openstack token issue -f json | jq -r '.id'`" \
-d '{"os-stop": null}' \
http://controller:8774/v2.1/servers/$SERVER_ID/action
[root@controller ~]# openstack server list -c ID -c Name -c Status
+--------------------------------------+--------------+------------------+
|                  ID                  |     Name     |      Status      |
+--------------------------------------+--------------+------------------+
|   46ec44a0-ad70-42b3-9d29-d7050c8500fc  | cirros-0.6.1 |     SHUTOFF      |
+--------------------------------------+--------------+------------------+
[root@controller ~]# curl -X POST \
-H "X-Auth-Token: `openstack token issue -f json | jq -r '.id'`" \
-H "Content-Type: application/json" \
-d '{"os-start": null}' \
http://controller:8774/v2.1/servers/$SERVER_ID/action
[root@controller ~]# openstack server list -c ID -c Name -c Status
+--------------------------------------+--------------+------------------+
| ID                                   | Name         | Status           |
+--------------------------------------+--------------+------------------+
| 46ec44a0-ad70-42b3-9d29-d7050c8500fc | cirros-0.6.1 | ACTIVE           |
+--------------------------------------+--------------+------------------+
```

最后，重建实例 cirros-0.6.1，命令和结果如下。

```
[root@controller ~]# IMAGE_ID=`openstack image show cirros-0.6.1 -f json | jq -r ".id"`
[root@controller ~]# echo $IMAGE_ID
55758bd0-031b-455d-aea8-ca5c9e9e19e0
[root@controller ~]# curl -s -X POST -H "Content-Type: application/json" \
-H "X-Auth-Token: `openstack token issue -f json | jq -r '.id'`" \
-d '{"rebuild": {"imageRef": "55758bd0-031b-455d-aea8-ca5c9e9e19e0"}}' \
http://controller:8774/v2.1/servers/$SERVER_ID/action | python3 -m json.tool
{
    "server": {
        "id": "46ec44a0-ad70-42b3-9d29-d7050c8500fc",
        "name": "cirros-0.6.1",
        "status": "REBUILD",
        "tenant_id": "d502ed6ca8304eb9a7eee463f5e8a924",
        "user_ id": "8d1170212a4c4d7f92d185 b08de629e7",
        "metadata" : {},
        "hostId": "b4933f8dde14227cf 20b7488edd8f762c386d051fbf409f6d3176093"
        "image": {
```

```
            "id": "55758bd0-031b-455d-aea8-ca5c9e9e19e0",
            "links": [
                {
                    "rel": "bookmark",
                   "href": "http://controller:8774/images/55758bd0-031b-
455d-aea8-ca5c9e9e19e0"
                }
            ]
        },
        "favor": {
            "id": "337ca314-4e57-45a3-a5 c6-2e18ce80a981",
            "links": [
                {
                    "rel": "bookmark"
                    "href": "http://controller:8774/flavors/337ca314-4e57-
45a3-a5c6-2e18ce80a981",
                }
            ]
        },
        "accessIPv4": " ",
        "accessIPv6": " ",
        "links": [
            {
                "rel": "self",
                "href": "http://controller:8774/v2.1/servers/46ec44a0-ad70-
42b3-9d29-d7050c8500fc"
            },
            {
                "rel": "bookmark"
                "href": "http://controller:8774/servers/46ec44a0-ad70-
42b3-9d29-d7050c8500fc"
            }
        ],
        "OS-DCF:diskConfig": "MANUAL",
        "progress": 0,
        "adminPass": "jvwvEY6kUtdY"
    }
}
```

任务 4.2 存储服务组件运维管理

存储服务组件运维管理涉及 Cinder 块存储服务运维管理、Swift 对象存储服务运维管理和 Manila 共享文件系统服务运维管理。

任务 4.2.1 Cinder 块存储服务运维管理

1. 创建镜像和网络

下载镜像文件，命令如下。

```
[root@controller ~]# wget \
http://download.cirros-cloud.net/0.6.1/cirros-0.6.1-x86_64-disk.img
```

创建名为 cirros-0.6.1 的镜像，命令和结果如下。

```
[root@controller ~]# openstack image create --disk-format qcow2 \
--container-format bare \
```

微课 4.5 Cinder 块存储服务运维管理

```
--progress --file cirros-0.6.1-x86_64-disk.img "cirros-0.6.1"
[=========================>] 100%
+---------------------+--------------------------------------------------------+
|        Field        |                         Value                          |
+---------------------+--------------------------------------------------------+
|   container_format  | bare                                                   |
|     created_at      | 2023-01-13T12:42:38Z                                   |
|     disk_format     | qcow2                                                  |
|        file         | /v2/images/c5a617f5-8b3e-4da8-b4a1-78953ea0453c/file   |
|         id          | c5a617f5-8b3e-4da8-b4a1-78953ea0453c                   |
|      min_disk       | 0                                                      |
|      min_ram        | 0                                                      |
|        name         | cirros-0.6.1                                           |
|        owner        | 06267a21b9494e31b2e30b1ee5f0fdc8                       |
|                     | os_hidden='False', owner_specified.openstack.md5=' ',  |
|     properties      | owner_specified.openstack.object='images/cirros-0.6.1', |
|                     | owner_specified.openstack.sha256=' '                   |
|      protected      | False                                                  |
|       schema        | /v2/schemas/image                                      |
|       status        | queued                                                 |
|        tags         |                                                        |
|     updated_at      | 2023-01-13T12:42:38Z                                   |
|     visibility      | shared                                                 |
+---------------------+--------------------------------------------------------+
```

查看镜像列表，命令和结果如下。

```
[root@controller ~]# openstack image list
+--------------------------------------+--------------+--------+
| ID                                   | Name         | Status |
+--------------------------------------+--------------+--------+
| c5a617f5-8b3e-4da8-b4a1-78953ea0453c | cirros-0.6.1 | active |
+--------------------------------------+--------------+--------+
```

创建网络 network-flat，网络类型为 flat，命令和结果如下。

```
[root@controller ~]# openstack network create --provider-network-type flat \
--provider-physical-network provider --external network-flat
+---------------------------+----------------------------------------------+
|           Field           |                    Value                     |
+---------------------------+----------------------------------------------+
|      admin_state_up       | UP                                           |
|  availability_zone_hints  |                                              |
|    availability_zones     |                                              |
|        created_at         | 2023-01-13T12:51:38Z                         |
|        description        |                                              |
|        dns_domain         | None                                         |
|            id             | 901341e2-4019-4281-aaf1-1429ef11f435         |
|    ipv4_address_scope     | None                                         |
|    ipv6_address_scope     | None                                         |
|        is_default         | False                                        |
|    is_vlan_transparent    | None                                         |
|            mtu            | 1500                                         |
|           name            | network-flat                                 |
|   port_security_enabled   | True                                         |
|        project_id         | 06267a21b9494e31b2e30b1ee5f0fdc8             |
|   provider:network_type   | flat                                         |
|  provider:physical_network | provider                                    |
```

```
|   provider:segmentation_id    | None                                      |
|        qos_polic_id           | None                                      |
|      revision_number          | 1                                         |
|      router:external          | External                                  |
|          segments             | None                                      |
|           shared              | False                                     |
|           status              | ACTIVE                                    |
|           subnet              |                                           |
|            tags               |                                           |
|         updated_at            | 2023-01-13T12:51:38Z                      |
+-------------------------------+-------------------------------------------+
```

查看网络列表，命令和结果如下。

```
[root@controller ~]# openstack network list
+------------------------------------+----------------------+-------+
|          ID                        |   Name               |Subnets|
+------------------------------------+----------------------+-------+


|901341e2-4019-4281-aaf1-1429ef11f435|network-flat          |      |
|a4ceaef7-b892-4897-bc14-5278b5c2d5c0|manila_service_network|      |
+------------------------------------+----------------------+------+
```

创建子网 network-flat-subnet，网络地址分配为 192.168.200.100～192.168.200.200，网关为 192.168.200.1，命令和结果如下。

```
[root@controller ~]# openstack subnet create --network network-flat \
--subnet-range 192.168.200.0/24 \
--allocation-pool start=192.168.200.100,end=192.168.200.200 \
--gateway 192.168.200.1 network-flat-subnet
+-----------------------------+-------------------------------------------+
|           Field             |                    Value                  |
+-----------------------------+-------------------------------------------+
|      allocation-pools       | 192.168.200.100-192.168.200.200           |
|            cidr             | 192.168.200.0/24                          |
|         created_at          | 2023-01-13T13:01:29Z                      |
|         description         |                                           |
|       dns_nameservers       |                                           |
|     dns_publish_fixed_ip    | None                                      |
|         enable_dhcp         | True                                      |
|         gateway_ip          | 192.168.200.1                             |
|         host_routes         |                                           |
|             id              | cf150d0f-6264-4005-a7f6-cea28c59bd46      |
|         ip_version          | 4                                         |
|      ipv6_address_mode      | None                                      |
|        ipv6_ra_mode         | None                                      |
|            name             | network-flat-subnet                       |
|         network_id          | 901341e2-4019-4281-aaf1-1429ef11f435      |
|         project_id          | 06267a21b9494e31b2e30b1ee5f0fdc8          |
|      revision_number        | 0                                         |
|         segment_id          | None                                      |
|        service_types        |                                           |
|        subnetpool_id        | None                                      |
|            tags             |                                           |
|         updated_at          | 2023-01-13T13:01:29Z                      |
+-----------------------------+-------------------------------------------+
```

查看子网列表，命令和结果如下。

```
[root@controller ~]# openstack subnet list
+--------------------+---------------+--------------------------+-------------+
|         ID         |     Name      |         Network          |   Subnet    |
+--------------------+---------------+--------------------------+-------------+
| cf150d0f-6264-4005-| network-flat- | 901341e2-4019-4281-aaf1- | 192.168.200.|
| a7f6-cea28c59bd46  |    subnet     |      1429ef11f435        |    0/24     |
```

2. 启动云主机

创建云主机类型 2V_1G_10G，命令如下。

```
[root@controller ~]# openstack flavor create 2V_1G_10G --vcpus 2 \
--ram 1024 --disk 10
```

使用先前创建的镜像、网络和云主机类型，启动云主机 cirros-test，命令和结果如下。

```
[root@controller ~]# openstack server create --image cirros-0.6.1 \
--flavor 2V_1G_10G --network network-flat cirros-test
+-----------------------------+----------------------------------------------------+
| Field                       | Value                                              |
+-----------------------------+----------------------------------------------------+
| OS-DCF:diskConfig           | MANUAL                                             |
| OS-EXT-AZ:availability_zone |                                                    |
| OS-EXT-SRV-ATTR:host        | None                                               |
| OS-EXT-SRV-ATTR:hypervisor_ | None                                               |
| hostname                    |                                                    |
| OS-EXT-SRV-ATTR:instance_name |                                                  |
| OS-EXT-sTs:power_state      | NOSTATE                                            |
| OS-EXT-STS:task_state       | scheduling                                         |
| OS-EXT-STS:vm_state         | building                                           |
| OS-SRV-USG:launched_at      | None                                               |
| OS-SRV-USG:terminated_at    | None                                               |
| accessIPv4                  |                                                    |
| accessIPv6                  |                                                    |
| addresses                   |                                                    |
| adminpass                   | Vvp4aVuv7gdE                                       |
| config_drive                |                                                    |
| created                     | 2023-01-13T13:16:07Z                               |
| flavor                      | 2V_1G_10G(9819c5bd-33d7-4f8a-b964-65488d335236)    |
| hostId                      |                                                    |
| id                          | f6850bc3-3606-4c3d-b14e-3764bf8f1549               |
| image                       | cirros-0.6.1(c5a617f5-8b3e-4da8-b4a1-78953ea0453c) |
| key_name                    | None                                               |
| name                        | cirros-test                                        |
| progress                    | 0                                                  |
| project_id                  | 06267a21b9494e31b2e30b1ee5f0fdc8                   |
| properties                  |                                                    |
| security_groups             | name='default'                                     |
| status                      | BUILD                                              |
| updated                     | 2023-01-13T13:16:07Z                               |
| user_id                     | 002279a051e7467a9abfcd6807effd71                   |
| volumes_attached            |                                                    |
+-----------------------------+----------------------------------------------------+
```

查看云主机列表，命令和结果如下。

```
[root@controller ~]# openstack server list
+----------------+--------+---------+------------+-----------+--------------+
|       ID       |  Name  | status  |  Networks  |   Image   |    Flavor    |
+----------------+--------+---------+------------+-----------+--------------+
```

```
  f6850bc3-3606-                         network-flat
|    4c3d-b14e-    |  cirros  |  ACTIVE  | =192.168.200 |  cirros-   |  2V_1G_10G   |
  3764bf8f1549        -test                  .123          0.6.1
+-----------------+--------+----------+------------+-----------+-------------+
```

3. 查看 Cinder 服务状态

查看 Cinder 服务状态，命令和结果如下。

```
[root@controller ~]# openstack volume service list
+-----------------+-----------+--------+----------+-------+------------------+
| Binary          | Host      | Zone   | Status   | State | Updated At       |
+-----------------+-----------+--------+----------+-------+------------------+
|                 |           |        |          |       | 2023-01-13T13:19:56 |
| cinder-scheduler| controller| nova   | enabled  | up    | .000000          |
|                 | compute01@|        |          |       | 2023-01-13T13:19:54 |
| cinder-volume   | lvm-1     | nova   | enabled  | up    | .000000          |
|                 |           |        |          |       | 2023-01-13T13:19:53 |
| cinder-backup   | compute01 | nova   | enabled  | up    | .000000          |
+-----------------+-----------+--------+----------+-------+------------------+
```

4. 创建卷

创建一个存储容量为 2GB 的卷（云硬盘），命令和结果如下。

```
[root@controller ~]# openstack volume create --size 2 SSD_2G
+---------------------------+--------------------------------------+
| Field                     | Value                                |
+---------------------------+--------------------------------------+
| attachments               | []                                   |
| availability_zone         | nova                                 |
| bootable                  | false                                |
| consistencygroup_id       | None                                 |
| created_at                | 2023-01-13T13:24:57.900784           |
| description               | None                                 |
| encrypted                 | False                                |
| id                        | 6917f023-1316-4807-b8ac-f8572d72770b |
| migration_status          | None                                 |
| multiattach               | False                                |
| name                      | SSD_2G                               |
| properties                |                                      |
| replication_status        | None                                 |
| size                      | 2                                    |
| snapshot_id               | None                                 |
| source_volid              | None                                 |
| status                    | creating                             |
| type                      | __DEFAULT__                          |
| updated_at                | None                                 |
| user_id                   | 002279a051e7467a9abfcd6807effd71     |
+---------------------------+--------------------------------------+
```

5. 查看卷

查看当前卷列表，命令和结果如下。

```
[root@controller ~]#openstack volume list
+---------------------------+------------+-------------+------+-------------+
| ID                        | Name       | Status      | Size | Attached to |
+---------------------------+------------+-------------+------+-------------+
| 6917f023-1316-4807-b8ac-  |            |             |      |             |
| f8572d72770b              | SSD_2G     | available   | 2    |             |
+---------------------------+------------+-------------+------+-------------+
```

查看SSD_2G卷的详细信息，命令和结果如下。

```
[root@controller ~]# openstack volume show SSD_2G
+--------------------------------+------------------------------------------------+
| Field                          | Value                                          |
+--------------------------------+------------------------------------------------+
| attachments                    | []                                             |
| availability_zone              | nova                                           |
| bootable                       | false                                          |
| consistencygroup_id            | None                                           |
| created_at                     | 2023-01-13T13:24:57.000000                     |
| description                    | None                                           |
| encrypted                      | False                                          |
| id                             | 6917f023-1316-4807-b8ac-f8572d72770b           |
| migration_status               | None                                           |
| multiattach                    | False                                          |
| name                           | SSD_2G                                         |
| os-vol-host-attr:host          | compute01@lvm-1#lvm-1                          |
| os-vol-mig-status-attr:migstat | None                                           |
| os-vol-mig-status-attr:name_id | None                                           |
| os-vol-tenant-attr:tenant_id   | 06267a21b9494e31b2e30b1ee5f0fdc8               |
| properties                     |                                                |
| replication_status             | None                                           |
| size                           | 2                                              |
| snapshot_id                    | None                                           |
| source_volid                   | None                                           |
| status                         | available                                      |
| type                           | __DEFAULT__                                    |
| updated_at                     | 2023-01-13T13:24:58.000000                     |
| user_id                        | 002279a051e7467a9abfcd6807effd71               |
+--------------------------------+------------------------------------------------+
```

6. 挂载卷

为云主机cirros-test挂载SSD_2G卷，命令和结果如下。

```
[root@controller ~]# openstack server add volume cirros-test SSD_2G
+--------------------------------+------------------------------------------------+
| Field                          | Value                                          |
+--------------------------------+------------------------------------------------+
| ID                             | 6917f023-1316-4807-b8ac-f8572d72770b           |
| Server ID                      | f6850bc3-3606-4c3d-b14e-3764bf8f1549           |
| Volume ID                      | 6917f023-1316-4807-b8ac-f8572d72770b           |
| Device                         | /dev/vdb                                       |
| Tag                            | None                                           |
| Delete On Termination          | False                                          |
+--------------------------------+------------------------------------------------+
```

查看当前卷列表，命令和结果如下。

```
[root@controller ~]# openstack volume list
+----------------------+----------+--------+------+------------------------------+
| ID                   | Name     | Status | Size | Attached to                  |
+----------------------+----------+--------+------+------------------------------+
| 6917f023-1316-4807-  | SSD_2G   | in-use | 2    | Attached to cirros-test on   |
| b8ac-f8572d72770b    |          |        |      | /dev/vdb                     |
+----------------------+----------+--------+------+------------------------------+
```

7. 扩容卷

将云主机cirros-test上挂载的SSD_2G卷移除，命令如下。

```
[root@controller ~]# openstack server remove volume cirros-test SSD_2G
```

查看当前卷列表，可看到该卷仍存在，命令和结果如下。

```
[root@controller ~]# openstack volume list
+--------------------------------------+----------+-----------+------+-------------+
| ID                                   | Name     | Status    | Size | Attached to |
+--------------------------------------+----------+-----------+------+-------------+
| 99c48f61-9f63-4c0c-80bc-3731b7127a3b | SSD_2G   | available | 2    |             |
+--------------------------------------+----------+-----------+------+-------------+
```

将 SSD_2G 卷由原来的 2GB 扩容为 3GB，并更名为 SSD_3G，命令如下。

```
[root@controller ~]# openstack volume set SSD_2G --size 3 --name SSD_3G
```

查看当前卷列表，命令和结果如下。

```
[root@controller ~]# openstack volume list
+------------------------+-----------+-----------+------+--------------------+
| ID                     | Name      | Status    | Size | Attached to        |
+------------------------+-----------+-----------+------+--------------------+
| 23326e1f-bf73-4e3c-8c25-
| 3631540691b0           | SSD_3G    | available | 3    |                    |
+------------------------+-----------+-----------+------+--------------------+
```

8．挂载卷并验证卷大小

为云主机 cirros-test 挂载扩容后的 SSD_3G 卷，命令和结果如下。

```
[root@controller ~]# openstack server add volume cirros-test SSD_3G
+---------------------------+--------------------------------------+
| Field                     | Value                                |
+---------------------------+--------------------------------------+
| ID                        | 23326e1f-bf73-4e3c-8c25-3631540691b0 |
| Server ID                 | f6850bc3-3606-4c3d-b14e-3764bf8f1549 |
| Volume ID                 | 23326e1f-bf73-4e3c-8c25-3631540691b0 |
| Device                    | /dev/vdb                             |
| Tag                       | None                                 |
| Delete On Termination     | False                                |
+---------------------------+--------------------------------------+
```

查看当前卷列表，命令和结果如下。

```
[root@controller ~]# openstack volume list
+------------------------+----------+-----------+------+--------------------+
| ID                     | Name     | Status    | Size | Attached to        |
+------------------------+----------+-----------+------+--------------------+
| 23326e1f-bf73-4e3c-8c25-                                | Attached to cirros-
| 3631540691b0           | SSD_3G   | in-use    | 3    | test on /dev/vdb   |
+------------------------+----------+-----------+------+--------------------+
```

列出所有的虚拟机，命令和结果如下。

```
[root@compute01 ~]# dnf -y install libvirt-client
[root@compute01 ~]# virsh list --all
Id    Name                    State
------------------------------------------------
2     instance-00000003       running
```

使用命令行切换到云主机 cirros-test 的控制台，命令如下。

```
[root@compute01 ~]# virsh console instance-00000003
```

使用用户名和密码登录云主机 cirros-test 的控制台，命令和结果如下。

```
[root@compute01 ~]# virsh console instance-00000003
Connected to domain 'instance-00000003'
Escape character is ^] (Ctrl + ])     # 按Enter键
```

```
login as 'cirros' user. default password: 'gocubsgo'. use 'sudo' for root.
cirros login: cirros
Password:
```

列出云主机cirros-test所有可用的块设备信息，验证SSD_2G卷扩容的大小，命令和结果如下。

```
$ lsblk
NAME  MAJ:MIN    RM  SIZE  RO  TYPE   MOUNTPOINTS
vda      252:0    0   10G   0   disk
|-vda1   252:1    0   10G   0   part  /
`-vda15  252:15   0   8M    0   part
vdb      252:16   0   3G    0   disk
```

任务4.2.2 Swift对象存储服务运维管理

1. 服务运维基础命令

（1）查看对象存储服务状态

查看对象存储服务状态，命令如下。

```
[root@controller ~]# swift stat
```

微课4.6 Swift对象存储服务运维管理

（2）创建容器

创建容器swift-wxic，命令如下。

```
[root@controller ~]# openstack container create swift-wxic
```

（3）查看容器

查看当前容器列表，命令如下。

```
[root@controller ~]# openstack container list
```

查看容器swift-wxic的详细信息，命令如下。

```
[root@controller ~]# openstack container show swift-wxic
```

（4）上传对象

创建wxic目录，命令如下。

```
[root@controller ~]# mkdir wxic
```

查看当前位置下的所有文件，命令和结果如下。

```
[root@controller ~]# ls
snap  wxic cirros-0.6.1-x86_64-disk.img
```

将cirros-0.6.1-x86_64-disk.img复制到wxic目录下，命令如下。

```
[root@controller ~]# mv cirros-0.6.1-x86_64-disk.img wxic/
```

将wxic/cirros-0.6.1-x86_64-disk.img文件上传到swift-wxic容器，命令如下。

```
[root@controller ~]# openstack object create swift-wxic \
wxic/cirros-0.6.1-x86_64-disk.img
```

（5）查看对象

查看swift-wxic容器内的上传对象，命令和结果如下。

```
[root@controller ~]# openstack object list swift-wxic
+-----------------------------------+
| Name                              |
+-----------------------------------+
| wxic/cirros-0.6.1-x86_64-disk.img |
+-----------------------------------+
```

查看上传至swift-wxic容器中的wxic/cirros-0.6.1-x86_64-disk.img文件的详细信息，命令如下。

```
[root@controller ~]# openstack object show swift-wxic \
wxic/cirros-0.6.1-x86_64-disk.img
```

（6）下载对象

进入/opt 目录，命令如下。

```
[root@controller ~]# cd /opt/
```

将 swift-wxic 容器中的 wxic/cirros-0.6.1-x86_64-disk.img 文件下载到/opt 目录下，命令如下。

```
[root@controller opt]# openstack object save swift-wxic wxic/cirros-0.6.1-x86_64-
disk.img
```

查看下载结果，命令如下。

```
[root@controller opt]# ls wxic/cirros-0.6.1-x86_64-disk.img
```

如果想要下载容器中的所有文件，则命令语法如下。

```
[root@controller ~]# openstack container save <容器名>
```

（7）删除对象

查看 swift-wxic 容器中的上传对象，命令和结果如下。

```
[root@controller ~]# openstack object list swift-wxic
+-----------------------------------+
|         Name                      |
+-----------------------------------+
| wxic/cirros-0.6.1-x86_64-disk.img |
+-----------------------------------+
```

删除 swift-wxic 容器中的上传对象 wxic/cirros-0.6.1-x86_64-disk.img，命令如下。

```
[root@controller ~]# openstack object delete swift-wxic wxic/cirros-0.6.1-x86_64-
disk.img
```

再次查看 swift-wxic 容器中的上传对象，命令如下。

```
[root@controller ~]# openstack object list swift-wxic
```

（8）删除容器

查看容器列表，命令和结果如下。

```
[root@controller ~]# openstack container list
+------------+
| Name       |
+------------+
| swift-wxic |
+------------+
```

删除 swift-wxic 容器，命令如下。

```
[root@controller ~]# openstack container delete swift-wxic
```

再次查看容器列表，检验上述操作的结果，命令如下。

```
[root@controller ~]# openstack container list
```

如果需要递归删除对象和容器，则需要加--recursive 参数，命令语法如下。

```
[root@controller ~]# openstack container delete <容器名> --recursive
```

2. 分片存储案例

（1）创建容器

创建一个名为 wxic 的容器，命令如下。

```
[root@controller ~]# openstack container create wxic
```

（2）上传镜像并分片存储

下载 cirros-0.6.1-x86_64-disk.img 镜像文件到本地，命令如下。

```
[root@controller ~]# wget \
http://download.cirros-cloud.net/0.6.1/cirros-0.6.1-x86_64-disk.img
```

将 cirros-0.6.1-x86_64-disk.img 镜像文件按 10MB 的大小分片存储在 wxic 容器中，命令如下。

```
[root@controller ~]# swift upload wxic -S 10M cirros-0.6.1-x86_64-disk.img
```

查看 wxic 容器中上传对象 cirros-0.6.1-x86_64-disk.img 的详细信息，命令如下。

```
[root@controller ~]# swift stat wxic cirros-0.6.1-x86_64-disk.img
```

查看 wxic 容器中的分片情况，命令和结果如下。

```
[root@controller ~]# swift list wxic_segments
cirros-0.6.1-x86_64-disk.img/1669128161.000000/21233664/10485760/00000000
cirros-0.6.1-x86_64-disk.img/1669128161.000000/21233664/10485760/00000001
cirros-0.6.1-x86_64-disk.img/1669128161.000000/21233664/10485760/00000002
```

数据分片分别存储在不同的存储设备中，以减小每台存储设备的数据访问压力，从而提高整个数据系统的性能。

任务 4.2.3　Manila 共享文件系统服务运维管理

微课 4.7　Manila 共享文件系统服务运维管理

1. Manila 服务命令

（1）通过帮助命令了解 Manila 服务的使用方法

共享文件系统服务的命令较多，但是命令行工具提供的帮助说明比较详细，读者可以参考其进行学习，由于回显命令过多，下面只给出帮助命令示例，读者可自行在已有环境下使用相关命令并查看结果。

```
[root@controller ~]# openstack help share
[root@controller ~]# openstack help share type create
[root@controller ~]# openstack help share create
[root@controller ~]# openstack help share access create
[root@controller ~]# openstack help share export location list
```

（2）服务状态查看命令

列出服务组件以验证每个进程是否成功启动，命令和结果如下。

```
[root@controller ~]# openstack share service list
+----+---------+-------------+--------+---------+-------+----------------------+
| ID | Binary  | status      |Networks| Image   | Flavor| Updated At           |
+----+---------+-------------+--------+---------+-------+----------------------+
| 1  | maniTa- | controller  | nova   | enabled | up    | 2023-05-08T09:08:17. |
|    |scheduler|             |        |         |       | 070795               |
| 2  | manila- | compute@lvm | nova   | enabled | up    | 2023-05-08T09:08:17. |
|    | share   |             |        |         |       | 906948               |
+----+---------+-------------+--------+---------+-------+----------------------+
```

2. 使用共享服务

（1）创建文件共享类型

创建名为 default_share_type 的文件共享类型，且不设置为默认文件共享类型，命令和结果如下。

```
[root@controller ~]# openstack share type create default_share_type False
+--------------------------------+------------------------------------------+
| Field                          | Value                                    |
+--------------------------------+------------------------------------------+
| id                             | 6cc57ed3-36ba-4d01-8829-15d1d495ac3a     |
| name                           | default_share_type                       |
| visibility                     | public                                   |
| is_default                     | True                                     |
| required_extra_specs           | driver_handles_share_servers : False     |
| optional_extra_specs           |                                          |
| description                    | None                                     |
+--------------------------------+------------------------------------------+
```

查询文件共享类型列表信息，命令和结果如下。

```
[root@controller ~]# openstack share type list
+--------------+--------+----------+-------+-----------+--------+----------+
| ID           | Name   |Visibility| Is    | Required  | Optional |Description|
|              |        |          |Default| Extra     | Extra    |          |
|              |        |          |       | Specs     | Specs    |          |
+--------------+--------+----------+-------+-----------+--------+----------+
| 6cc57ed3-36ba-| default| public  | True  | driver_   |        | None     |
| 4d01-8829-   | _      |          |       | handles_  |        |          |
| 15d1d495ac3a | share_ |          |       | share_    |        |          |
|              | type   |          |       | servers : |        |          |
|              |        |          |       | False     |        |          |
+--------------+--------+----------+-------+-----------+--------+----------+
```

（2）创建共享文件目录

创建目录大小为 10GB 的共享文件目录 share01，命令和结果如下。

```
[root@controller ~]# openstack share create --name share01 --share-type
default_share_type NFS 10
+------------------------------------+------------------------------------------+
| Field                              | Value                                    |
+------------------------------------+------------------------------------------+
| access_rules_status                | active                                   |
| availability_zone                  | None                                     |
| create_share_from_snapshot_support | False                                    |
| created_at                         | 2023-01-18T12:22:30.420775               |
| description                        | None                                     |
| has_replicas                       | False                                    |
| host                               |                                          |
| id                                 | a62e282a-a4b8-4657-b3b3-a476d6ac527d     |
| is_public                          | False                                    |
| is_soft_deleted                    | False                                    |
| metadata                           | {}                                       |
| mount_snapshot_support             | False                                    |
| name                               | share01                                  |
| progress                           | None                                     |
| project_id                         | 0d393503f5d34b9c86303090f41c9a56         |
| replication_type                   | None                                     |
| revert_to_snapshot_support         | False                                    |
| scheduled_to_be_deleted_at         | None                                     |
| share_group_id                     | None                                     |
| share_network_id                   | None                                     |
| share_proto                        | NFS                                      |
| share_server_id                    | None                                     |
| share_type                         | 6cc57ed3-36ba-4d01-8829-15d1d495ac3a     |
| share_type_name                    | default_share_type                       |
| size                               | 10                                       |
| snapshot_id                        | None                                     |
| snapshot_support                   | False                                    |
| source_share_group_snapshot_member_| None                                     |
| id                                 |                                          |
| status                             | creating                                 |
| task_state                         | None                                     |
| user_id                            | 18246e0db9d845aa979c3b7e90ac3c0b         |
| volume_type                        | default_share_type                       |
+------------------------------------+------------------------------------------+
```

查询所创建的共享文件目录列表信息，命令和结果如下。

```
[root@controller ~]# openstack share list
+-----------+--------+-----+------+---------+-------+---------+---------+-------+
|           |        |     | Share|         |  Is   | Share   |         | Availa|
| ID        |Name    |Size | Proto|Status   | Public|Type     | Host    |bility |
|           |        |     |      |         |       | Name    |         | Zone  |
+-----------+--------+-----+------+---------+-------+---------+---------+-------+
| a62e282a- |        |     |      |         |       | default | compute01|      |
| a4b8-4657-|share01 | 10  | NFS  |available| False | _share_ | @lvm#lvm-| nova |
| b3b3-a476d|        |     |      |         |       | type    | single-  |      |
| 6ac527d   |        |     |      |         |       |         | pool     |      |
+-----------+--------+-----+------+---------+-------+---------+---------+-------+
```

（3）挂载共享文件目录

开放share01文件共享目录对OpenStack内部管理网段的使用权限，授予其读写访问权限，命令和结果如下。

```
[root@controller ~]# openstack share access create share01 ip 192.168.100.0/24
--access-leve rw
+----------------------+----------------------------------------------------------+
|Field                 | Value                                                    |
+----------------------+----------------------------------------------------------+
| id                   | 9d347fb9-46c2-436c-99bc-a9408ef7eb14                     |
| share_id             | a62e282a-a4b8-4657-b3b3-a476d6ac527d                     |
| access_level         | rw                                                       |
| access_to            | 192.168.100.0/24                                         |
| access_type          | ip                                                       |
| state                | queued_to_apply                                          |
| access_key           | None                                                     |
| created_at           | 2023-01-18T12:38:49.631018                               |
| updated_at           | None                                                     |
| properties           |                                                          |
+----------------------+----------------------------------------------------------+
```

查看share01共享文件目录的权限及开放网段，命令和结果如下。

```
[root@controller ~]# openstack share access list share01
+----------------+-------+--------+--------+--------+--------+----------+---------+
|                | Access| Access | Access |        | Access |          | Updatde |
| ID             | Type  | To     | Level  | State  | Key    |Created At| At      |
+----------------+-------+--------+--------+--------+--------+----------+---------+
|                |       | 192.   |        |        |        |          |         |
| 9d347fb9-46c2- |       | 168.   |        |        |        | 2023-01- |         |
| 436c-99bc-a940 | ip    | 100.0/ | rw     | active | None   | 18T12:38:| None    |
| 8ef7eb14       |       | 24     |        |        |        | 49.631018|         |
+----------------+-------+--------+--------+--------+--------+----------+---------+
```

查看share01共享文件目录的访问路径，命令和结果如下。

```
[root@controller ~]# openstack share show share01 | awk '/path/ {print $5}'
192.168.100.20:/var/lib/manila/mnt/share-726614bd-89e8-4ffb-b2d5-387d0d7b3173
# 或者
[root@controller ~]# openstack share show share01 | egrep 'path' | awk '{print $5}'
192.168.100.20:/var/lib/manila/mnt/share-726614bd-89e8-4ffb-b2d5-387d0d7b3173
```

此访问路径在Skyline界面左侧的“文件存储→共享”中的“基本信息”处可以看到，如图4-8所示。

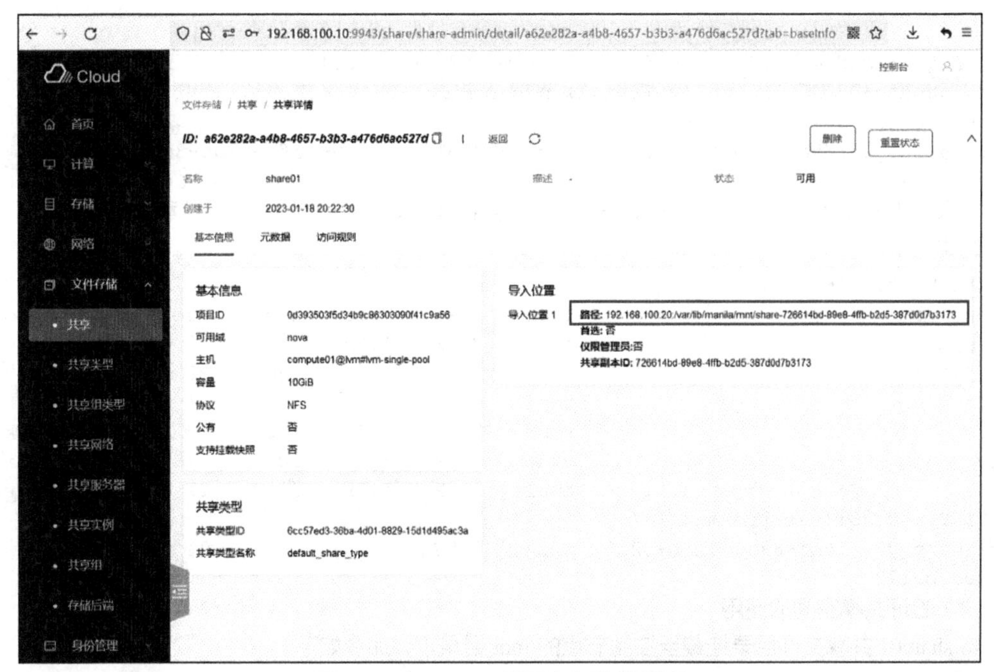

图 4-8　在 Skyline 界面中查看访问路径

将 share01 共享文件目录挂载至云主机 Ubuntu-22.04 的/mnt 目录下，命令和结果如下。

```
[root@controller ~]# openstack server list
+----------------+---------+-------+-------------------+-----------+------------+
| ID             | Name    |status | Networks          | Image     | Flavor     |
+----------------+---------+-------+-------------------+-----------+------------+
| 44c5d960-dfef- |         |       |                   |           |            |
| 488a-8981-8ec9 | Ubuntu- |ACTIVE | int-net-10.0.0.236,| ubuntu-  | 2V_4G_20G  |
| d8401227       | 22.04   |       | 192.168.200.200   | 22.04     |            |
+----------------+---------+-------+-------------------+-----------+------------+
[root@controller ~]# ssh root@192.168.200.200
welcome to ubuntu 22.04.1 LTS (GNU/Linux 5.15.0-50-generic x86_64)

* Documentation:     https://help.ubuntu.com
* Management:        https://landscape.canonical.com
* Support:           https://ubuntu.com/advantage

System information as of Wed Jan 18 15:03:32 UTC 2023
System Load:         0.0                    Processes:              111
Usage of:            8.2% of 19.20GB        Users logged in:        1
Memory usage:        7%                     IPv4 address for eth0:  10.0.0.236
Swap usage:          0%

* Strictly confined kubernetes makes edge and IoT secure. Learn how MicroK8s
just raised the bar for easy, resilient and secure K8s cluster deployment.

https://ubuntu.com/engage/secure-kubernetes-at-the-edge

91 updates can be applied immediately.
55 of these updates are standard security updates.
To see these additional updates run: apt list -upgradable
```

```
Last login: Wed Jan 18 14:50:59 2023 from 192.168.200.10
[root@ubuntu-2204 ~]# apt -y install nfs-common
[root@ubuntu-2204 ~]# mount -t nfs
192.168.100.20:/var/lib/manila/mnt/share-726614bd-89e8-4ffb-b2d5-387d0d7b3173
/mnt/
```

在云主机 Ubuntu-22.04 上查询挂载信息，可以看到 share01 共享文件目录被挂载至/mnt 目录下，命令和结果如下。

```
[root@ubuntu-2204 ~]# df -Th
Filesystem                              Type   Size  Used  Avail Use%  Mounted on
tmpfs                                   tmpfs  393M  1.1M  392M  1%    /run
/dev/vda1                               ext4   20G   1.6G  18G   9%    /
tmpfs                                   tmpfs  2.0G  0     2.0G  0%    /dev/shm
tmpfs                                   tmpfs  5.0M  0     5.0M  0%    /run/lock
/dev/vda15                              vfat   105M  5.3M  100M  5%    /boot/efi
tmpfs                                   tmpfs  393M  4.0K  393M  1%    /run/user/0
192.168.100.20:/var/lib/manila/
mnt/share-726614bd-89e8-4ffb-b2d5-      nfs4   9.8G  0     9.3G  0%    /mnt
387d0d7b3173
```

（4）验证共享空间的使用

将 share01 共享文件目录挂载至控制节点的/mnt 目录下，命令如下。

```
[root@controller ~]# dnf -y install nfs-utils
[root@controller ~]# mount -t nfs \
192.168.100.20:/var/lib/manila/mnt/share-726614bd-89e8-4ffb-b2d5-387d0d7b3173 /mnt/
```

查看挂载信息，命令和结果如下。

```
[root@controller ~]# dt -hT /mnt/
Filesystem
Type    SizeUsedAvail   Use%Mounted on
192.168.100.20:/var/lib/manila/mnt/share-726614bd-89e8-4ffb-b2d5-387d0d7b3173
nfs4    9.8G    0   9.3G0%   /mnt
```

在云主机 Ubuntu-22.04 和控制节点的/mnt 目录下分别创建不同的文件，命令如下。

```
[root@controller mnt]# touch controller.txt
[root@ubuntu-2204 mnt]# touch server-ubuntu-22.04.txt
```

分别在云主机 Ubuntu-22.04 和控制节点上查看/mnt 目录下的内容，命令和结果如下。

```
[root@controller ~]# /mnt
[root@controller ~]# ls
controller.txt  lost+found  server-ubuntu-22.04.txt
[root@ubuntu-2204 ~]# /mnt
[root@ubuntu-2204 ~]# ls
controller.txt  lost+found  server-ubuntu-22.04.txt
```

可以发现，任何一个节点创建的文件都能被挂载了共享文件目录 share01 的节点访问，下面查看共享文件目录 share01 在集群中的实际位置，命令和结果如下。

```
[root@controller ~]# openstack share export location list share01
+----------------------+------------------------------------------+--------------+
| ID                   | Path                                     | Preferred    |
+----------------------+------------------------------------------+--------------+
| a4fa58bf-cd99-4919-  | 192.168.100.20:/var/lib/manila/mnt/share | False        |
| be0a-d89f4264dfd6    | -726614bd-89e8-4ffb-b2d5-387d0d7b3173    |              |
+----------------------+------------------------------------------+--------------+
[root@compute01 ~]# ls \
/var/lib/manila/mnt/share-726614bd-89e8-4ffb-b2d5-387d0d7b3173
controller.txt  lost+found  server-ubuntu-22.04.txt
```

查询集群中共享的可用区、所有共享实例的信息、指定共享实例的导出位置信息、共享存储资源池列表，命令和结果如下。

```
[root@controller ~]# openstack share availability zone list
```

Id	Name	Created At	Updated At
1cec6c8b-70a6-4d2f-8b25-0d5fd0b423c2	nova	2023-01-18T08:25:46.302604	None

```
[root@controller ~]# openstack share instance list
```

ID	Share ID	Host	Status	Availability Zone	Share Network ID	Share Server ID	Share Type ID
726614bd-89e8-4ffb-b2d5-387d0d7b3173	a62e282a-a4b8-4657-b3b3-a476d6ac527d	compute01@lvm#lvm-single-pool	available	nova	None	None	6cc57ed3-36ba-4d01-8829-15d1d495ac3a

```
# 共享实例 ID 以实际查询的结果为准
[root@controller ~]# openstack share instance export location list 726614bd-89e8-4ffb-b2d5-387d0d7b3173
```

ID	Path	Is Admin Only	Preferred
a4fa58bf-cd99-4919-be0a-d89f4264dfd6	192.168.100.20:/var/lib/manila/mnt/share-726614bd-89e8-4ffb-b2d5-387d0d7b3173	False	False

```
[root@controller ~]# openstack share pool list
```

Name	Host	Backend	Pool
compute01@lvm#lvm-single-pool	compute01	lvm	lvm-single-pool

任务 4.3 高级服务组件运维管理

高级服务组件运维管理涉及 Heat 编排服务运维管理、Ceilometer 监控服务运维管理和 Cloudkitty 计费服务运维管理。

任务 4.3.1 Heat 编排服务运维管理

微课 4.8 Heat 编排服务运维管理

1. Heat 的基础命令和堆栈的语法

查看当前系统中 Heat 的构建信息，命令如下。

```
[root@controller ~]# openstack orchestration build info
```

查看 Heat 组件的服务状态，命令如下。

```
[root@controller ~]# openstack orchestration service list
```

查看当前支持的所有资源类型，命令如下。

```
[root@controller ~]# openstack orchestration resource type list
```

查看资源类型的详细信息，如OS::Nova::Flavor类型，命令如下。

```
[root@controller ~]# openstack orchestration resource type show OS::Nova::Flavor
```

查看当前支持的Heat模板版本列表，命令如下。

```
[root@controller ~]# openstack orchestration template version list
```

查看当前OpenStack Yoga版本中最新的2021-04-16支持的功能字段列表，命令如下。

```
[root@controller ~]# openstack orchestration template function list
heat_template_version.2021-04-16
```

验证模板文件是否有语法错误，命令如下。

```
[root@controller ~]# openstack orchestration template validate -t Heat-ServerCreate.yaml
```

创建堆栈的语法如下。

```
openstack stack create
  [-e <environment>]
  [-s <files-container>]
  [--timeout <timeout>]
  [--pre-create <resource>]
  [--enable-rollback]
  [--parameter <key=value>]
  [--parameter-file <key=file>]
  [--wait]
  [--poll SECONDS]
  [--tags <tag1,tag2...>]
  [--dry-run]
  -t <template>
  <stack-name>
```

查询堆栈资源列表的语法如下。

```
openstack stack resource list
  [--sort-column SORT_COLUMN]
  [--sort-ascending | --sort-descending]
  [--long]
  [-n <nested-depth>]
  [--filter <key=value>]
  <stack>
```

查询堆栈文件映射列表的语法如下。

```
openstack stack file list <NAME or ID>
```

查看堆栈详细事件的语法如下。

```
openstack stack event show <stack> <resource> <event>
```

显示堆栈环境的语法如下。

```
openstack stack environment show <NAME or ID>
```

导出堆栈的语法如下。

```
openstack stack export [--output-file <output-file>] <stack>
```

查询堆栈列表的语法如下。

```
openstack stack list
  [--sort-column SORT_COLUMN]
  [--sort-ascending | --sort-descending]
  [--deleted]
  [--nested]
  [--hidden]
  [--property <key=value>]
```

```
  [--tags <tag1,tag2...>]
  [--tag-mode <mode>]
  [--limit <limit>]
  [--marker <id>]
  [--sort <key>[:<direction>]]
  [--all-projects]
  [--short]
  [--long]
```

查询堆栈输出列表的语法如下。

```
openstack stack output list
   [--sort-column SORT_COLUMN]
   [--sort-ascending | --sort-descending]
   <stack>
```

放弃堆栈的语法如下。

```
openstack stack abandon [--output-file <output-file>] <stack>
openstack stack cancel
  [--sort-column SORT_COLUMN]
  [--sort-ascending | --sort-descending]
  [--wait]
  [--no-rollback]
 <stack>
[<stack> ...]
```

恢复堆栈的语法如下。

```
openstack stack resume
  [--sort-column SORT_COLUMN]
  [--sort-ascending | --sort-descending]
  [--wait]
 <stack>
[<stack> ...]
```

创建堆栈快照的语法如下。

```
openstack stack snapshot create [--name <name>] <stack>
```

恢复堆栈快照的语法如下。

```
openstack stack snapshot restore <stack> <snapshot>
```

2. 利用 Heat 模板创建云主机并挂载卷

（1）创建云主机类型的模板文件

编写一个创建云主机类型的模板文件 Heat-FlavorCreate.yaml，创建名为 flavor 的资源堆栈来实现云主机类型的创建，命令和结果如下。

```
[root@controller ~]# cat Heat-FlavorCreate.yaml
heat_template_version: 2021-04-16
description: Template for creating a '2V_4G_20G' flavor.
resources:
 2V_4G_20G:
  type: OS::Nova::Flavor
  properties:
   name: 2V_4G_20G
   flavorid: 9943
   ram: 4096
   vcpus: 2
   disk: 20
   ephemeral: 0
outputs:
 flavor_name:
```

```
   description: Name of the new flavor
   value: 2V_4G_20G
 flavor_id:
   description: ID of the new flavor
   value: { get_resource: 2V_4G_20G }
 flavor_show:
   value: { get_attr: [2V_4G_20G, show] }
 flavor_is_public:
   value: { get_attr: [2V_4G_20G, is_public] }
[root@controller ~]# openstack stack create -t Heat-FlavorCreate.yaml flavor
+--------------------------------------+--------------------------------------------+
|Field                                 |Value                                       |
+--------------------------------------+--------------------------------------------+
| id                                   |14a1e3ff-60e9-4185-a140-c8ccb84af4b2        |
| stack_name                           |flavor                                      |
| description                          |Template for creating a '2V_4G_20G'' flavor.|
| creation_time                        |2023-01-20T04:33:19Z                        |
| updated_time                         |None                                        |
| stack_status                         |CREATE_IN_PROGRESS                          |
| stack_status_reason                  |Stack CREATE started                        |
+--------------------------------------+--------------------------------------------+
[root@controller ~]# openstack flavor list
+-------+-------------+----------+------+-----------+---------+-----------+
| ID    | Name        | RAM      | Disk | Ephemeral | vCPUs   | Is Public |
+-------+-------------+----------+------+-----------+---------+-----------+
| 9943  | 2V_4G_20G   | 4096     | 20   | 0         | 2       | True      |
+-------+-------------+----------+------+-----------+---------+-----------+
```

查看 flavor 资源堆栈的输出信息并以表格形式将其输出，命令和结果如下。

```
[root@controller ~]# openstack stack show flavor -c outputs -f table
+-----------+----------------------------------------------------------------------+
| Field     | Value                                                                |
+-----------+----------------------------------------------------------------------+
| outputs   | - description: ID of the new flavor                                  |
|           | output_key: flavor_id                                                |
|           |   output_value :  '9943'                                             |
|           | - description: Name of the new flavor                                |
|           |   output_key: flavor_name                                            |
|           |   output_value: 2V_4G_20G                                            |
|           | - description: No description given                                  |
|           |   output_key: flavor_show                                            |
|           |   output_value:                                                      |
|           |     OS-FLV-DISABLED:disabled: false                                  |
|           |     OS-FLV-EXT-DATA:ephemeral: 0                                     |
|           |     description: null                                                |
|           |     disk: 20                                                         |
|           |     extra_specs: {}                                                  |
|           |     id:  '9943'                                                      |
|           |     links:                                                           |
|           |     - href: http://192.168.100.10:8774/v2.1/flavors/9943             |
|           |       rel: self                                                      |
|           |     - href: http://192.168.100.10:8774/flavors/9943                  |
|           |       rel: bookmark                                                  |
|           |     name: 2V_4G_20G                                                  |
|           |     os-flavor-access:is_public: true                                 |
|           |     ram: 4096                                                        |
```

```
|           |     rxtx_factor: 1.0                                  |
|           |     swap: 0                                           |
|           |     vcpus: 2                                          |
|           |   - description: No description given                |
|           |     output_key: flavor_is_public                      |
|           |     output_value: true                                |
+-----------+-------------------------------------------------------+
```

（2）创建对象存储容器的模板文件

编写一个创建对象存储容器的模板文件 Heat-ContainerCreate.yaml，创建名为 Container 的资源堆栈来实现容器的创建，命令和结果如下。

```
[root@controller ~]# cat Heat-ContainerCreate.yaml
heat_template_version: "2021-04-16"
description: "This template is used to create a swift container and upload a local
image to it."
resources:
 swift_container:
  type: "OS::Swift::Container"
  properties:
   name: "glance"
   # 删除容器时会删除容器中的所有对象
   PurgeOnDelete: True
   # 允许公开访问
   "X-Container-Read": ".r:*"
outputs:
 container_name:
  description: Name of the swift container
  value: { get_attr: [swift_container, HeadContainer] }
 container_url:
  description: URL of the swift container
  value: { get_attr: [swift_container, WebsiteURL] }
[root@controller ~]# openstack stack create -t Heat-ContainerCreate.yaml container
+---------------------+-----------------------------------------------------+
| Field               | Value                                               |
+---------------------+-----------------------------------------------------+
| id                  | 5d77f3aa-7bfb-4b51-8753-87eecca3ca5f                |
| stack_name          | container                                           |
|                     | This template is used to create a swift container and |
| description         | upload a local image to it.                         |
| creation_time       | 2023-01-20T05:03:12Z                                |
| updated_time        | None                                                |
| stack_status        | CREATE_IN_PROGRESS                                  |
| stack_status_reason | Stack CREATE started                                |
+---------------------+-----------------------------------------------------+
[root@controller ~]# openstack container list
+--------------------+
|    Name            |
+--------------------+
|    glance          |
+--------------------+
```

查看 Container 资源堆栈的输出信息中的容器公开访问链接，并上传 cirros-0.6.1-x86_64-disk.img 镜像到 glance 容器中，命令和结果如下。

```
[root@controller ~]# openstack stack output show container container_url
+---------------------+-----------------------------------------------------+
```

```
| Field               | Value                                                  |
+---------------------+--------------------------------------------------------+
| description         | URL of the swift container                             |
| output_key          | container_url                                          |
| output_value        | http://192.168.100.10:8080/v1/AUTH_95ea70f0b429418797  |
|                     | 6e9dd56f605c94/glance                                  |
+---------------------+--------------------------------------------------------+
[root@controller ~]# openstack object create glance
/root/cirros-0.6.1-x86_64-disk.img
+-----------------------------------+-----------+----------------------------------+
| object                            | container | etag                             |
+-----------------------------------+-----------+----------------------------------+
| /root/cirros-0.6.1-x86_64-disk.img | glance    | 0c839612eb3f2469420f2ccae990827F |
+-----------------------------------+-----------+----------------------------------+
```

（3）创建和上传镜像的模板文件

编写一个创建和上传镜像的模板文件 Heat-ImageCreate.yaml，创建名为 image 的资源堆栈来实现镜像的创建和上传，命令和结果如下。

```
[root@controller ~]# cat Heat-ImageCreate.yaml
heat_template_version: 2021-04-16
description: Create an image using swift URL
resources:
 image:
  type: OS::Glance::WebImage
  #type: OS::Glance::Image
  properties:
   name: "cirros-0.6.1"
   disk_format: qcow2
   container_format: bare
   min_disk: 2
   min_ram: 512
   visibility: public
   os_version: 0.6.1
   tags: [wxic-cloud]
   location:
http://192.168.100.10:8080/v1/AUTH_95ea70f0b4294187976e9dd56f605c94/gl
ance/cirros-0.6.1-x86_64-disk.img
outputs:
 image_id:
  description: Image ID
  value: { get_resource: image }
 image_show:
  value: { get_attr: [image, show] }
 image_location:
  description: "镜像的位置"
value: { get_property: [image, location] }
[root@controller ~]# openstack stack create -t Heat-ImageCreate.yaml image
+---------------------+--------------------------------------------------------+
| Field               | Value                                                  |
+---------------------+--------------------------------------------------------+
| id                  | d5e93c03-ae6d-48a2-8d6e-d1d2772eb66e                   |
| stack_name          | image                                                  |
| description         | Create an image using swift URL                        |
| creation_time       | 2023-01-20T05:10:10Z                                   |
| updated_time        | None                                                   |
```

```
| stack_status          |  CREATE_IN_PROGRESS                                  |
| stack_status_reason |  Stack CREATE started                                |
+---------------------+------------------------------------------------------+
[root@controller ~]# openstack image list
+--------------------------------------+----------------+--------------+
| ID                                   |  Name          |  Status      |
+--------------------------------------+----------------+--------------+
|  95c20d72-8ae4-451e-ac8b-032358ac8f61 |  cirros-0.6.1  |  active      |
+--------------------------------------+----------------+--------------+
```

（4）创建内部私有网络的模板文件

编写一个创建内部私有（Private）网络和子网，以及连接外部公有（Public）网络的路由（Public-router），并绑定内部私有网络端口的模板文件（Heat-NetworkCreate.yaml），创建名为 private-net 的资源堆栈来实现内部私有网络的使用，命令和结果如下。

```
[root@controller ~]# cat Heat-NetworkCreate.yaml
heat_template_version: "2021-04-16"
description: "此模板文件用于创建内部私有网络和子网，以及创建连接外部公有网络的路由，并绑定内部私有网络端口"
resources:
 private-net:
  type: "OS::Neutron::Net"
  properties:
   name: "private"
   port_security_enabled: true
  private-subnet:
   type: "OS::Neutron::Subnet"
   properties:
    network_id:
     get_resource: private-net
    cidr: "10.0.0.0/24"
    allocation_pools: [{start: "10.0.0.100",end: "10.0.0.200"}]
    ip_version: 4
    enable_dhcp: true
    name: "private-subnet"
    gateway_ip: "10.0.0.1"
 Public-router:
  type: "OS::Neutron::Router"
  properties:
   admin_state_up: true
   external_gateway_info:
    network: Public
    enable_snat: true
   name: "Public-router"
   distributed: true
 router_interface:
  type: "OS::Neutron::RouterInterface"
  properties:
   router_id: { get_resource: Public-router }
   subnet_id: { get_resource: private-subnet }
outputs:
 int_net_id:
  description: ID of the internal network
  value: { get_resource: private }
```

```
 int_subnet_id:
  description: ID of the internal subnet
  value: { get_resource: private-subnet }
 ext_router_id:
  description: ID of the external router
  value: { get_resource: Public-router }
 router_interface_id:
  description: ID of the router interface
  value: { get_resource: router_interface }
 int_subnet_cidr:
  description: CIDR of the internal subnet
  value: { get_property: [private-subnet, cidr] }
 int_subnet_gateway:
  description: Gateway IP of the internal subnet
  value: { get_property: [private-subnet, gateway_ip] }
 ext_network:
  description: Name of the external network
  value: { get_property: [Public-router, external_gateway_info, network] }
 int_subnet_allocation_pools:
  description: Allocation pools of the internal subnet
  value: { get_property: [private-subnet, allocation_pools] }
 ext_router_name:
  description: Name of the external router
  value: { get_property: [Public-router, name] }
[root@controller ~]# openstack stack create -t Heat-NetworkCreate.yaml private-net
+---------------------+------------------------------------------------------------+
| Field               | Value                                                      |
+---------------------+------------------------------------------------------------+
| id                  | A670fa63-049f-48a4-8007-049c7a3d91b5                       |
| stack_name          | private-net                                                |
|                     | "此模板文件用于创建内部私有网络和子网,以及创建连接外部公有网络的 |
| description         | 路由，并绑定内部私有网络端口"                                  |
| creation_time       | 2023-01-20T05:33:50Z                                       |
| updated_time        | None                                                       |
| stack_status        | CREATE_IN_PROGRESS                                         |
| stack_status_reason | Stack CREATE started                                       |
+---------------------+------------------------------------------------------------+
[root@controller ~]# openstack network list
+------------------------+--------------------------+--------------------------+
| ID                     | Name                     | Status                   |
+------------------------+--------------------------+--------------------------+
| 3641c6bf-90c1-4aa7-    |                          | d97147d1-87c4-4ec2-      |
| 83e3-cd3674811a28      | private                  | a290-6916f2e5f426        |
| 6680f655-4806-4ea2-    |                          |                          |
| b86d-0d1fce159f23      | manila_service_network   |                          |
| d46023dd-60a0-4eeb-    |                          | 8f1a332c-4770-469f-      |
| 8c8c-b391842d390a      | Public                   | a75d-5021e0fbf140        |
+------------------------+--------------------------+--------------------------+
[root@controller ~]# openstack subnet list
+-------------------------+----------+---------------------+-------------------+
| ID                      | Name     | Status              | Subnet            |
+-------------------------+----------+---------------------+-------------------+
| 8f1a332c-4770-469f-a75d-| pub-     | d46023dd-60a0-4eeb- | 192.168.200.0/24  |
| 5021e0fbf140            | subnet   | 8c8c-b391842d390a   |                   |
```

```
|  d97147d1-87c4-4ec2-a290-  | private-  | 3641c6bf-90c1-4aa7-   | 10.0.0.0/24       |
|  6916f2e5f426              | subnet    | 83e3-cd3674811a28     |                   |
+----------------------------+-----------+-----------------------+-------------------+
[root@controller ~]# openstack router list
+---------------+--------+--------+-------+----------------+------------+---------+
| ID            | Name   | RAM    | State | Project        | Distributed| HA      |
+---------------+--------+--------+-------+----------------+------------+---------+
|  cb8569ae-8cb2-|        |        |       | 95ea70f0b429   |            |         |
|  4c46-b8e5-    | Pub7ic-| ACTIVE | UP    | 4187976e9dd5   | True       | False   |
|  cc39d47eaca1  | router |        |       | 6f605c94       |            |         |
+---------------+--------+--------+-------+----------------+------------+---------+
```

查看 private-net 资源堆栈的输出信息，从中可知创建路由时绑定的内部私有网络的端口设备 ID，命令和结果如下。

```
[root@controller ~]# openstack stack output show private-net router_interface_id
+--------------------+------------------------------------------------------------+
| Field              |  Value                                                     |
+--------------------+------------------------------------------------------------+
|  description       |  ID of the router interface                                |
|  output_key        |  router_interface_id                                       |
|  output_value      |  cb8569ae-8cb2-4c46-b8e5-cc39d47eaca1:subet_id=d97147d     |
|                    |  1-87c4-4ec2-a290-6916f2e5f426                             |
+--------------------+------------------------------------------------------------+
```

private-net 资源堆栈输出信息与 Dashboard 的内容是对应的，Dashboard 的内容如图 4-9 所示。

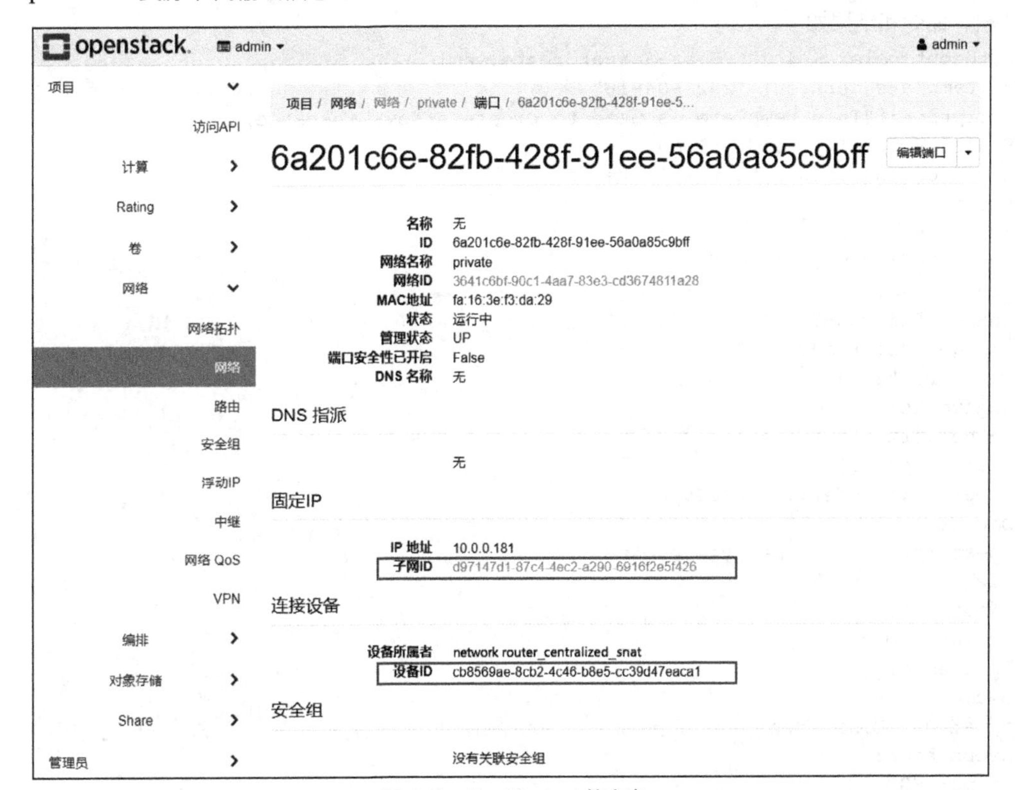

图 4-9　Dashboard 的内容

通过在 OpenStack Dashboard 中“项目→编排→堆栈”中查看信息，可以看到创建的 private-net 资源堆栈的拓扑信息如图 4-10 所示。

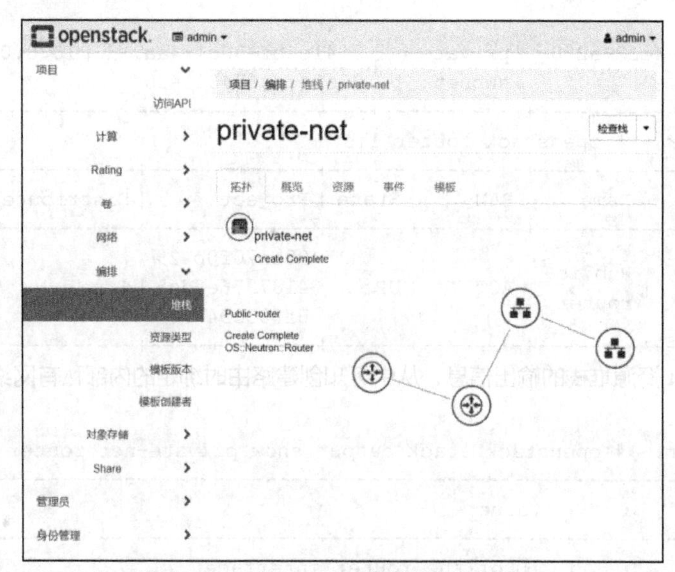

图4-10 private-net资源堆栈的拓扑信息

（5）创建云主机并挂载卷的模板文件

编写一个创建云主机cirros-0.6.1，并在云主机上挂载一块固态盘（Solid State Disk，SSD）类型的5GB大小的卷的模板文件Heat-ServerCreate.yaml，创建名为server的资源堆栈来按照自定义需求创建云主机，命令和结果如下。

```
[root@controller ~]# cat Heat-ServerCreate.yaml
heat_template_version: "2021-04-16"
description: "This template is used to create a Nova instance,
a Cinder VolumeType and Volume,
and attach the volume to the instance."
resources:
 create_instance:
  type: "OS::Nova::Server"
  properties:
   name: "cirros-0.6.1"
   image: "cirros-0.6.1"
   flavor: "2V_4G_20G"
   networks:
   - network: "private"
 Volume_Type_SSD:
  type: OS::Cinder::VolumeType
  properties:
   description: volume type create
   is_public: true
   #metadata:
   name: SSD
   projects: []
 volume:
  type: OS::Cinder::Volume
  properties:
   name: SSD_5G
   size: 5
   volume_type: {get_resource: Volume_Type_SSD}
 Attachment_instance:
  type: OS::Cinder::VolumeAttachment
```

```
  properties:
   instance_uuid: {get_resource: create_instance}
   # 可选参数，表示挂载路径，生效优先级低
   mountpoint: "/dev/vdb"
   volume_id: { get_resource: volume }
outputs:
 instance_name:
  description: The name of the created instance
  value: {get_attr: [create_instance, name]}
 volume_type:
  description: The volume type of the created volume
  value: {get_resource: Volume_Type_SSD}
 Attachment_instance_show:
  description: Show the details of the created attachment
  value: { get_attr: [Attachment_instance, show] }
[root@controller ~]# openstack stack create -t Heat-ServerCreate.yaml server
+----------------------+--------------------------------------------------------+
| Field                |  Value                                                 |
+----------------------+--------------------------------------------------------+
| id                   |  25b1ce26-9bbd-4dad-b205-a8ac19f881ee                  |
| stack_name           |  server                                                |
|                      |  This template is used to create a Nova instance, a cinder
| description          |  volumeType and volume, and attach the volume to the i |
|                      |  nstance.
| creation_time        |  2023-01-20T05:53:17Z                                  |
| updated_time         |  None                                                  |
| stack_status         |  CREATE_IN_PROGRESS                                    |
| stack_status_reason  |  stack CREATE started                                  |
+----------------------+--------------------------------------------------------+
[root@controller ~]# openstack server list
+------------------+-------------+--------+------------+------------+----------+
| ID               | Name        | State  | Networks   |  Image     | Flavor   |
+------------------+-------------+--------+------------+------------+----------+
| 2bf47abe -25ed-  |             |        | private=   | cirros-    |          |
| 4242-a721-       | cirros-0.6.1| ACTIVE | 10.0.0.149 | 0.6.1      | 2V_4G_20G|
| 84f72bde36f2     |             |        |            |            |          |
+------------------+-------------+--------+------------+------------+----------+
[root@controller ~]# openstack volume type list
+---------------------------------------+------------------+--------------+
| ID                                    | Name             | Is Public    |
+---------------------------------------+------------------+--------------+
| 004a3aa9-a7cc-4bb9-9122-ff117adbb2a5  | SSD              |True          |
| 4d528fa1-17cf-475f-be8a-2bf2832ec62a  | __DEFAULT__      |True          |
+---------------------------------------+------------------+--------------+
[root@controller ~]# openstack volume list
+-----------------------+----------+--------+------+-------------------------+
| ID                    |  Name    | Status | Size |  Attached to            |
+-----------------------+----------+--------+------+-------------------------+
| 158f4840-6b28-4393-8e41|         |        |      | Attached to cirros-0.6.1|
| -5d877f5d886f         |  SSD_5G  | in-use |  5   | on /dev/vdb             |
+-----------------------+----------+--------+------+-------------------------+
```

（6）创建绑定浮动 IP 地址的模板文件

编写一个为刚才创建的云主机 cirros-0.6.1 绑定浮动 IP 地址的模板文件 Heat-Floating_Ip_Association.yaml，创建名为 floating_ip 的资源堆栈来实现浮动 IP 地址的绑定，该绑定类似于公有云中公网 IP 地址的绑定，

命令和结果如下。

```
[root@controller ~]# openstack server list
+--------------------+---------+---------+----------------+--------------+----------+
| ID                 | Name    | Status  | Networks       | Image        | Flavor   |
+--------------------+---------+---------+----------------+--------------+----------+
| 2bf47abe-25ed-4242-| cirros- | ACTIVE  | private=       | cirros-0.6.1 | 2V_4G_20G |
| a721-84f72bde36f2  | 0.6.1   |         | 10.0.0.149     |              |          |
+--------------------+---------+---------+----------------+--------------+----------+
[root@controller ~]# openstack port list -f value | grep 10.0.0.149 |awk '{print $1}'
  3a1c4b21-015d-42f0-87f0-e898676ef490
[root@controller ~]# openstack port list |grep 10.0.0.149|awk '{print $2}'
  3a1c4b21-015d-42f0-87f0-e898676ef490
[root@controller ~]# cat Heat-Floating_Ip_Association.yaml
  heat_template_version: "2021-04-16"
  description: "This template is used to associate a floating IP to an \
  existing Nova instance and allocate a specific IP address."
  resources:
   floating_ip:
    type: OS::Neutron::FloatingIP
    properties:
     floating_network: "Public"
     # 定义申请指定的浮动 IP 地址
     floating_ip_address: "192.168.200.55"
     floating_subnet: "pub-subnet"
  floating_ip_association:
   type: "OS::Neutron::FloatingIPAssociation"
   properties:
    fixed_ip_address: 10.0.0.149
    floatingip_id: { get_resource: floating_ip }
    port_id: "3a1c4b21-015d-42f0-87f0-e898676ef490"
  outputs:
   floating_ip_address:
    description: Allocated floating IP address
    value: { get_attr: [floating_ip, floating_ip_address] }
[root@controller ~]# openstack stack create -t Heat-Floating_Ip_Association.yaml
floating_ip
+---------------------+------------------------------------------------------------+
| Field               | Value                                                      |
+---------------------+------------------------------------------------------------+
| id                  | 11b9e1ce-6f54-4031-ae29-a79afeac8131                       |
| stack_name          | floating_ip                                                |
| description         | This template is used to associate a floating IP to an existing |
|                     | Nova instance and allocate a specific IP address.          |
| creation_time       | 2023-01-20T06:24:40Z                                       |
| updated_time        | None                                                       |
| stack_status        | CREATE_IN_PROGRESS                                         |
| stack_status_reason | stack CREATE started                                       |
+---------------------+------------------------------------------------------------+
[root@controller ~]# openstack server list
+--------------------+---------+---------+------------------+----------+-----------+
|         ID         |  Name   | Status  |     Networks     |  Image   |  Flavor   |
+--------------------+---------+---------+------------------+----------+-----------+
| 2bf47abe-25ed-4242-| cirros- | ACTIVE  | private=10.0.0.  | cirros-  | 2V_4G_20G |
|  a721-84f72bde36f2 |  0.6.1  |         | 149,192.168.     |  0.6.1   |           |
|                    |         |         |   200.55         |          |           |
```

```
+-------------------------------+-----------+------------+-------------------+---------
----+-----------+
[root@controller ~]# openstack floating ip list
+------------+---------+-----------+-----------+----------+------------+
|     ID     | Floating|  Fixed IP |   Port    | Floating |   Project  |
|            |    IP   |  Address  |           |  Network |            |
|            | Address |           |           |          |            |
+------------+---------+-----------+-----------+----------+------------+
|  34ae71e9- |         |           | 3a1c4b21- | d46023dd-|            |
| 51cf-4193- | 192.168.| 10.0.0.149| 015d-42f0-| 60a0-4eeb-| 95ea70f0b4294|
| 8455-2e5628| 200.55  |           | 87f0-e898676| 8c8c-63918| 187976e9dd56f|
|   544070   |         |           |   ef490   | 42d390a  |   605c94   |
+------------+---------+-----------+-----------+----------+------------+
```

通过使用以下命令，查看 floating_ip 资源堆栈的所有输出信息，包括申请到的浮动 IP 地址，命令和结果如下。

```
[root@controller ~]# openstack stack output show floating_ip --all
+--------------------+--------------------------------------------------------+
| Field              | Value                                                  |
+--------------------+--------------------------------------------------------+
| floating_ip_address | {                                                     |
|                    |   "output_key": "floating_ip_address",                 |
|                    |   "description": "Allocated floating IP address",      |
|                    |   "output_value": "192.168.200.55"                     |
|                    | }                                                      |
+--------------------+--------------------------------------------------------+
[root@controller ~]# openstack stack show floating_ip -c outputs
+--------------------+--------------------------------------------------------+
| Field              | Value                                                  |
+--------------------+--------------------------------------------------------+
| outputs            | - description: Allocated floating IP address           |
|                    |   output_key: floating_ip_address                      |
|                    |   output_value: 192.168.200.55                         |
+--------------------+--------------------------------------------------------+
```

（7）Dashboard 验证

云主机 cirros-0.6.1 创建结果如图 4-11 所示，通过使用 Heat 模板文件，可以实现自动化的创建操作，使得整个过程更加便捷和高效。

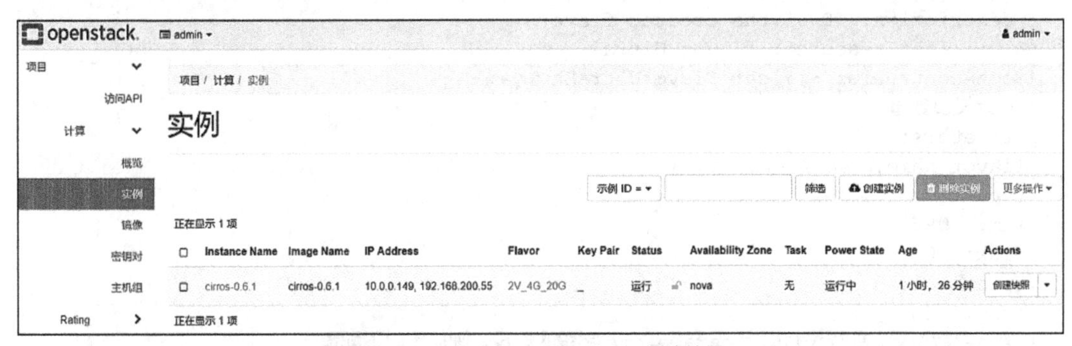

图 4-11　云主机 cirros-0.6.1 创建结果

3．模板文件的进阶编写及运用

之前编写的模板文件中的参数都是固定的，无法达到参数可变的效果，下面以创建云主机类型为例，通过使用 parameters 字段，在执行模板文件时将参数作为变量传入。

编写模板文件 Heat-FlavorUltra.yaml、parameters 字段变量文件 FlavorUltraEnv.yaml，创建名为 FlavorUltra 的资源堆栈来实现云主机类型的创建，命令和结果如下。

```
[root@controller ~]# cat Heat-FlavorUltra.yaml
heat_template_version: "2021-04-16"
description: "This template is used to create a Nova flavor."
parameters:
 flavor_name:
  type: string
  description: Name of the flavor to be created
 vcpus:
  type: number
  description: Number of virtual CPUs for the flavor
 ram:
  type: number
  description: Amount of RAM for the flavor in MB
 disk:
  type: number
  description: Amount of local disk storage for the flavor in GB
 ephemeral:
  type: number
  description: Amount of ephemeral disk storage for the flavor in GB
 swap:
  type: number
  description: Amount of swap space for the flavor in MB
 resources:
  flavor_Ultra:
   type: "OS::Nova::Flavor"
   properties:
    name: {get_param: flavor_name}
    vcpus: {get_param: vcpus}
    ram: {get_param: ram}
    disk: {get_param: disk}
    ephemeral: {get_param: ephemeral}
    swap: {get_param: swap}
    is_public: true
  outputs:
   flavor_id:
    description: ID of the created flavor
  value: {get_resource: flavor_Ultra}
  [root@controller ~]# cat FlavorUltraEnv.yaml
  # 自定义参数值
  parameters:
   flavor_name:  s6.xlarge.2
   vcpus: 2
   ram: 4096
   disk: 20
   ephemeral: 5
   swap: 256
  # 默认参数设置，如果没有以上 parameters 字段的配置，则应用以下配置
  parameter_defaults:
   flavor_name: c7.2xlarge.2
   vcpus: 8
   ram: 16384
   disk: 40
```

```
    ephemeral: 10
    swap: 512
[root@controller ~]# openstack stack create -t Heat-FlavorUltra.yaml \
-e FlavorUltraEnv.yaml FlavorUltra
+---------------------+------------------------------------------------------+
| Field               | Value                                                |
+---------------------+------------------------------------------------------+
| id                  | 19c0a480-5e11-4dba-af92-60508ea4ce70                 |
| stack_name          | FlavorUltra                                          |
| description         | This template is used to create a Nova flavor.      |
| creation_time       | 2023-01-20T07:40:53Z                                 |
| updated_time        | None                                                 |
| stack_status        | CREATE_IN_PROGRESS                                   |
| stack_status_reason | stack CREATE started                                 |
+---------------------+------------------------------------------------------+
[root@controller ~]# openstack flavor list
+---------------------+-------------+------+------+-----------+-------+-----------+
|        ID           |    Name     | RAM  | Disk | Ephemeral | vCPUs | Is Public |
+---------------------+-------------+------+------+-----------+-------+-----------+
|       9943          | 2V_4G_20G   | 4096 |  20  |     0     |   2   |   True    |
|  e13546bf-32f0-4379-|             |      |      |           |       |           |
|   80db-cc64f19497df | s6.xlarge.2 | 4096 |  20  |     5     |   2   |   True    |
+---------------------+-------------+------+------+-----------+-------+-----------+
```

查看 FlavorUltra 堆栈的所有文件列表和环境变量信息，命令和结果如下。

```
[root@controller ~]# openstack stack file list FlavorUltra
files:
  file:///root/FlavorUltraEnv.yaml: '{"parameters": {"flavor_name": "s6.xlarge.2",
   "vcpus":
2, "ram":
4096
, "disk":
20, "ephemeral":
5, "swap":
256}
, "parameter_
   defaults": {"flavor_name": "c7.2xlarge.2", "vcpus": 8, "ram": 16384, "disk": 40,
   "ephemeral": 10, "swap": 512}}'
[root@controller ~]# openstack stack environment show FlavorUltra
parameters:
    disk: 20
    ephemeral: 5
    flavor_ame: s6. xlarge.2
    ram: 4096
    swap: 256
    vcpus: 2
resource_registry:
    resources: {}
parameter_defaults:
    disk: 40
    ephemeral: 10
    flavor_name: c7.2xlarge.2
    ram: 16384
    swap: 512
    vcpus: 8
```

查看 FlavorUltra 堆栈的所有事件信息和堆栈是否符合预期状态，命令和结果如下。

```
[root@controller ~]# openstack stack event list FlavorUltra
2025-01-20 07:40:53Z [FlavorUltra]: _CREATE_IN_PROGRESS Stack CREATE started
2025-01-20 07:40:53Z [Flavorultra.flavor_ultra]: CREATE_IN_PROGRESS state changed
2023-01-20 07:40:54Z [Flavorultra.flavor_ultra]: CREATE_COMPLETE state changed
2023-01-20  07:40:54Z  [FlavorUltra]:  CREATE_COMPLETE  Stack  CREATE  completed
successfully
[root@controller ~]# openstack stack check FlavorUltra
+--------------------+-----------+-----------+-------------+------------+
|        ID          | Stack Name |Stack Status| Creation Time | Updated Time |
+--------------------+-----------+-----------+-------------+------------+
|  19c0a480-5e11-4dba- |            |CHECK_COMPLE |   2023-01-    |            |
|  af92-60508ea4ce70  | Flavorultra |     TE      | 20T07:40:53Z  |    None    |
+--------------------+-----------+-----------+-------------+------------+
```

除了通过使用前面编写的 parameters 字段变量文件 FlavorUltraEnv.yaml 的形式创建堆栈，还可以在创建堆栈时使用 parameter 参数，这样可以在执行命令时快速传递参数，而无须提前编写 env 文件。这对于短时间内的测试或者堆栈使用次数较少的情况是非常方便的，但是在生产环境中还是尽量使用 env 文件来管理参数，命令如下，结果如图 4-12 所示。

```
[root@controller ~]# openstack stack create -t Heat-FlavorUltra.yaml --parameter \
"flavor_name=Heat-FlavorUltra;vcpus=1;ram=2048;disk=10;ephemeral=5;swap=51
2" Heat-FlavorUltra
[root@controller ~]# openstack flavor list
```

```
root@controller:~# openstack stack create -t Heat-FlavorUltra.yaml --parameter \
> "flavor_name=Heat-FlavorUltra;vcpus=1;ram=2048;disk=10;ephemeral=5;swap=512" Heat-FlavorUltra
+---------------------+-----------------------------------------------+
| Field               | Value                                         |
+---------------------+-----------------------------------------------+
| id                  | fa54dcb4-0136-4da6-ac94-72b7bbf7a2d5          |
| stack_name          | Heat-FlavorUltra                              |
| description         | This template is used to create a Nova flavor. |
| creation_time       | 2023-01-20T07:47:31Z                          |
| updated_time        | None                                          |
| stack_status        | CREATE_IN_PROGRESS                            |
| stack_status_reason | Stack CREATE started                          |
+---------------------+-----------------------------------------------+
root@controller:~# openstack flavor list
+--------------------------------------+------------------+------+------+-----------+-------+-----------+
| ID                                   | Name             |  RAM | Disk | Ephemeral | vCPUs | Is Public |
+--------------------------------------+------------------+------+------+-----------+-------+-----------+
| 5bdbab4d-72ab-4070-8b3f-fbdaa9fde4be | Heat-FlavorUltra | 2048 |   10 |         5 |     1 | True      |
| 9943                                 | 2V_4G_20G        | 4096 |   20 |         0 |     2 | True      |
| e13546bf-32f0-4379-80db-cc64f19497df | s6.xlarge.2      | 4096 |   20 |         5 |     2 | True      |
+--------------------------------------+------------------+------+------+-----------+-------+-----------+
```

图 4-12　创建和查询堆栈的结果

将当前堆栈的状态创建为快照，可以方便地备份和恢复堆栈状态。下面演示如何创建堆栈 Heat-FlavorUltra 的快照 snap-flavor，命令如下，结果如图 4-13 所示。

```
[root@controller ~]# openstack stack snapshot create --name snap-flavor
Heat-FlavorUltra
[root@controller ~]# openstack stack snapshot list  Heat-FlavorUltra
```

```
root@controller:~# openstack stack snapshot create --name snap-flavor Heat-FlavorUltra
+----------------+--------------------------------------+
| Field          | Value                                |
+----------------+--------------------------------------+
| ID             | 07835cd5-98a9-4697-b511-13f02e51346f |
| name           | snap-flavor                          |
| status         | IN_PROGRESS                          |
| status_reason  | None                                 |
| data           | None                                 |
| creation_time  | 2023-01-20T11:47:44Z                 |
+----------------+--------------------------------------+
root@controller:~# openstack stack snapshot list  Heat-FlavorUltra
+--------------------------------------+-------------+----------+------------------------------------+----------------------+
| id                                   | name        | status   | status_reason                      | creation_time        |
+--------------------------------------+-------------+----------+------------------------------------+----------------------+
| 07835cd5-98a9-4697-b511-13f02e51346f | snap-flavor | COMPLETE | Stack SNAPSHOT completed successfully | 2023-01-20T11:47:44Z |
+--------------------------------------+-------------+----------+------------------------------------+----------------------+
```

图 4-13　创建堆栈快照的结果

任务 4.3.2　Ceilometer 监控服务运维管理

微课 4.9　Ceilometer 监控服务运维管理

1. 查看与服务相关的监控数据

查看当前存在的实例列表，命令和结果如下，可以发现实例 cirros-0.6.1 目前是关闭状态，启动该实例之后 Ceilometer 便可以监控相关资源数据。

```
[root@controller ~]# openstack server list -f json
[
 {
  "ID": "46ec44a0-ad70-42b3-9d29-d7050c8500fc",
  "Name": "cirros-0.6.1",
  "Status": "SHUTOFF",
  "Networks": {
   "int-net": [
    "10.0.0.185",
    "192.168.200.23"
   ]
  },
  "Image": "cirros-0.6.1",
  "Flavor": "2V_4G_20G"
 }
]
[root@controller ~]# openstack server start cirros-0.6.1
```

查看可用于监控的资源列表，命令如下，结果如图 4-14 所示。

```
[root@controller ~]# openstack metric resource list
```

```
root@controller:~# openstack metric resource list
| id                                   | type                       | project_id                       | user_id                          |
| revision_start                   | revision_end | creator                                                           |
| b0dcffff-2015-5876-bbf3-a90067d55f84 | instance_disk              | d502ed6ca8304eb9a7eee463f5e8a924 | 8d1170212a4c4d7f92d185b08de629e7 |
| 2023-01-29T05:45:45.228970+00:00 | None         | ba25cf17e4d249dfad554595e8d818e8:ac0a72482c1f40d79f54fa16e6dbba1e |
| 2e41a69c-af08-5dd7-a3ae-2a64393265d4 | instance_network_interface | d502ed6ca8304eb9a7eee463f5e8a924 | 8d1170212a4c4d7f92d185b08de629e7 |
| 2023-01-29T05:45:45.250755+00:00 | None         | ba25cf17e4d249dfad554595e8d818e8:ac0a72482c1f40d79f54fa16e6dbba1e |
| 46ec44a0-ad70-42b3-9d29-d7050c8500fc | instance                   | d502ed6ca8304eb9a7eee463f5e8a924 | 8d1170212a4c4d7f92d185b08de629e7 |
| 2023-01-29T05:45:45.279909+00:00 | None         | ba25cf17e4d249dfad554595e8d818e8:ac0a72482c1f40d79f54fa16e6dbba1e |
```

图 4-14　查看可用于监控的资源列表

仅查看监控实例类型的资源，命令如下，结果如图 4-15 所示。

```
[root@controller ~]# openstack metric resource list --type instance
[root@controller ~]# openstack metric resource show 46ec44a0-ad70-42b3-9d29-
d7050c8500fc
```

```
root@controller:~# openstack metric resource list --type instance
| id                                   | type     | project_id                       |
 user_id                          | original_resource_id                 | started_at                       |
 ended_at | revision_start                   | revision_end | creator                                                           |
 flavor_id                            | image_ref                            | host       | display_name | server_group | flavor_name |
 launched_at | created_at | deleted_at | availability_zone |
| 46ec44a0-ad70-42b3-9d29-d7050c8500fc | instance | d502ed6ca8304eb9a7eee463f5e8a924 |
 8d1170212a4c4d7f92d185b08de629e7 | 46ec44a0-ad70-42b3-9d29-d7050c8500fc | 2023-01-29T05:45:45.279903+00:00 |
 None     | 2023-01-29T05:45:45.279909+00:00 | None         | ba25cf17e4d249dfad554595e8d818e8:ac0a72482c1f40d79f54fa16e6dbba1e |
 337ca314-4e57-45a3-a5c6-2e18ce80a981 | 55758bd0-031b-455d-aea8-ca5c9e9e19e0 | controller | cirros-0.6.1 | None         | 2v_4G_20G   |
 None        | None       | None       | None              |
root@controller:~# openstack metric resource show 46ec44a0-ad70-42b3-9d29-d7050c8500fc
| Field                 | Value                                                             |
| created_by_project_id | ac0a72482c1f40d79f54fa16e6dbba1e                                  |
| created_by_user_id    | ba25cf17e4d249dfad554595e8d818e8                                  |
| creator               | ba25cf17e4d249dfad554595e8d818e8:ac0a72482c1f40d79f54fa16e6dbba1e |
| ended_at              | None                                                              |
| id                    | 46ec44a0-ad70-42b3-9d29-d7050c8500fc                              |
| metrics               | cpu: 64851946-ddcc-4536-8e78-cdb092792b4c                         |
|                       | memory.usage: 24e57da9-852f-46e7-942b-f1a290cd2016                |
| original_resource_id  | 46ec44a0-ad70-42b3-9d29-d7050c8500fc                              |
| project_id            | d502ed6ca8304eb9a7eee463f5e8a924                                  |
| revision_end          | None                                                              |
| revision_start        | 2023-01-29T05:45:45.279909+00:00                                  |
| started_at            | 2023-01-29T05:45:45.279903+00:00                                  |
| type                  | instance                                                          |
| user_id               | 8d1170212a4c4d7f92d185b08de629e7                                  |
```

图 4-15　仅查看监控实例类型的资源

查看CPU和memory.usage的资源监控数据，命令和结果如下。

```
[root@controller ~]# openstack metric measures show \
64851946-ddcc-4536-8e78-cdb092792b4c
+---------------------------+-------------+----------------+
| timestamp                 | granularity | value          |
+---------------------------+-------------+----------------+
| 2023-01-29T13:45:00+08:00 |  300.0      | 8620000000.0 |
……
+---------------------------+-------------+----------------+
 [root@controller ~]# openstack metric measures show \
24e57da9-852f-46e7-942b-f1a290cd2016
+---------------------------+-------------+----------------+
| timestamp                 | granularity | value          |
+---------------------------+-------------+----------------+
| 2023-01-29T13:45:00+08:00 |  300.0      | 169.52734375 |
……
+---------------------------+-------------+----------------+
```

仅查看所有网络接口实例类型的可用资源，命令如下，结果如图4-16所示。

```
[root@controller ~]# openstack metric resource list --type instance_network_interface
[root@controller ~]# openstack metric resource show \
2e41a69c-af08-5dd7-a3ae-2a64393265d4
```

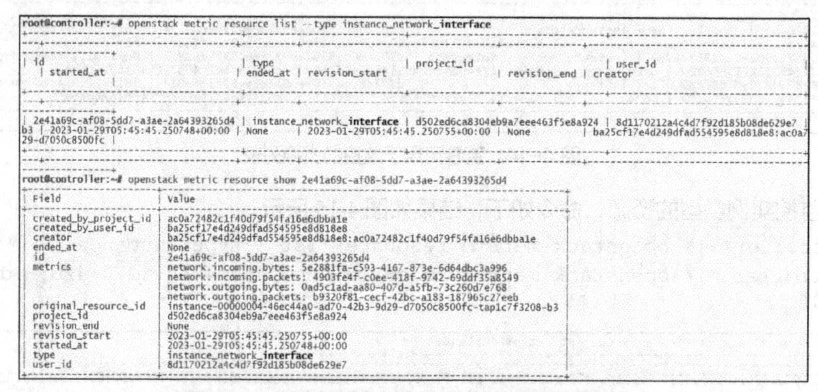

图4-16 仅查看所有网络接口实例类型的可用资源

查看实例接收和发送字节数的详细数据，命令和结果如下。

```
[root@controller ~]# openstack metric measures show \
5e2881fa-c593-4167-873e-6d64dbc3a996
+---------------------------+-------------+----------+
| timestamp                 | granularity | value    |
+---------------------------+-------------+----------+
| 2023-01-29T13:45:00+08:00 |  300.0      | 2834.0   |
……
+---------------------------+-------------+----------+
[root@controller ~]# openstack metric measures show \
0ad5c1ad-aa80-407d-a5fb-73c260d7e768
+---------------------------+-------------+----------+
| timestamp                 | granularity | value    |
+---------------------------+-------------+----------+
| 2023-01-29T13:45:00+08:00 |  300.0      | 3056.0   |
……
+---------------------------+-------------+----------+
```

2. 查看与镜像相关的监控数据

查看当前镜像列表，将镜像 cirros-0.6.1 下载并保存为本地 wxic-cirros.qcow2 文件，使用下载的镜像文件创建一个新的镜像 wxic-cirros，命令和结果如下。

```
[root@controller ~]# openstack image list -f json
[
 {
  "ID": "55758bd0-031b-455d-aea8-ca5c9e9e19e0",
  "Name": "cirros-0.6.1",
  "Status": "active"
 }
]
[root@controller ~]# openstack image save --file wxic-cirros.qcow2 cirros-0.6.1
[root@controller ~]# openstack image create --disk-format qcow2  \
--progress --public --file wxic-cirros.qcow2 "wxic-cirros"
```

列出所有镜像类型的可用资源，命令如下，结果如图 4-17 所示。

```
[root@controller ~]# openstack metric resource list --type image
[root@controller ~]# openstack metric resource show \
2e1cb57c-698d-40ad-b0f4-d5ec50155198
```

```
root@controller:~# openstack metric resource list --type image
+--------------------------------------+-------+----------------------------------+---------+--------------------------------------+
| id                                   | type  | project_id                       | user_id | original_resource_id                 |
 started_at                       | ended_at | revision_start                   | revision_end | creator
                                  | name     | container_format | disk_format |
+--------------------------------------+-------+----------------------------------+---------+--------------------------------------+
| 2e1cb57c-698d-40ad-b0f4-d5ec50155198 | image | d502ed6ca8304eb9a7eee463f5e8a924 | None    | 2e1cb57c-698d-40ad-b0f4-d5ec50155198 |
 2023-01-29T05:58:19.971464+00:00 | None     | 2023-01-29T05:58:19.971471+00:00 | None    | ba25cf17e4d249dfad554595e8d818e8:ac
0a72482c1f40d79f54fa16e6dbba1e | wxic-cirros | bare             | qcow2       |
+--------------------------------------+-------+----------------------------------+---------+--------------------------------------+
root@controller:~# openstack metric resource show 2e1cb57c-698d-40ad-b0f4-d5ec50155198
+-----------------------+--------------------------------------------------------------------+
| Field                 | Value                                                              |
+-----------------------+--------------------------------------------------------------------+
| created_by_project_id | ac0a72482c1f40d79f54fa16e6dbba1e                                   |
| created_by_user_id    | ba25cf17e4d249dfad554595e8d818e8                                   |
| creator               | ba25cf17e4d249dfad554595e8d818e8:ac0a72482c1f40d79f54fa16e6dbba1e  |
| ended_at              | None                                                               |
| id                    | 2e1cb57c-698d-40ad-b0f4-d5ec50155198                               |
| metrics               | image.download: a974c893-79af-48b6-a6ae-4c7d795432d8               |
|                       | image.serve: b689b611-b485-452d-805b-77ba7cbf613f                  |
|                       | image.size: d86a2e77-25d0-48ae-acc2-6c90cd9fa063                   |
| original_resource_id  | 2e1cb57c-698d-40ad-b0f4-d5ec50155198                               |
| project_id            | d502ed6ca8304eb9a7eee463f5e8a924                                   |
| revision_end          | None                                                               |
| revision_start        | 2023-01-29T05:58:19.971471+00:00                                   |
| started_at            | 2023-01-29T05:58:19.971464+00:00                                   |
| type                  | image                                                              |
| user_id               | None                                                               |
+-----------------------+--------------------------------------------------------------------+
```

图 4-17　列出所有镜像类型的可用资源

查看镜像的监控数据，命令和结果如下。

```
[root@controller ~]# openstack metric measures show \
d86a2e77-25d0-48ae-acc2-6c90cd9fa063
+---------------------------+-------------+------------+
| timestamp                 | granularity | value      |
+---------------------------+-------------+------------+
| 2023-01-29T13:55:00+08:00 | 300.0       | 21233664.0 |
+---------------------------+-------------+------------+
```

此处只演示了查看 image.size 资源的监控数据的方法，按照此方法还可以查看 image.download、image.serve 的监控数据。

3. 查看与卷相关的监控数据

创建 10GB 大小的测试卷 disk01，命令如下。

```
[root@controller ~]# openstack volume create --size 10 disk01
```

仅查看与卷相关的可用资源，命令如下，结果如图 4-18 所示。

```
[root@controller ~]# openstack metric resource list --type volume
[root@controller ~]# openstack metric resource show \
6626f004-03a8-4a2f-9581-bd9c0dff30c4
```

```
root@controller:~# openstack metric resource list --type volume
| id                                   | type   | project_id                       | user_id                          |
     | ended_at | revision_start                   | revision_end | creator
   | image_id | instance_id |
| 6626f004-03a8-4a2f-9581-bd9c0dff30c4 | volume | d502ed6ca8304eb9a7eee463f5e8a924 | 8d1170212a4c4d7f92d185b08de629e7 |
00:00 | None     | 2023-01-29T06:07:41.238727+00:00 | None         | ba25cf17e4d249dfad554595e8d818e8:ac0a72482c1f40d79f
5d | None     | None        |
root@controller:~# openstack metric resource show 6626f004-03a8-4a2f-9581-bd9c0dff30c4
+-----------------------+-------------------------------------------------------------------+
| Field                 | Value                                                             |
+-----------------------+-------------------------------------------------------------------+
| created_by_project_id | ac0a72482c1f40d79f54fa16e6dbba1e                                  |
| created_by_user_id    | ba25cf17e4d249dfad554595e8d818e8                                  |
| creator               | ba25cf17e4d249dfad554595e8d818e8:ac0a72482c1f40d79f54fa16e6dbba1e |
| ended_at              | None                                                              |
| id                    | 6626f004-03a8-4a2f-9581-bd9c0dff30c4                              |
| metrics               | volume.size: b9ae838c-4aae-4cb8-a724-e0f0de8dafed                 |
| original_resource_id  | 6626f004-03a8-4a2f-9581-bd9c0dff30c4                              |
| project_id            | d502ed6ca8304eb9a7eee463f5e8a924                                  |
| revision_end          | None                                                              |
| revision_start        | 2023-01-29T06:07:41.238727+00:00                                  |
| started_at            | 2023-01-29T06:07:41.238720+00:00                                  |
| type                  | volume                                                            |
| user_id               | 8d1170212a4c4d7f92d185b08de629e7                                  |
+-----------------------+-------------------------------------------------------------------+
```

图 4-18　仅查看与卷相关的可用资源

查看卷的监控数据，可以发现此时监控到的数值为 10GB，与实际情况一致，命令和结果如下。

```
[root@controller ~]# openstack metric measures show \
b9ae838c-4aae-4cb8-a724-e0f0de8dafed
+---------------------------+-------------+-------+
| timestamp                 | granularity | value |
+---------------------------+-------------+-------+
| 2023-01-29T14:05:00+08:00 |  300.0      | 10.0  |
+---------------------------+-------------+-------+
```

创建一个 20GB 大小的测试卷 disk02，查看其资源详情，命令如下，结果如图 4-19 所示。

```
[root@controller ~]# openstack metric resource list --type volume
[root@controller ~]# openstack metric resource show \
848594cb-94d6-4298-90d9-35371e292a0e
```

```
root@controller:~# openstack metric resource list --type volume
| id                                   | type   | project_id                       |
 user_id                          | original_resource_id                 | started_at
     | ended_at | revision_start                   | revision_end | creator
                                                  | display_name | volume_type
| image_id | instance_id |
| 6626f004-03a8-4a2f-9581-bd9c0dff30c4 | volume | d502ed6ca8304eb9a7eee463f5e8a924 |
 8d1170212a4c4d7f92d185b08de629e7 | 6626f004-03a8-4a2f-9581-bd9c0dff30c4 | 2023-01-29T06:07:41.238720+
00:00 | None     | 2023-01-29T06:07:41.238727+00:00 | None         | ba25cf17e4d249d
fad554595e8d818e8:ac0a72482c1f40d79f54fa16e6dbba1e | disk01       | ff8dd8e8-6595-4d5f-b10e-142f5c4f1e5d
| None     | None        |

| 848594cb-94d6-4298-90d9-35371e292a0e | volume | d502ed6ca8304eb9a7eee463f5e8a924 |
 8d1170212a4c4d7f92d185b08de629e7 | 848594cb-94d6-4298-90d9-35371e292a0e | 2023-01-29T06:09:33.872151+
00:00 | None     | 2023-01-29T06:09:33.872158+00:00 | None         | ba25cf17e4d249d
fad554595e8d818e8:ac0a72482c1f40d79f54fa16e6dbba1e | disk02       | ff8dd8e8-6595-4d5f-b10e-142f5c4f1e5d
| None     | None        |
root@controller:~# openstack metric resource show 848594cb-94d6-4298-90d9-35371e292a0e
+-----------------------+-------------------------------------------------------------------+
| Field                 | Value                                                             |
+-----------------------+-------------------------------------------------------------------+
| created_by_project_id | ac0a72482c1f40d79f54fa16e6dbba1e                                  |
| created_by_user_id    | ba25cf17e4d249dfad554595e8d818e8                                  |
| creator               | ba25cf17e4d249dfad554595e8d818e8:ac0a72482c1f40d79f54fa16e6dbba1e |
| ended_at              | None                                                              |
| id                    | 848594cb-94d6-4298-90d9-35371e292a0e                              |
| metrics               | volume.size: 22b61553-9765-43db-96d1-4d7045985fc6                 |
| original_resource_id  | 848594cb-94d6-4298-90d9-35371e292a0e                              |
| project_id            | d502ed6ca8304eb9a7eee463f5e8a924                                  |
| revision_end          | None                                                              |
| revision_start        | 2023-01-29T06:09:33.872158+00:00                                  |
| started_at            | 2023-01-29T06:09:33.872151+00:00                                  |
| type                  | volume                                                            |
| user_id               | 8d1170212a4c4d7f92d185b08de629e7                                  |
+-----------------------+-------------------------------------------------------------------+
```

图 4-19　查看测试卷 disk02 的资源详情

查看卷的监控数据，可以发现此时监控到的数值已经为 20GB 了，命令和结果如下。

```
[root@controller ~]# openstack metric measures show \
22b61553-9765-43db-96d1-4d7045985fc6
+-------------------------------+-------------+-------+
| timestamp                     | granularity | value |
+-------------------------------+-------------+-------+
| 2023-01-29T14:05:00+08:00 |  300.0      | 20.0  |
+-------------------------------+-------------+-------+
```

任务 4.3.3 Cloudkitty 计费服务运维管理

1. 了解 Cloudkitty 服务命令

微课 4.10 Cloudkitty 计费服务运维管理

列出 Cloudkitty 可用的映射类型，包括“flat”（单一值）和“rate”（按照时间或数量计费），命令和结果如下。

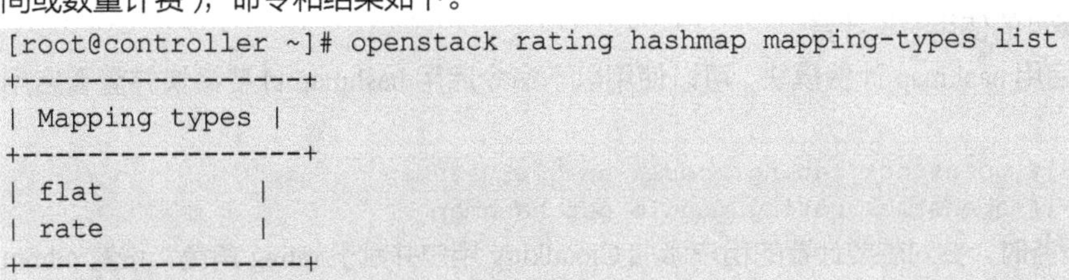

```
[root@controller ~]# openstack rating hashmap mapping-types list
+---------------+
| Mapping types |
+---------------+
| flat          |
| rate          |
+---------------+
# 以上命令等价于以下命令
[root@controller ~]# cloudkitty hashmap mapping-types list
```

列出所有租户列表，命令和结果如下。

```
[root@controller ~]# cloudkitty report tenant list
+----------------------------------+
| Tenant ID                        |
+----------------------------------+
| d502ed6ca8304eb9a7eee463f5e8a924 |
+----------------------------------+
# 查看该 ID 可以发现，该 ID 对应的租户其实就是 admin 项目，因为实验环境中还没有创建其他租户
[root@controller ~]# openstack project list  -f json
[
 {
  "ID": "ac0a72482c1f40d79f54fa16e6dbba1e",
  "Name": "service"
 },
 {
  "ID": "d502ed6ca8304eb9a7eee463f5e8a924",
  "Name": "admin"
 }
]
```

查看用于计费度量的列表，其中的元数据可以用于计费，命令和结果如下。

```
[root@controller ~]# openstack rating info metric list
+-----------------------------+--------+----------------------------------+
|Metric                       |Unit    | Metadata                         |
+-----------------------------+--------+----------------------------------+
|instance                     |instance|['flavor_name', 'flavor_id', 'vcpus'] |
|image.size                   | MiB    |['disk_format', 'container_format'] |
|volume.size                  | GiB    |['volume_type']                   |
|network.outgoing.bytes.rate  | MB     |['instance_id']                   |
|network.incoming.bytes.rate  | MB     |['instance_id']                   |
|ip.floating                  | ip     |['state']                         |
```

```
|radosgw.objects.size       | GiB   |[]                                 |
+---------------------------+-------+-----------------------------------+
# 以上命令等价于以下命令
[root@controller ~]# cloudkitty info metric list
```

查看当前的计费信息，命令和结果如下。

```
[root@controller ~]# openstack rating summary get
+--------------------+-----------+------+-------------------+------------------+
|     Tenant ID      | Resource  | Rate |    Begin Time     |    End Time      |
|                    |   Type    |      |                   |                  |
+--------------------+-----------+------+-------------------+------------------+
| 0d393503f5d34b9c8630 |   ALL   |  0   |2023-01-01T00:00:00|2023-02-01T00:00:00|
|    3090f41c9a56    |           |      |                   |                  |
+--------------------+-----------+------+-------------------+------------------+
```

2. 使用 Cloudkitty 服务

（1）根据云主机类型的使用时长计费

云主机默认没有启用 hashmap 计费模块，可以使用以下命令启用 hashmap 计费模块并查看启用状态。

```
[root@controller ~]# openstack rating module enable hashmap
[root@controller ~]# openstack rating module get hashmap
```

使用 Cloudkitty 服务时，要对想要计费的租户添加 Cloudkitty 用户并赋予 rating 角色，设置 admin 租户的命令如下。

```
[root@controller ~]# openstack role add --user cloudkitty --project admin rating
```

根据云主机实例类型 ID 和正常运行时间的计费规则对云主机进行计费，创建一个 instance_uptime_flavor_id 组，命令和结果如下。

```
[root@controller ~]# openstack rating hashmap \
group create instance_uptime_flavor_id
[root@controller ~]# openstack rating hashmap group list
+---------------------------+--------------------------------------+
| Name                      | Group ID                             |
+---------------------------+--------------------------------------+
| instance_uptime_flavor_id | ce59e66f-1ada-4d32-a33e-19c99a86083c |
+---------------------------+--------------------------------------+
```

创建服务匹配规则 instance，命令和结果如下。

```
[root@controller ~]# openstack rating hashmap service create instance
+----------+--------------------------------------+
| Name     | Service ID                           |
+----------+--------------------------------------+
| instance | e0f6e1f2-7d5e-42fb-8280-8e335ee21abc |
+----------+--------------------------------------+
```

创建字段匹配规则 flavor_id，命令如下。

```
[root@controller ~]# openstack rating hashmap field create \
e0f6e1f2-7d5e-42fb-8280-8e335ee21abc flavor_id
```

在 instance_uptime_flavor 组中创建映射 2V_4G_20G 云主机类型的实例成本为 1 元，命令和结果如下。

```
[root@controller ~]# openstack flavor list
+--------------------+-----------+------+------+-----------+-------+-----------+
|        ID          |   Name    | RAM  | Disk | Ephemeral | vCPUs | Is Public |
+--------------------+-----------+------+------+-----------+-------+-----------+
| 98e69fab-13e5-4e01-| 2V_4G_20G | 4096 |  20  |     0     |   2   |   True    |
|  abed-66c62a167193 |           |      |      |           |       |           |
```

```
+--------------------+------------+------+------+----------+------+----------+
[root@controller ~]# openstack rating hashmap mapping create 1 \
--field-id bb9ad708-69e2-44e9-ad9d-74d7620b3fa0 \
--value 98e69fab-13e5-4e01-abed-66c62a167193 \
-g ce59e66f-1ada-4d32-a33e-19c99a86083c -t flat
#查询费用信息，计费规则为刚创建时 Rate 的值为 0，经过一定时间后将开始计费（默认计费周期为 1h）
#更改计费周期可以通过修改 cloudkitty.conf 文件的 period 参数来实现，计费周期默认为 3600s，即 1h
[root@controller ~]# openstack rating summary get
```

（2）根据使用卷的大小进行计费

创建卷的计费规则，创建 volume_thresholds 组，命令和结果如下。

```
[root@controller ~]# openstack rating hashmap group create volume_thresholds
[root@controller ~]# openstack rating hashmap group list
+------------------------------+--------------------------------------+
| Name                         | Group ID                             |
+------------------------------+--------------------------------------+
| volume_thresholds            | 5020af51-cfe9-40e7-9173-a0c8395383e5 |
| instance_uptime_flavor_id    | ce59e66f-1ada-4d32-a33e-19c99a86083c |
+------------------------------+--------------------------------------+
```

创建服务匹配规则 volume.size，命令和结果如下。

```
[root@controller ~]# openstack rating hashmap service create volume.size
+--------------+--------------------------------------+
| Name         | Service ID                           |
+--------------+--------------------------------------+
| volume.size  | 2a958378-8aee-4b9e-9865-20a03fa253a4 |
+--------------+--------------------------------------+
```

设置费率，1GB 的费用为 0.05 元，命令如下。

```
[root@controller ~]# openstack rating hashmap mapping create 0.05 \
-s 2a958378-8aee-4b9e-9865-20a03fa253a4 \
-g 5020af51-cfe9-40e7-9173-a0c8395383e5 -t flat
```

创建优惠服务，若使用的卷大小超过 20GB，则应用 2%的折扣。如果要对特定的项目租户进行计费，则应使用参数-p，命令如下。

```
[root@controller ~]# openstack rating hashmap threshold create 20 0.98 \
-s 2a958378-8aee-4b9e-9865-20a03fa253a4 \
-g 5020af51-cfe9-40e7-9173-a0c8395383e5 -t rate
```

设置若使用卷大小超过 40GB，则应用 5%的折扣，命令如下。

```
[root@controller ~]# openstack rating hashmap threshold create 40 0.95 \
-s 2a958378-8aee-4b9e-9865-20a03fa253a4 \
-g 5020af51-cfe9-40e7-9173-a0c8395383e5 -t rate
```

创建大小不同的卷，测试不同计费规则下卷的费用，命令如下。

```
[root@controller ~]# openstack volume create --size 10 disk01
[root@controller ~]# openstack volume create --size 20 disk02
[root@controller ~]# openstack volume create --size 40 disk03
```

经过几个小时后，查看已经产生的费用，目前产生的费用为 80.0124 元，命令和结果如下。

```
[root@controller ~]# openstack rating summary get
+--------------------+-----------+------+---------------------+---------------------+
|     Tenant ID      | Resource  | Rate |     Begin Time      |      End Time       |
|                    |   Type    |      |                     |                     |
+--------------------+-----------+------+---------------------+---------------------+
| d502ed6ca8304eb9a7ee |   ALL   | 80.0 | 2023-01-01T00:00:00 | 2023-02-01T00:00:00 |
|   e463f5e8a924     |           | 124  |                     |                     |
+--------------------+-----------+------+---------------------+---------------------+
```

查看检索计费过程中产生的数据帧，命令如下。

```
[root@controller ~]# openstack rating dataframes get
```

获取过去 30 天的资源使用情况，命令和结果如下。

```
[root@controller ~]# openstack usage show
Usage from 2023-12-22 to 2024-01-20 on project
95ea70f0b4294187976e9dd56f605c94:
+-------------------+-----------------------------------------------+
| Field             | Value                                         |
+-------------------+-----------------------------------------------+
| Project           | 95ea70f0b4294187976e9dd56f605c94              |
| Servers           | 8                                             |
| RAM MB-Hours      | 69035.8                                       |
| CPU Hours         | 33.71                                         |
| Disk GB-Hours     | 337.09                                        |
+-------------------+-----------------------------------------------+
```

项目小结

本项目首先介绍了云服务组件的定义和基本概念；在此基础上讲解了云基础服务组件的运维管理，如对认证服务、镜像服务、网络服务和计算服务的运维管理，针对存储服务组件的运维管理，如块存储服务、对象存储服务和共享文件系统服务的运维管理，以及对高级服务组件的运维管理，如对编排服务、监控服务和计费服务的运维管理。

拓展知识

OpenStack 网络服务的运维命令思维导图

在项目 4 中，介绍了 OpenStack 的云基础服务组件、存储服务组件和高级服务组件，其实 OpenStack 所提供的服务组件远不止这些。围绕这些服务组件的运维命令的数量繁多，要想熟记这些命令，除了需要不断实践之外，还可借助思维导图。下面以 Neutron 网络服务为例，为读者提供与其运维命令相应的思维导图，如图 4-20 所示。

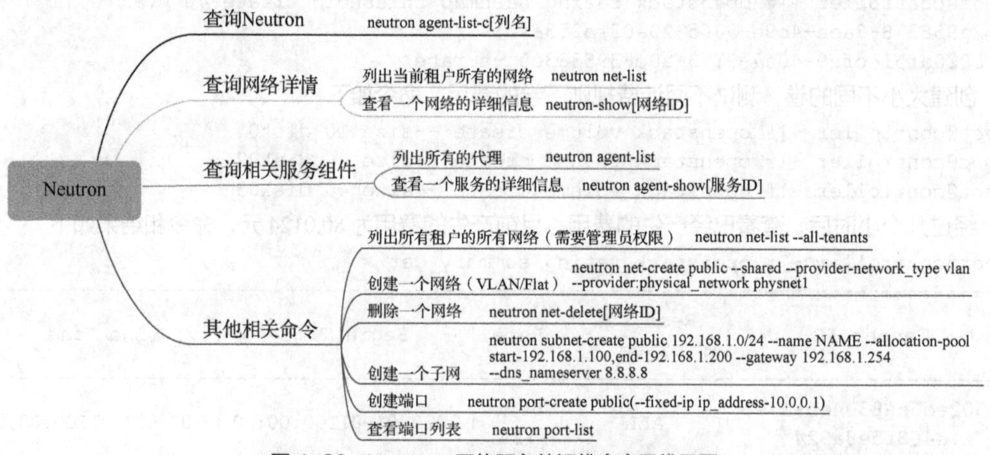

图 4-20　Neutron 网络服务的运维命令思维导图

知识巩固

1. 单选

（1）下列不属于 Keystone 主要功能的是（　　）。

A．身份认证　　B．用户管理　　C．文件管理　　D．用户授权

（2）下列属于基础服务的是（　　）。

A．Heat　　B．Glance　　C．Zun　　D．Swift

（3）（　　）网络中的云主机实例能与位于同一网络的云主机实例进行通信，并且可以跨多个节点。

A．Flat　　B．VLAN　　C．VXLAN　　D．GRE

2. 填空

（1）Neutron 网络类型可分为____________、____________、____________、____________和____________。

（2）Cloudkitty 当前的计费模型有 3 个，分别是 noop、_______和 pyscripts。

3. 简答

（1）简述 Keystone 的工作流程。

（2）说明 Flat 网络类型和 VLAN 网络类型的不同之处。

（3）简述公有云服务的计费模式。

拓展任务

openEuler 22.09 部署 NFS 服务

NFS 是当前主流异构平台的共享文件系统之一，主要应用在 UNIX 环境中。它最早是由 Sun Microsystems 开发的，能够支持在不同类型的系统之间通过网络进行文件共享，并允许一个系统在网络上与其他系统共享目录和文件。通过使用 NFS，用户和程序可以像访问本地文件一样访问远端系统中的文件，每个计算机的节点也能够像使用本地资源一样方便地使用网络中的资源。换言之，NFS 可支持用户在不同类型的计算机、操作系统、网络架构和传输协议运行环境中远程访问及共享网络文件。

微课 4.11　openEuler 22.09 部署 NFS 服务

任务步骤

1. 规划要搭建的 NFS 服务节点

搭建 NFS 服务节点的规划如表 4-1 所示。

表 4-1　搭建 NFS 服务节点的规划

IP 地址	主机名	节点
192.168.100.10	nfs-server	NFS-SERVER 节点
192.168.100.20	nfs-client	NFS-CLIENT 节点

2. 配置 NFS-SERVER 节点

修改主机名，命令如下。

```
[root@localhost ~]# hostnamectl set-hostname nfs-server
[root@localhost ~]# exec bash
[root@nfs-server ~]#
```

安装 NFS 服务并设置开机自启动，命令如下。

```
[root@nfs-server ~]# dnf -y install nfs-utils
[root@nfs-server ~]# systemctl enable --now nfs-server
```

查看 NFS 的版本信息，“+”表示支持，“-”表示不支持，命令和结果如下。

```
[root@nfs-server ~]# cat /proc/fs/nfsd/versions
-2 +3 +4 +4.1 +4.2
```

安装 NFS 服务后，远程过程调用（Remote Procedure Call，RPC）服务已经启用了对 NFS 的端口映射列表，查看 RPC 服务本地端口映射详情，命令如下，结果如图 4-21 所示。

```
[root@nfs-server ~]# rpcinfo -p localhost
```

```
[root@nfs-server ~]# rpcinfo -p localhost
   program vers proto   port  service
    100000    4   tcp    111  portmapper
    100000    3   tcp    111  portmapper
    100000    2   tcp    111  portmapper
    100000    4   udp    111  portmapper
    100000    3   udp    111  portmapper
    100000    2   udp    111  portmapper
    100024    1   udp  38518  status
    100024    1   tcp  34971  status
    100005    1   udp  20048  mountd
    100005    1   tcp  20048  mountd
    100005    2   udp  20048  mountd
    100005    2   tcp  20048  mountd
    100005    3   udp  20048  mountd
    100005    3   tcp  20048  mountd
    100003    3   tcp   2049  nfs
    100003    4   tcp   2049  nfs
    100227    3   tcp   2049  nfs_acl
    100021    1   udp  42678  nlockmgr
    100021    3   udp  42678  nlockmgr
    100021    4   udp  42678  nlockmgr
    100021    1   tcp  44661  nlockmgr
    100021    3   tcp  44661  nlockmgr
    100021    4   tcp  44661  nlockmgr
[root@nfs-server ~]#
```

图 4-21　RPC 服务本地端口映射详情

启用 NFS 4.2 并禁用其他版本，命令如下。

```
[root@nfs-server ~]# nfsconf --set nfsd udp n
[root@nfs-server ~]# nfsconf --set nfsd tcp y
[root@nfs-server ~]# nfsconf --set nfsd vers2 n
[root@nfs-server ~]# nfsconf --set nfsd vers3 n
[root@nfs-server ~]# nfsconf --set nfsd vers4 y
[root@nfs-server ~]# nfsconf --set nfsd vers4.1 y
[root@nfs-server ~]# nfsconf --set nfsd vers4.2 y
```

因为启用 NFS 4.0 后无须使用 RPC 服务，所以此时需要禁用 RPC 服务，命令如下。

```
[root@nfs-server ~]# systemctl mask --now rpc-statd.service
[root@nfs-server ~]# systemctl mask --now rpcbind.service
[root@nfs-server ~]# systemctl mask --now rpcbind.socket
```

设置防火墙放行 NFS 服务，命令和结果如下。

```
[root@nfs-server ~]# firewall-cmd --add-service=nfs --permanent
success
[root@nfs-server ~]# firewall-cmd --reload
success
```

关闭 SELinux，命令如下。

```
[root@nfs-server ~]# setenforce 0
[root@nfs-server ~]# sed -i \
's/^SELINUX=.*/SELINUX=permissive/g' /etc/selinux/config
```

创建 NFS 共享目录，命令如下。

```
[root@nfs-server ~]# mkdir -p /wxic/nfs-share
```

修改 NFS 配置文件/etc/exports，命令和结果如下。

```
[root@nfs-server ~]# cat << WXIC > /etc/exports
/wxic/nfs-share 192.168.100.0/24(rw,no_subtree_check,no_root_squash)
WXIC
```

该配置文件的内容可分为 3 部分，第一部分为本地 nfs-server 端的共享目录；第二部分为允许访问的主机（可以是 IP 地址或者一个网段范围）；第三部分为权限选项，其详细列举如表 4-2 所示。

表 4-2　第三部分权限选项的详细列举

选项	描述
rw	允许在 NFS 卷上读取和写入请求
ro	仅允许在 NFS 卷上读取请求
sync	同步模式，表示把内存中的数据实时写入磁盘（默认）
async	非同步模式，表示把内存中的数据定期写入磁盘
secure	只允许客户端使用 TCP 进行访问（默认）
insecure	允许客户端使用不安全的用户数据报协议（User Datagram Protocol，UDP）进行访问
wdelay	开启写入延迟，确保 NFS 服务器的数据一致性（默认）
no_wdelay	关闭写入延迟，提高 NFS 服务器的性能，但可能会导致数据丢失
subtree_check	启用子树检查（默认）
no_subtree_check	禁用子树检查，这具有轻微的安全隐患，但在某些情况下可以提高服务器的性能
root_squash	限制 root 用户对共享目录的权限，使其只具有普通用户的权限
no_root_squash	使 root 用户对共享的目录有最高的权限控制
all_squash	将所有 uid 和 gid 映射到匿名用户
no_all_squash	保留客户端用户的身份和权限，no_all_squash 选项仅在与 root_squash 或 no_root_squash 选项一起使用时才有效（默认）
sec	指定在 NFS 共享上使用系统级别的安全机制，不提供加密传输和数据完整性验证，不适合在不受信任的网络上使用。其中，选项 sec=krb5i 或 sec=krb5p 可以提供加密传输和数据完整性验证
anonuid	限定使用者的 uid
anongid	限定使用者的 gid

重启 NFS 服务，命令如下。

```
[root@nfs-server ~]# systemctl restart nfs-server
```

使 NFS 配置文件生效，命令和结果如下。

```
[root@nfs-server ~]# exportfs -rv
exporting 192.168.100.0/24:/wxic/nfs-share
[root@nfs-server ~]# exportfs
/wxic/nfs-share
        192.168.100.0/24
```

查看本地 NFS 服务的文件共享列表，命令和结果如下。

```
[root@nfs-server ~]# showmount -e localhost
Export list for localhost:
/wxic/nfs-share 192.168.100.0/24
```

3. 配置nfs-client节点

修改主机名，命令如下。

```
[root@localhost ~]# hostnamectl set-hostname nfs-client
[root@localhost ~]# exec bash
[root@nfs-client ~]#
```

安装NFS客户端工具，命令如下。

```
[root@nfs-client ~]# dnf -y install nfs-utils
```

4. 挂载使用

当前环境下NFS服务的服务端与客户端的简易架构如图4-22所示。

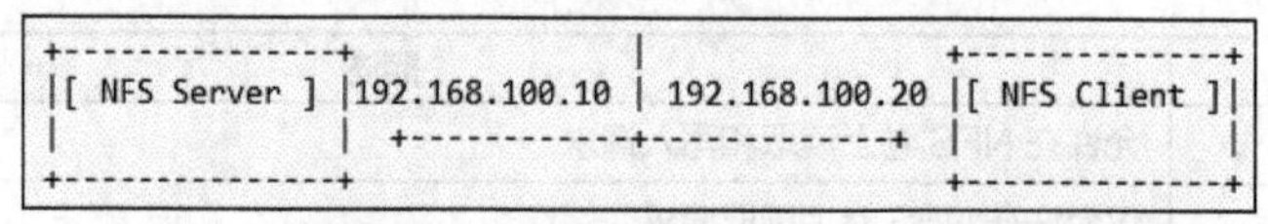

图4-22 当前环境下NFS服务的服务端与客户端的简易架构

在客户端将NFS共享目录挂载至/mnt目录，并查看挂载目录的大小及使用率等，命令和结果如下。

```
[root@nfs-client ~]# mount -t nfs 192.168.100.10:/wxic/nfs-share /mnt/
[root@nfs-client ~]# df -hT
Filesystem                    Type     Size  Used  Avail  Use%   Mounted ON
tmpfs                         tmpfs    1.2G  1.6M  1.2G   1%     /run
/dev/sda4                     xfs      115G  7.9G  108G   7%     /
……
192.168.100.10:/wxic/nfs-share  nfs4  115G  8.6G  107G   8%  /mnt
```

如果需要在系统启动时自动挂载，则需要配置/etc/fstab文件，命令和配置内容如下。

```
[root@nfs-client ~]# vi /etc/fstab
# 在末尾添加，设置NFS共享目录开机自动挂载
192.168.100.10:/wxic/nfs-share  /mnt   nfs    defaults      0      0
```

设置开机自动挂载的相关参数说明，如表4-3所示。

表4-3 开机自动挂载的相关参数说明

参数	说明
192.168.100.10:/wxic/nfs-share	nfs-server节点的IP地址和共享目录路径
/mnt	本地挂载点路径，这个路径必须在本地存在
nfs	挂载类型
defaults	默认选项，包括读写权限和其他的默认设置
0	是否对文件系统进行备份（0即否）
0	是否检查文件系统（0即否）

在客户端上创建文件夹和文件以验证NFS服务，命令和结果如下。

```
[root@nfs-client ~]# cd /mnt/
[root@nfs-client mnt]# mkdir wxic-dir
[root@nfs-client mnt]# touch wxic-file
[root@nfs-client mnt]# ll
total 0
drwxr-xr-x  3 root root  39 Mar 10 14:30 ./
drwxr-xr-x 18 root root 279 Oct 26 01:28 ../
drwxr-xr-x  2 root root   6 Mar 10 14:30 wxic-dir/
-rw-r--r--  1 root root   0 Mar 10 14:30 wxic-file
```

在 nfs-server 端查看共享目录中文件和目录的详细信息，命令和结果如下。

```
[root@nfs-server ~]# ll /wxic/nfs-share/
total 0
drwxr-xr-x 3 root root 39 Mar 10 14:30 ./
drwxr-xr-x 3 root root 23 Mar 10 10:27 ../
drwxr-xr-x 2 root root  6 Mar 10 14:30 wxic-dir/
-rw-r--r-- 1 root root  0 Mar 10 14:30 wxic-file
```

至此，NFS 服务部署完成。

项目5
云基础架构平台管理

学习目标

【知识目标】

1. 学习云平台的管理策略。
2. 学习云平台的监控和日志分析工具。

【技能目标】

1. 掌握云平台资源的规划方法。
2. 掌握常见云平台监控管理工具的使用。
3. 具备云平台故障排查的能力。

【素养目标】

1. 培养耐心细致的品质。
2. 培养解决问题的能力。
3. 培养创新思维与创造力。

项目概述

在熟悉 OpenStack 组件架构、熟练使用 OpenStack 各组件、了解 OpenStack 云服务中各组件之间的关系后，小张能够根据需求完成公司业务环境的部署工作。现在公司需要对云基础架构平台进行规划和管理，包括部门间、项目间的安全策略上的资源规划，对云基础架构平台的资源使用状况、运行状态进行监控，以及对突发故障提供应急解决方法。因此，接下来小张需要在了解云平台管理策略的前提下，通过操控常见的云平台监控和日志分析工具，对 OpenStack 云基础架构平台进行运维管理。

知识准备

5.1 云平台管理策略

云平台安全管理不仅是防范数据泄露和规避技术风险的有效措施，更是构建可信、可靠云计算环境的基石。

5.1.1 安全管理

在制定云平台管理策略时，在关注风险识别和合规要求的同时，自然会提到云平台安全策略。通过认识潜在的安全风险和遵守相关法规标准，我们不仅能够确保云平台的稳健运作，更能够有效地制定相应的安全策略，保障数据和应用程序的安全性，促进业务的可持续发展。

1. 什么是云平台安全策略

云平台安全策略是由公司在云运营过程中的一些正式准则定义的，它能够对所有有关云资产安全的决策进行指导。云平台安全策略的具体功能如下。

（1）能够判断数据类型能否迁移至云上。

（2）能够帮助团队应对每种数据类型可能带来的风险。

（3）能够决定负载迁移至云上的责任分配。

（4）能够确定可对数据进行访问或迁移的授权用户。

（5）能够对法律条款和当前状态进行合规审查。

（6）能够正确应对威胁、黑客攻击和数据泄露。

（7）能够制定风险优先级规则。

云平台安全策略是公司安全项目的重要组成部分。云平台安全策略能够保证信息的完整性和私密性，帮助团队快速做出正确的决定。

2. 对云平台安全进行管理的必要性

尽管云计算能够带来很多收益，但是云计算也具有一些安全隐患，举例如下。

（1）第三方设置中缺乏安全控制。

（2）在多云环境中可见性差。

（3）数据容易被窃取和滥用。

（4）易遭受分布式拒绝服务（Distributed Denial of Service，DDoS）攻击。

（5）攻击会从一个环境快速扩散至另一个环境。

云计算的安全隐患可能会对每个企业部门及其网络中的所有设备产生广泛的影响。因此，确保保护措施的强大性、多样性和广泛性显得尤为关键。一个可靠的云平台安全策略应当具备消除上述所有安全隐患的特征。

3. 云平台安全管理策略

云平台安全管理策略是一个综合性的安全管理体系，是指在利用云计算服务时，为确保用户和数据的安全而采取的各种措施和管理方法。在云计算环境中，为保障云平台的安全性和可靠性，生产企业通常会制定一系列的安全管理措施和方法。通过建立综合的云平台安全管理策略，可以有效地降低云平台的安全风险，保障云服务的安全性和稳定性。以下是一些常见的云平台安全管理策略。

（1）安全框架。建立全面、系统的安全框架，确保云平台的所有组件和服务都受到充分的保护。

（2）访问控制。通过多因素身份认证、强密码策略、细分权限和日志审计等手段，对云平台中的资源实行访问控制策略，限制用户对云平台资源的访问权限，确保只有授权用户可以访问云平台资源，防止未经授权的访问和操作。

（3）安全合规。通过遵守相关法规和规范（如《通用数据保护条例》《ISO/IEC 27001:2022 信息安全-网络安全-隐私保护-信息安全管理体系要求》等），确保云平台符合安全标准和规范，以及确保云平台的安全策略符合安全法规和应用开发规范（包括审计规范、报告规范和合规性测试）。同时，定期进行安全检测和审计，推广最佳的安全实践和标准，并对云供应商的合规性进行确认，以确保云供应商符合安全标准。

（4）数据加密。使用数据加密技术，对云平台中的敏感数据进行加密存储，保障数据的机密性和完整性，防止数据泄露和篡改，确保数据传输和存储的安全，大幅降低因数据泄露导致的安全风险。同时，也可在不同的数据存储级别上实施适当的加密策略。

（5）事件响应。建立事件响应体系，及时对云平台安全事件做出响应，追踪、申报漏洞，并通过严格的安全审查流程来处理问题。

（6）安全监控与审计。运用实时监控工具，加强网络交通分析，识别并阻止攻击行为，及时防御网络威胁。建立安全监控与审计机制，对云平台的操作和访问进行审计，监控云平台安全状况，及时发现和处理安全事件，保障云平台的安全性和可靠性。

（7）网络安全与应用安全。采用防火墙、入侵检测和防御等技术，实施多层次的安全措施，如网络隔离、安全大门、入侵检测和防火墙，防止网络攻击和恶意软件的入侵，确保用户安全地访问云平台，保障云平台的网络安全。对于运行在云平台上的应用程序，需要进行内部和外部的安全评估及测试，及时发现和修复漏洞，保障应用程序的安全性和可靠性，同时采用安全实践来保护应用程序。

（8）安全培训。对云平台用户、管理员和开发人员进行安全培训，加强其安全意识，加深其对云平台安全的认识和理解，提高其安全防范意识和应急响应能力，减少人为因素对云平台安全的影响。

综上所述，云平台安全管理策略需要从多个方面进行考虑和实施，以保障云平台的安全性和可靠性。

5.1.2 备份监控与调整

云平台备份监控与调整是指对云平台中的备份数据进行监控和调整，以保障备份数据的完整性和可用性。在对云平台实施有效监控的基础上，做好数据备份和恢复准备工作，设计能够应对意外事故的应急计划和故障恢复程序；制订灾备和恢复计划，保障云平台的可用性和可靠性，防止因灾害或故障导致的数据丢失和业务中断；定期备份和恢复云数据，并监督云平台的完整性以确保数据不被篡改。

1. OpenStack 平台备份监控方式

在 OpenStack 云平台中，对资源的监控与计量是确保云平台稳定运行的标准配置。尤其在公有云平台中，对资源的监控与计量不仅可以向业务使用者展现资源的使用情况，还可以成为按需计费模式下的计费依据。但是确保监控数据的准确性、实时性，以及对海量监控数据进行处理、存储和索引等工作都是具有挑战性的。

OpenStack 社区由 Ceilometer 项目来实现对 OpenStack 集群资源的监控与计量的功能。Ceilometer 项目从 OpenStack Folsom 开始发布，经过不断的迭代，功能逐渐丰富，能够对集群资源进行监控、计量与警报。借助 OpenStack 的 RESTful API 及消息队列，Ceilometer 项目可以非常好地与 OpenStack 中的其他项目相结合，实现分租户的自动化资源监控和计量。然而，由于 Ceilometer 在运行性能上的局限性，OpenStack 社区对 Ceilometer 项目进行了功能的拆分，Ceilometer 项目主要实现资源数据的采集，由 Gnocchi 项目负责对资源进行计量和数据存储，由 Aodh 项目负责对风险进行警报。

2. 云平台备份与调整策略

云平台备份与调整都非常重要，它们可以保证云平台的稳定性和可用性。云平台备份可以帮助用户在突发情况下恢复数据和应用，以保证数据的完整性和可用性。应根据业务流程的需要，尽可能多地创建备份，同时要确保备份的完整性和安全性。在制定云平台备份策略时，应该考虑备份的频率、备份的类型、备份数据的存储位置和恢复策略等因素。

云平台调整是指根据实际需求，对云平台进行优化和改进，以便更好地满足用户需求。调整的目的是提高云平台的性能、扩展云平台的计算和存储资源、改进运行不佳的应用程序等。云平台调整需要根据不同的需求制定不同的策略，例如，如果需要改进应用程序的性能，则可以从代码优化、部署优化，以及增加计算资源等方面对云平台进行调整。以下是一些常见的云平台备份与调整的策略。

（1）备份运行状态。需要定期备份运行状态，以确保备份的完整性和可用性。此外，还需要关注数据备份的速度、存储容量，以及备份过程中出现的错误和异常。如果发现备份失败或备份数据损坏，则需要及时根据业务需求和数据量修复、重新备份或调整备份策略。

同时，云平台环境可能会因为各种因素需要进行修改和调整，如升级软件版本、更改配置、扩展容量等。在进行环境调整之前，需要做好备份，在调整后需要对环境进行测试和验证，确保环境调整成功和服务正常运行。

（2）调整备份策略。需要根据业务需求和数据量调整备份策略，如调整备份频率、备份时段、备份存储周期等，以确保备份数据的最新性和有效性。如定期备份，制订备份计划，定期对云平台中的数据进行备份，确保备份数据的及时性和完整性；多地备份，将备份数据存储在多个地点，以防止因地震、火灾等自然灾害或人为因素导致的数据丢失；自动备份，使用自动备份工具，定期对云平台中的数据进行备份，减少人工操作的错误和漏洞。

（3）备份恢复测试。需要定期对备份的数据进行恢复测试，以确保备份数据的可用性和恢复速度，确保在出现数据丢失或系统故障时快速地恢复数据并避免业务停顿。

云平台的容灾备份是确保业务连续性的重要手段。在灾难发生时，容灾备份能够快速被调用来恢复服务，以保证服务的可用性和稳定性。容灾备份同样需要定期进行测试，以确保备份数据的完整性和快速恢复。

灾难恢复是云服务的较极端的备份和调整措施。在发生灾难时，需要快速响应，进行应急处理，确保业务恢复。因此，在云平台的备份和调整中，灾难恢复应当被充分考虑，制订灾难恢复计划，设计应急流程，确保在灾难发生时能够快速恢复服务和数据。

总之，云平台备份与调整是保障备份数据完整性和可用性的重要措施，为了保证云服务的持续稳定运行，需要定期进行云平台监控和调整。有了备份，可以保证数据的完整性和可用性；通过调整，可以提高云平台的性能和可靠性，为用户提供更优质的服务。

3. 云平台监控策略

云平台监控是指对云平台中的基础设施、应用程序、安全、数据、日志等进行全面监控，及时发现问题，并对其进行处理和优化。云平台监控对于云服务商和云平台用户来说都非常重要。

为了保证云平台能够持续稳定地运行，需要在云平台中配置监控系统，监控系统会对云平台的各项服务进行实时监控，如果出现问题会及时报警并向相关人员发送通知。云平台的监控类型通常包括以下几种。

（1）基础设施监控。监控云平台的基础设施，如物理主机的磁盘、CPU 及内存等部件的使用状况、网络带宽、I/O 吞吐量等指标。

（2）系统监控。监控云平台的系统资源利用情况和性能指标，包括 CPU、内存、磁盘等系统资源的利用情况，以及网络流量、连接数等系统性能指标。监控这些资源的利用情况和性能指标可以通过系统自带的监控工具或第三方监控工具实现。

（3）应用程序监控。监控云平台中运行的应用程序的状态、运行信息及资源利用情况等。可以通过部署应用程序的监控组件、监控容器等工具实现。

（4）安全监控。监控云平台的安全事件，包括对云平台的攻击、未授权访问等恶意行为进行检测，如网络入侵、端口扫描、Web 应用漏洞检测等，并及时发出警报。

（5）数据监控。监控云平台中数据的访问、存储、备份和恢复等情况，包括访问次数、访问权限等信息。例如，数据库的磁盘使用率、数据库拥塞程度等，可以通过数据库监控工具实现。

（6）日志监控。对云平台中的操作、访问、安全事件等进行日志记录和监控，以及进行日志分析，并对其中的重要事件自动发送通知，及时发现问题并进行处理。

针对不同的云平台类型和使用场景，选择合适的监控策略，并加以持续优化，能够提高云平台的管理效率和服务质量。同时，通过对云平台进行全面的监控，能够有效地提高云平台的稳定性和可用性，为用户提供更优质的云服务。

5.2 常见云平台监控系统和日志分析工具

在实现云平台高效运维与故障快速定位的过程中，监控系统和日志分析工具扮演着至关重要的角色，它们为管理员提供了实时的系统视图和深入的问题诊断方法。

5.2.1 Zabbix 监控系统

生产环境下，作为系统运维人员，需要会使用监控系统查看服务器状态及网站流量指标，利用监控系统的数据来了解网站发布的结果和健康状态。Zabbix 监控系统可以完成以下任务。

（1）通过一个友好的界面展示整个网站的所有服务器的状态。

（2）在 Web 前端方便地查看监控数据。

（3）回溯寻找事故发生时系统的问题和报警情况。

1. Zabbix 介绍及架构

Zabbix 是一种企业级分布式开源监控解决方案，能够监控众多网络参数和服务器的健康度、完整性。Zabbix 具有灵活的警报机制，允许用户为任何事件信息配置基于邮件的警报，这样用户可以快速响应服务器的问题。Zabbix 基于存储的数据提供出色的报表和数据可视化功能，这些功能使得 Zabbix 成为容量规划的理想选择。

Zabbix 支持主动轮询（Polling）和被动捕获（Trapping）。Zabbix 所有的报表、统计数据和配置参数都可以通过基于 Web 的前端界面进行访问。基于 Web 的前端界面确保使用者可以在任何地方访问监控的网络状态和服务器健康状况，对 Zabbix 进行适当的配置后，它可以在监控 IT 基础设施方面发挥重要作用，对于拥有少量服务器的小型组织和拥有大量服务器的大企业均适用。

Zabbix 主要由以下 5 个组件构成，其架构如图 5-1 所示，对这 5 个组件的说明如下。

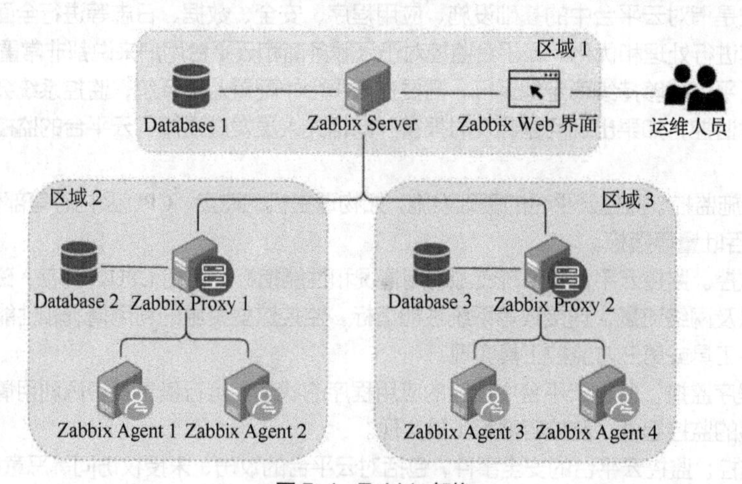

图 5-1　Zabbix 架构

（1）Zabbix Server。Zabbix Server 是 Zabbix 的核心组件，Zabbix Server 内部存储了所有的配置信息、统计信息和操作信息。Zabbix Agent 会向 Zabbix Server 报告可用性、完整性及其他统计信息。

（2）Zabbix Web 界面。Zabbix Web 界面也是 Zabbix 的一部分，通常和 Zabbix Server 位于一台物理设备上，但是在特殊情况下也可以将它们分开配置。Zabbix Web 界面主要提供了直观的监控信息，以方便系统运维人员监控和管理。

（3）Database。Database 内存储了配置信息、统计信息等 Zabbix 的相关内容。

（4）Zabbix Proxy。可以根据具体生产环境选择使用或者放弃使用 Zabbix Proxy。如果使用了 Zabbix Proxy，则其会替代 Zabbix Server 采集数据信息，可以很好地分担 Zabbix Server 的负载。Zabbix Proxy 通常应用于架构过大、Zabbix Server 负载过重，或者企业设备跨机房、跨网段，Zabbix Server 无法与 Zabbix Agent 直接通信的场景。

（5）Zabbix Agent。Zabbix Agent 通常部署在被监控目标上，用于主动监控本地资源和应用程序，并将监控的数据发送给 Zabbix Server。

2. Zabbix 多重功能

Zabbix 是一种高度成熟且完善的网络监控解决方案，它包含多种功能，使用了灵活的通知机制，可以通过邮件、短信和即时通信等方式向管理员发送通知并记录事件。Zabbix 的主要功能如下。

（1）可以监控多种系统和服务，包括网络设备、物理服务器、虚拟服务器、云平台和应用程序等。

（2）提供了多种监控方式，包括简单网络管理协议（Simple Network Management Protocol，SNMP）、Java 管理扩展（Java Management Extensions，JMX）、智能平台管理接口（Intelligent Platform Management Interface，IPMI）、SSH、Telnet 和 ICMP 等；支持可用性检查及自定义检查，可以按照自定义的时间间隔采集需要的数据，通过 Zabbix Server/Proxy 和 Zabbix Agent 来执行数据采集操作。

（3）可以监控多种指标，如 CPU 使用率、内存使用率、磁盘空间、网络流量，以及自定义指标等。

（4）支持事件触发器，可以在某个期限内检测到指标异常时触发警报，可以参考后端数据库定义灵活的警报阈值，设置高度可配置化的触发器警报，支持包含远程执行命令的自动操作，可以根据递增计划、接收者、媒介类型等自定义发送警报通知，同时可以使用宏变量使警报通知变得更加高效。

（5）可以设置监控周期，按指定时间间隔对系统和服务进行周期性检查，同时提供实时监控功能。

（6）Zabbix 不仅有内置图形功能可以将监控项实时绘制成图形，还提供了可视化界面，可以组合多个监控项（item）到单个视图中，可以创建自定义图表、表格、网络拓扑图和地图等，还可以以仪表盘样式展示自定义聚合图形并使用幻灯片演示监控数据。

（7）Zabbix API 为 Zabbix 提供可编程接口，用于批量操作、第三方软件集成和其他用途。使用者可以通过 Zabbix API 访问和控制 Zabbix 的所有功能。

（8）配置简单，Zabbix 的 Web 前端基于 PHP，使用者可以从任何地方访问 Zabbix 的 Web 前端界面，定制自己的操作方式，同时可以通过审计日志来查看操作。

（9）Zabbix Agent 强大且易于扩展，Zabbix Agent 支持 Linux 和 Windows 操作系统，它部署在被监控对象上。Zabbix Agent 支持通过网络发现设备，并且在发现设备后可以自动注册、自动发现文件系统和网络接口。

（10）强大的模板应用功能，使用者可以将模板用于监控设备，被监控设备一旦添加到数据库中，就会被自动采集数据以用于监控。同时模板可以分组检查、相互关联和继承已关联模板的属性。

（11）适应更加复杂的应用环境，使用 Zabbix Proxy 可以轻松实现分布式远程监控。

总而言之，Zabbix 是一种强大的开源监控平台，可以为企业和组织提供多样化的监控服务和报告功能。因其具有广泛的适用性和定制性，故在许多企业和组织中得到了广泛应用。

3. Zabbix 常用监控架构

在实际监控架构中，Zabbix 根据网络环境、监控规模等分为 3 种架构：server-client、server-proxy-client、master-node-client。对这 3 种架构的说明如下。

（1）server-client 架构。该架构是 Zabbix 的最简单的架构，监控主机和被监控主机之间不经过任何代理，直接在 Zabbix Server 和 Zabbix Agent 之间进行数据交互。该架构适用于网络比较简单、设备比较少的监控环境。

（2）server-proxy-client 架构。Zabbix Proxy 是 Zabbix Server 和 Zabbix Agent 之间沟通的“桥梁”，Zabbix Proxy 本身没有前端，且并不长期存放数据，只是将 Zabbix Agent 发来的数据暂时存放，而后提交给 Zabbix Server。该架构经常被用来与 master-node-client 架构做比较。该架构一般适用于跨机房、跨网络的中型网络架构的监控环境。

（3）master-node-client 架构。该架构是 Zabbix 最复杂的监控架构，适用于跨网络、跨机房、设备较多的大型环境。每个节点同时是一个 Zabbix Server，节点下面可以接 Zabbix Proxy，也可以直接接 Zabbix Agent。节点有自己的配置文件和数据库，其要做的是将配置信息和监控数据向主节点（master）同步，master 的故障或损坏会破坏节点下架构的完整性。

在不同的架构基础上，Zabbix 监控系统的运行是通过 Zabbix Agent 的数据来完成的。Zabbix Agent 需要安装到被监控的主机上，它负责定期收集各项数据，并将其发送到 Zabbix Server，Zabbix Server 将数据存储到数据库中，Zabbix Web 界面根据数据在前端进行展现和绘图。

5.2.2 ELK 日志分析工具

日志分析是系统运维和开发人员解决系统故障的主要手段。日志主要包括系统日志、应用程序日志和安全日志等。系统运维和开发人员可以通过日志了解服务器软硬件信息来检查配置过程中的错误及错误发生的原因。经常分析日志可以了解服务器的负载、安全性，并能够及时采取措施纠正错误。

通常，日志被分散地存储在不同的设备上。如果管理数十或数百台服务器时，还在使用依次登录每台机器的传统方法查阅日志，则这样做既烦琐又效率低下。这时用户可以使用集中化的日志管理，如开源的 syslog，它能够对所有服务器上的日志进行收集并汇总。集中化管理日志后，日志的统计和检索又成为一件比较麻烦的事情，一般使用 grep、awk 和 wc 等 Linux 命令对日志实现统计和检索，但是在面对查询、排序和统计等要求更高，或者机器数量更加庞大的情况时，使用这样的方法难免有些力不从心。

开源实时日志分析工具 ELK 能够完美地解决上述问题。

1. ELK 日志分析工具简介

随着系统架构的不断升级，系统架构由单体转变为分布式、微服务、网格系统等，用户访问产生的日志量也在不断增加，用户急需一个可以快速、准确查询和分析日志的工具。

一个完整的日志分析工具需要具有以下几个主要特点。

（1）收集：能够收集多种来源的日志数据（如系统的错误日志和业务的访问日志）。

（2）传输：能够稳定地把日志数据传输到日志分析平台。

（3）存储：能够存储日志数据。

（4）分析：能够支持可视化分析。

（5）警告：能够提供错误报告和监控机制。

ELK 日志分析工具的出现为用户提供了一种完整的解决方案，它能够使多种开源软件之间互相配合使用、完美衔接，从而高效地满足了多种应用场景的需求。ELK 是目前的一种主流的日志分析工具，是 Elasticsearch（ES）开源生态中提供的一套完整的日志收集、分析及展示的解决方案。ELK 是 Elasticsearch、Logstash 和 Kibana 这 3 个产品的首字母缩写。

ELK 实现的原理是通过 Logstash 收集日志数据，将日志数据存储到 Elasticsearch 中，并使用 Kibana

来查询和分析日志数据。ELK 可以收集多种来源的日志数据，并将它们转换为易于查询和分析的数据格式。ELK 执行架构如图 5-2 所示。

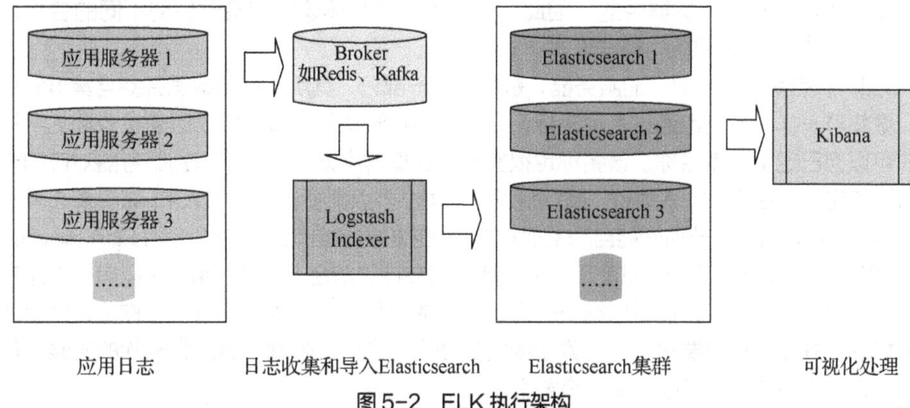

图 5-2　ELK 执行架构

Elasticsearch 是一个开源分布式实时数据分析搜索引擎，可以用来存储和分析大量的日志数据，以便后续的查询和分析，它建立在全文搜索引擎库 Apache Lucene 的基础上，同时隐藏了 Apache Lucene 的复杂性。Elasticsearch 将所有的功能打包成一个独立的动画片，其具有索引副本机制、RESTful 风格接口、多数据源、自动搜索等特点。

Logstash 是一个完全开源的工具，是一个日志收集器，主要用于日志收集，它可以通过多种方式对数据进行处理，并将其输出给 Elasticsearch。

Kibana 也是一个完全开源的数据可视化工具，Kibana 可以将 Elasticsearch 中的数据转换为可视化的图表和仪表盘，为 Logstash 和 Elasticsearch 提供图形化的日志分析。通过 Web 界面，Kibana 可以汇总、分析和搜索重要数据日志。

使用 ELK 可以实时监控应用程序、服务器、网络等相关信息，并自动收集日志数据以进行后续分析。它能够应对多种复杂场景，面对数据量大、时间跨度长等问题，利用 ELK 进行多维度查询可以看到业务链路的全景视图并做到及时反应，为运维人员提供便利。ELK 对查询进行了优化，它可以构建复杂的搜索和分析查询，并用直观、易理解的可视化方式展示数据。同时，它提供了多种操作接口和工具，方便集成和自动化使用。

总之，ELK 是非常强大的一种日志收集方案，它能够帮助企业和组织收集、分析和可视化大量的日志数据，以便迅速识别和解决问题，提高运维效率。

2. Elasticsearch

（1）Elasticsearch 概述

Elasticsearch 是一个开源的、高扩展的具有分布式多用户能力的数据分析搜索引擎，它可以近乎实时地检索、存储数据；其扩展性很好，可以扩展到上百台服务器，处理 PB 级别的数据。Elasticsearch 是一个基于 Apache Lucene 的搜索引擎，使用 Java 开发并使用 Apache Lucene 作为其核心来实现所有索引和搜索的功能，但是它的目的是通过简单的 RESTful API 来隐藏 Apache Lucene 的复杂性，从而让全文搜索变得简单。

Elasticsearch 作为 Apache 许可条款下的开放源代码发布，是流行的数据分析搜索引擎，它被用于云计算中，能够满足实时搜索、稳定、可靠、快速、实用、方便的要求。

（2）Elasticsearch 核心概念

① 接近实时：Elasticsearch 是一个接近实时的数据分析搜索引擎，这意味着从索引一个文档直到这个文档能够被索引到仅有短暂延迟（通常是 1s）。

② 集群：一个集群由一个或多个节点组织在一起，它们共同持有完整的数据，并一起提供索引和

搜索功能。其中一个节点为主节点，主节点通过选举产生，提供跨节点的联合索引和搜索功能，一个集群可以只有一个节点。集群有一个唯一性标识的名称，默认是 Elasticsearch。集群名称很重要，每个节点都是基于集群名称加入集群中的，因此，必须确保不同环境中的集群使用不同的名称。在配置 Elasticsearch 时，建议优先配置为集群模式。

③ 节点：节点就是一台单一的服务器，是集群的一部分，其用于存储数据并参与集群的索引和搜索功能。像集群一样，节点也通过名称来标识，节点名称默认是在节点启动时随机分配的字符名。当然，用户可以自己定义节点名称。该名称也很重要，在集群中该名称用于识别服务器对应的节点。节点可以通过指定集群名称来加入集群。默认情况下，每个节点都被设置为加入 Elasticsearch 集群中。如果启动了多个节点，且多个节点间能自动发现对方，则它们将会自动组建一个名为 Elasticsearch 的集群。

④ 索引：一个索引就是一个拥有几分相似特征的文档的集合。例如，用户可以有一个用户数据的索引、一个产品目录的索引，还可以有一个订单数据的索引。一个索引用一个名称（必须全部是小写字母）来标识，并且当用户要对相应的索引中的文档进行索引、收缩、更新和删除的时候，都要用到这个名称。在一个集群中，可以定义多个索引。

⑤ 类型：在一个索引中，用户可以定义一种或多种类型。一种类型是索引的一个逻辑上的分类分区，其寓意完全由用户来定义。通常，用户会为具有一组共同字段的文档定义一种类型。假设运营一个博客平台并且将所有的数据存储到一个索引中，在这个索引中，用户可以为用户数据定义一种类型，为博客数据定义一种类型，也可以为评论数据定义另一种类型。

⑥ 文档：一个文档是一个可被索引的基础信息单元。例如，用户可以拥有一个用户的文档或某一个产品的文档；文档以 JSON 格式来表示，JSON 是一种通用的互联网数据交互类型。同时，在一个索引/类型内，用户可以存储任意数量的文档。

虽然文档在物理上位于一个索引内，这意味着它存储在特定的位置或文件夹中，但是从逻辑上讲，文档必须在一个索引内才能被有效地索引和分配一种类型。

⑦ 分片和副本：在实际情况下，索引存储的数据可能超过单个节点的硬件限制。如一个文件需要 1TB 空间，那么这个文件可能不适合存储在单个节点的磁盘上，或者从单个节点对其进行搜索太慢了。为了解决这个问题，Elasticsearch 提供将索引分成多个分片的功能。当创建索引时，可以定义想要分片的数量。每一个分片就是一个全功能的独立的索引，可以位于集群中的任何节点上。

每个索引可以被分成多个分片。一个索引也可以被复制 0 次（意思是没有复制）或多次。一旦复制了，每个索引就有了主分片（作为复制源的原来的分片）和复制分片（副本，主分片的复制）之别。主分片和副本的数量可以在创建索引的时候指定。

在创建索引之后，用户可以在任何时候动态地改变副本的数量，但是用户不能改变主分片的数量。

（3）Elasticsearch 的分片和副本

Elasticsearch 的分片可以进行水平分割、横向扩展，以增大存储量，实现分布式并行跨分片操作，提高性能和吞吐量。在实际应用中，通过开启副本提高可用性，可应对分片或者节点故障，出于这个原因，副本要在不同节点上。为了提高 I/O 性能，增大吞吐量，搜索可以并行在所有副本上执行。

默认情况下，Elasticsearch 中的每个索引被分为 5 个主分片和 1 个副本，这意味着，在用户的集群中至少有两个节点的情况下，用户的索引将会被分为 5 个主分片和 5 个副本（1 个完全复制），这样每个索引总共被分为 10 个分片。

3. Logstash

（1）Logstash 概述

Logstash 是一个用于传输、处理和存储流数据的开源工具，它由 Ruby 语言编写，采用基于消息（Message-based）的简单架构，运行在 Java 虚拟机（Java Virtual Machine，JVM）上。不同于分离的代理端（Agent）或主机端（Server），Logstash 可配置单一的代理端与其他开源软件结合，以实现不同的功能。它可以实现数据传输、格式处理、格式化输出。它可以从各种数据源中收集日志和事件数据，并对其进行过滤、转换和增强，最终将它们输出。

（2）Logstash 主要组件

Logstash 的组件在 ELK 架构中可以独立部署，提供了很好的集群扩展性，主要组件如下。

① Shipper：日志收集者，负责监控本地日志文件的变化，及时把日志文件的最新内容收集起来。通常，远程代理端只需要运行这个组件即可。

② Indexer：日志存储者，负责接收日志并写入本地文件。

③ Broker：日志仓库，负责连接多个 Shipper 和多个 Indexer。

④ Search and Storage：允许对事件进行搜索和存储。

⑤ Web Interface：基于 Web 的展示界面。

在此基础上，Logstash 与 Elasticsearch 和 Kibana 配合使用，来构建完整的日志分析解决方案，实现日志的收集、存储和可视化等功能。

4. Kibana

（1）Kibana 概述

Kibana 是一个针对 Elasticsearch 的开源的数据分析及可视化平台，它是 Elastic Stack 中的一个组件，与 Elasticsearch、Logstash 和 Beats 一同构成了强大的数据分析解决方案。

Kibana 提供了丰富的数据可视化、仪表盘和监控功能，让海量数据更容易理解。它支持多种数据源的可视化和检索，通过搜索、查看存储在 Elasticsearch 索引中的数据，能够使用各种图表进行高级数据分析及展示。例如，使用 Elasticsearch、Timelion 等工具可以帮助用户快速地发现数据中的模式、趋势和异常。Kibana 操作简单，用户基于浏览器的用户界面就可以快速创建仪表盘以实时显示 Elasticsearch 的查询动态。

安装并设置 Kibana 非常简单，用户无须编写代码，在几分钟内就可以完成 Kibana 安装并设置启动 Elasticsearch 监测。

（2）Kibana 主要功能

Kibana 可帮助用户全面洞察数据，发现数据中的价值和趋势，并通过可视化方式为用户提供清晰的解释和指引。Kibana 的主要功能如下。

① 数据可视化：支持多种可视化图表，包括柱状图、饼图、折线图、热力图、地图、仪表盘等，可以直观地呈现数据，并支持自定义样式和布局。

② 数据检索：Kibana 支持通过查询和过滤来检索 Elasticsearch 中的数据，可以快速定位用户感兴趣的数据，并支持保存查询和过滤器。

③ 插件扩展：Kibana 提供了丰富的插件机制，可以通过插件扩展 Kibana 的功能，如 Timelion 插件可以帮助用户对时间序列数据进行可视化。

5. ELK 应用场景

（1）日志分析需求

在需要私有化部署的系统中，大部分系统仅提供系统本身的业务功能，如用户管理、财务管理、客户管理等。但是系统本身仍然需要进行日志、应用指标的收集，如请求速率、主机磁盘、内存使用量的收集等。能够方便地进行分布式系统日志的查看、应用指标的监控和警报也是系统稳定运行的一个重要保证。

为了使私有化部署的系统更健壮，同时不增加额外的部署运维工作量，通常可以选择基于 ELK 的开箱即用的日志和应用指标收集方案。在私有化部署的环境中，日志的收集和使用有以下几个特点。

① 需要能快速部署系统。由于用户的数量较多，系统应能被快速部署，监控系统本身的运维压力也应较小。

② 部署组件要简单，且健壮性强。由于部署环境较为复杂，每个组件自身应该是健壮的，同时组件之间的交互应尽量简单，避免复杂的网络拓扑。

③ 功能性优于稳定性。由于日志和应用指标信息在宿主主机及应用上是有副本的，所以即使监控系统的数据丢失了，影响也不大。但是如果系统能提供更多强大的功能，则对于分析数据是很有帮助的。

④ 性能要求不高。由于在私有化部署环境中，对接系统的容量和复杂度可控，因此，可以使用单机部署，查询慢一些也没关系。

日志的收集还需要满足以下几个需求。

① 能收集分布式的日志，并集中式地对日志进行查看。

② 能收集机器的基本信息，如CPU、磁盘等的信息，并进行监控。

③ 最好能收集应用的数据，如导入数据的条目数，并进行监控。

④ 最好能实现异常指标的警报功能。

（2）警报应用方案

为满足日志分析和警报的需求，通常情况下可以利用Zabbix、Open-Falcon等运维监控工具进行系统基础组件的监控，同时利用自定义指标进行数据的监控和警报。

ELK也能实现警报，可以利用Elasticsearch、Logstash、Kibana作为整体的监控基础组件，同时使用Elastic新推出的Beats系列作为采集工具。Elasticsearch的原生警报是付费功能，可以使用开源项目ElastAlert实现警报。ElastAlert是Yelp公司（美国的类似大众点评网的公司）开发的基于Python和Elasticsearch的警报系统。

ElastAlert可以配置多种警报类型，举例如下。

① 某条件连续触发N次（frequency类型）。

② 某指标出现的频率提高或者降低（spike类型）。

③ N分钟未检测到某指标（flatline类型）等。

每个警报的配置核心其实都是一个Elasticsearch的查询语句，通过查询语句返回的条目数来判断是否触发警报。

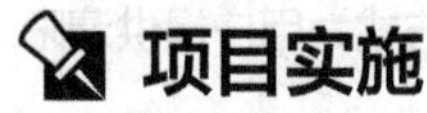

任务5.1 云平台资源规划

云平台资源规划对于优化资源配置、保障业务高效运行具有重要意义。本任务将逐步介绍部门资源隔离、部门资源共享和企业云平台资源规划的最佳实践等内容，旨在构建一个既能保障部门间资源隔离与安全性，又能促进有效资源共享，并借鉴最佳实践优化企业级云平台资源规划的策略体系。

任务5.1.1 部门资源隔离

1. 基于项目的资源隔离

微课5.1 部门资源隔离

云平台基于项目的资源隔离是一种常见的资源隔离方法，可以确保不同项目的资源彼此独立，避免资源冲突和信息泄露。一组隔离的资源和对象，由一组关联的用户进行管理。在旧版本里，也用租户来表示用户。根据配置的需求，一个项目对应一个组织、一个公司或者一个使用用户等，项目中可以有多个用户，项目中的用户可以在该项目中创建、管理虚拟资源。具有admin角色的用户可以创建项目，并将项目相关信息保存到数据库中。默认情况下，OpenStack有以下两个独立的项目。

① admin：为admin角色的用户创建的项目。

② services：与安装的各项服务相关联。

（1）创建项目

在云平台上创建不同的项目，将同一项目的相关资源放入该项目中。

使用命令创建项目，命令如下。

```
[root@controller ~]# openstack project create wxicproject
```

查看所有项目，命令如下。

```
[root@controller ~]# openstack project list
```

查看项目的详细信息，命令如下。

```
[root@controller ~]# openstack project show wxicproject
```

启用项目，命令如下。

```
[root@controller ~]# openstack project set --enable wxicproject
```

（2）创建用户并将其加入项目中

在项目中分配资源，如计算实例、存储、网络等，确保每个项目都拥有自己独立的计算实例、存储和网络资源。

创建“wxicuser”用户并将其加入“wxicproject”项目中，命令如下。

```
[root@controller ~]# openstack user create --project wxicproject --password 000000 wxicuser
```

每个项目都可以设置特定的访问权限，以确保项目资源只被授权的用户或团队访问，将项目中的用户激活，命令如下。

```
[root@controller ~]# openstack user set --enable wxicuser
```

使用命令查看所有用户，命令如下。

```
[root@controller ~]# openstack user list
```

使用命令将“wxicuser”用户添加为“admin”角色，命令如下。

```
[root@controller ~]# openstack role add --project wxicproject --user wxicuser admin
```

（3）测试所创建的项目能否使用

项目和项目内用户创建完毕后，为每个项目设置资源使用配额，以确保同一项目内的不同用户或应用程序不会过度使用资源，从而影响其他项目的资源使用。Skyline 平台登录界面如图 5-3 所示。

图 5-3　Skyline 平台登录界面

登录 Skyline 平台后，在进入的界面中单击右上角的“管理平台”按钮，如图 5-4 所示。

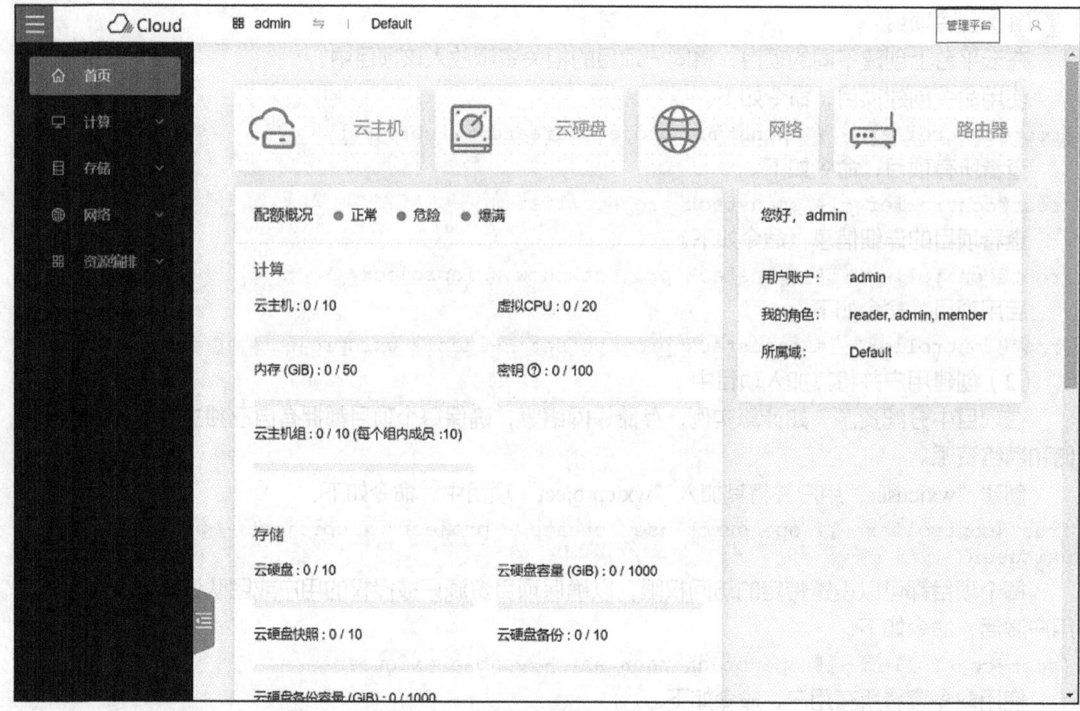

图5-4　单击“管理平台”按钮

进入管理平台界面，选择左侧导航栏中的“身份管理→项目”选项，单击项目名称“wxicproject”，进入该项目，如图5-5所示。

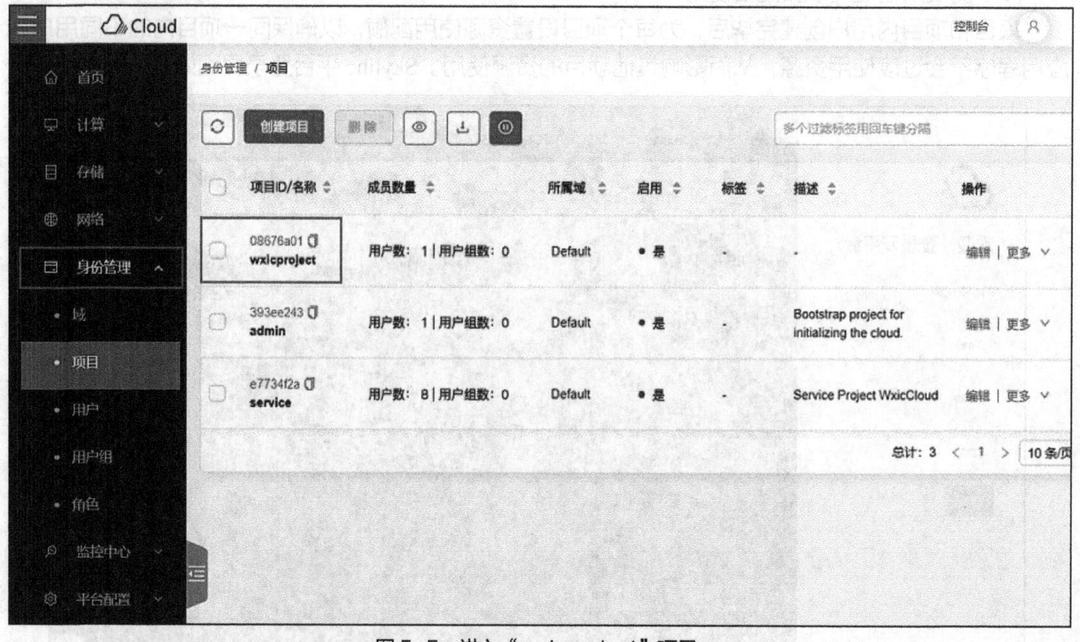

图5-5　进入“wxicproject”项目

在Skyline项目用户界面中，可以看到创建的项目用户数为1，说明该项目创建成功，并有可用用户，如图5-6所示。

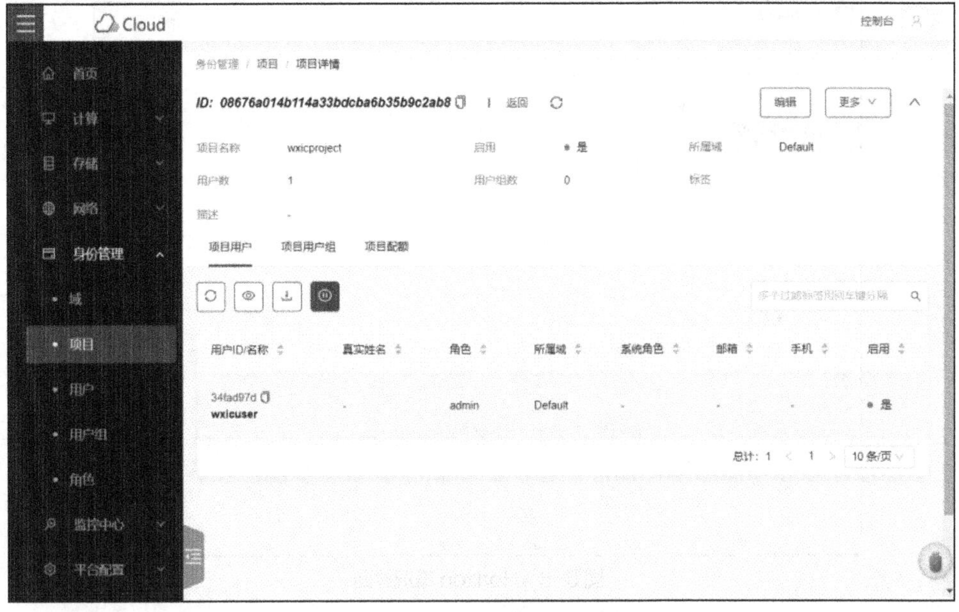

图 5-6　Skyline 项目用户界面

查看完成后，单击界面右上角的“用户”图标，在其下拉列表中选择“退出登录”选项，进入项目内的新创建用户的 Skyline 平台界面，重新登录平台。

登录后，在 Skyline 首页中可以查看“wxicuser”用户在当前“wxicproject”项目内的资源情况，如图 5-7 所示。

图 5-7　Skyline 首页（1）

项目组内的用户可以登录 OpenStack 云平台，跳转到 Horizon 概况界面，对资源使用情况进行查看，如图 5-8 所示。

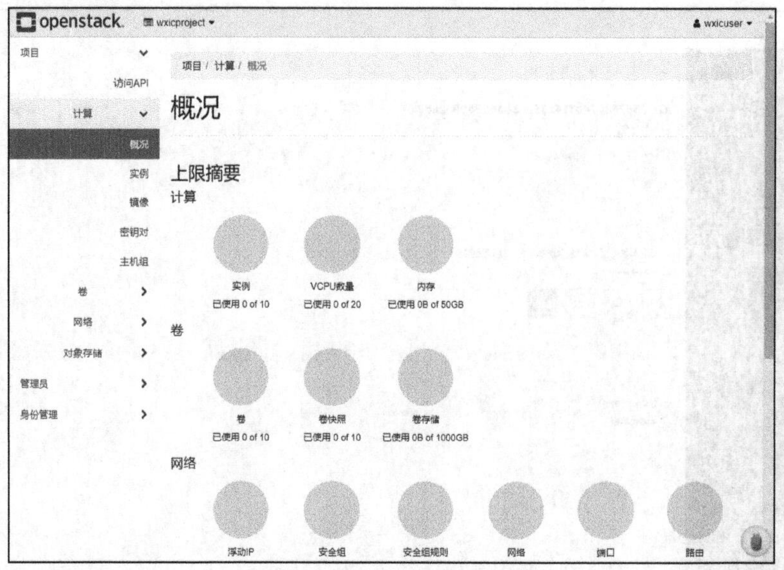

图 5-8　Horizon 概况界面

2. 基于安全组的资源隔离

云平台基于安全组的资源隔离是防止不同云服务之间相互干扰及对内部和外部的攻击的重要方法。安全组是云平台中一种实现安全隔离和访问控制的网络级别的安全管理模型。

openstack security group 命令用于管理 OpenStack 中的安全组，包括查看、创建、删除和修改等操作。

（1）创建“wxicproject”项目对应的安全组

在云平台上创建项目“wxicproject”对应的安全组“wxicsecurity”，命令如下。

```
[root@controller ~]# openstack security group create --project wxicproject wxicsecurity
```

使用命令查看创建的安全组，命令如下。

```
[root@controller ~]# openstack security group list
```

使用 Skyline 平台检测创建的安全组，此时，安全组概览界面如图 5-9 所示。

图 5-9　安全组概览界面（1）

通过 Skyline 平台，可以发现项目可以设置多个安全组，安全组所对应的项目不同。

（2）配置安全组规则（UDP、TCP、ICMP）

在安全组中添加安全组规则以确定哪些流量可以进入或离开该安全组，确保资源只被指定的流量访问，如阻止所有不必要的访问，或者允许限定源 IP 地址的流量等。

添加流量入口方向规则，命令如下。

```
[root@controller ~]# openstack security group rule create --ingress --protocol udp wxicsecurity
[root@controller ~]# openstack security group rule create --ingress --protocol tcp wxicsecurity
[root@controller~]# openstack security group rule create --ingress --protocol icmp wxicsecurity
```

添加流量出口方向规则，命令如下。

```
[root@controller ~]# openstack security group rule create --egress --protocol udp wxicsecurity
[root@controller ~]# openstack security group rule create --egress --protocol tcp wxicsecurity
[root@controller ~]# openstack security group rule create --egress --protocol icmp wxicsecurity
```

使用安全组命令，检测安全组规则，查看根据 IP 创建的规则，命令如下。

```
[root@controller ~]# openstack security group rule list
```

使用 Skyline 平台检测（可以使用“admin”用户检测，也可以使用“wxicuser”用户检测）。此处使用“wxicuser”用户登录 Skyline 平台，打开“安全组”检测其创建的规则，此时，安全组概览界面如图 5-10 所示。

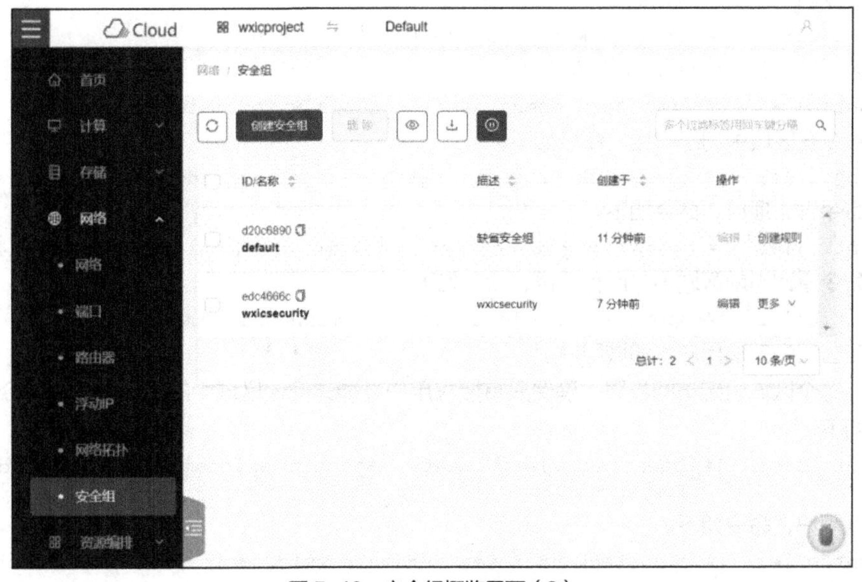

图 5-10 安全组概览界面（2）

将安全组绑定到不同的实例、子网或网络，可保护它们不被未授权的访问或攻击。通过创建规则、增加规则方法、绑定不同端口，进行资源的进出管控。在管理平台界面中，选择左侧导航栏中的“安全组”选项，再选择右侧的“wxicsecurity”选项，查看目前设置的安全组规则，如图 5-11 所示。

3. 基于区域的资源隔离

云平台基于区域的资源隔离是一种资源隔离方法，其将不同的资源放置在不同的区域中以实现资源的隔离和保护。

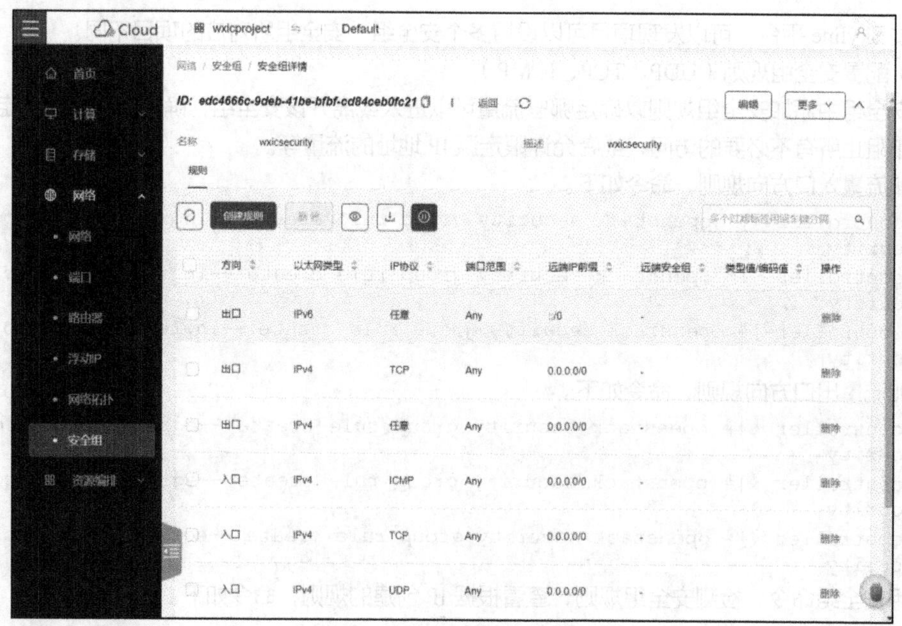

图 5-11　安全组规则概览界面

（1）创建区域

在云平台上创建不同的区域，将同一区域的相关资源放入该区域中。

使用命令创建区域“wxicdomain”，命令如下。

```
[root@controller ~]# openstack domain create wxicdomain
```

激活区域“wxicdomain”，命令如下。

```
[root@controller ~]# openstack domain set --enable wxicdomain
```

使用命令在区域中创建项目，命令如下。

```
[root@controller ~]# openstack project create --domain wxicdomain wxicproject
```

使用命令激活项目，命令如下。

```
[root@controller ~]# openstack project set --domain wxicdomain --enable wxicproject
```

使用命令查询当前区域中的所有项目，命令如下。

```
[root@controller ~]# openstack project list --domain wxicdomain
```

（2）在区域中创建用户（即域用户）

可以为每个区域设定访问权限，确保只有授权用户才能访问该区域中的资源。使用命令创建域用户，命令如下。

```
[root@controller ~]# openstack user create --domain wxicdomain --password 000000
wxicuser
```

激活域用户，命令如下。

```
[root@controller ~]# openstack user set --domain wxicdomain --enable wxicuser
```

查询创建的域用户，命令如下。

```
[root@controller ~]# openstack user list --domain wxicdomain
```

使用命令将创建的域用户“wxicuser”添加到项目“wxicproject”中，命令如下。

```
[root@controller  ~]#  openstack  role  add  --project-domain  wxicdomain  --project
wxicproject --user-domain wxicdomain --user wxicuser admin
```

使用 Skyline 平台，选择左侧导航栏中的“身份管理→项目”选项，单击“更多”按钮，选择下拉列表中的“管理用户”选项，在弹出的“管理用户”对话框中，选择域“wxicdomain”，以勾选的方式添加域用户“wxicuser”到“wxicproject”中，单击▶图标，如图 5-12 所示。

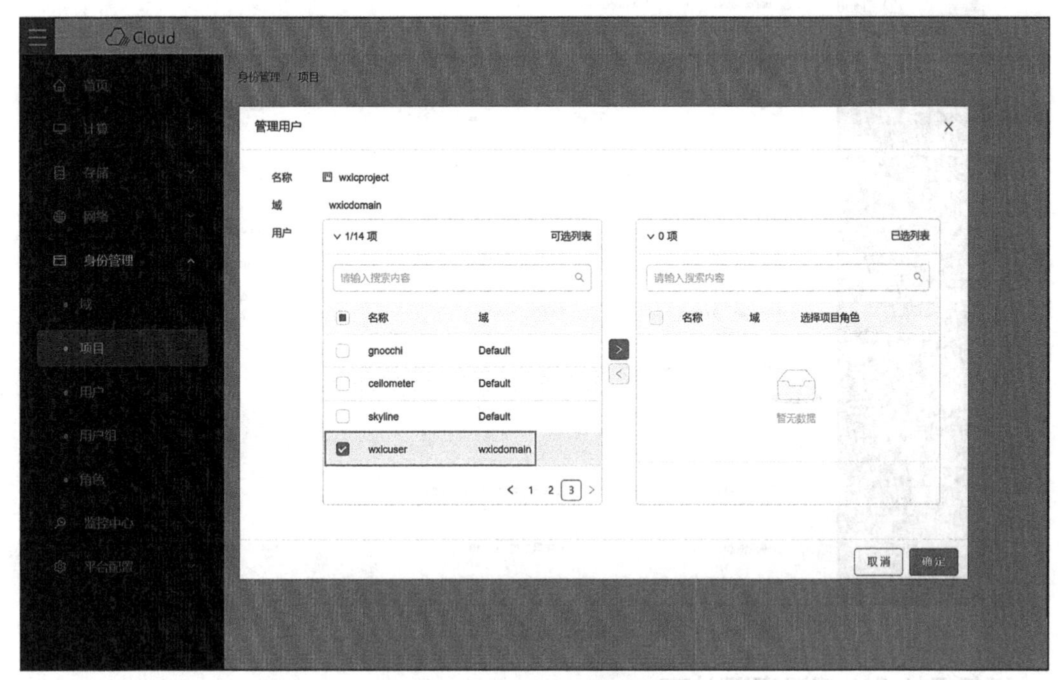

图 5-12 “管理用户”对话框（1）

添加域用户后，在“选择项目角色”下拉列表中选择“admin”选项，单击“确定”按钮即可完成操作，如图 5-13 所示。

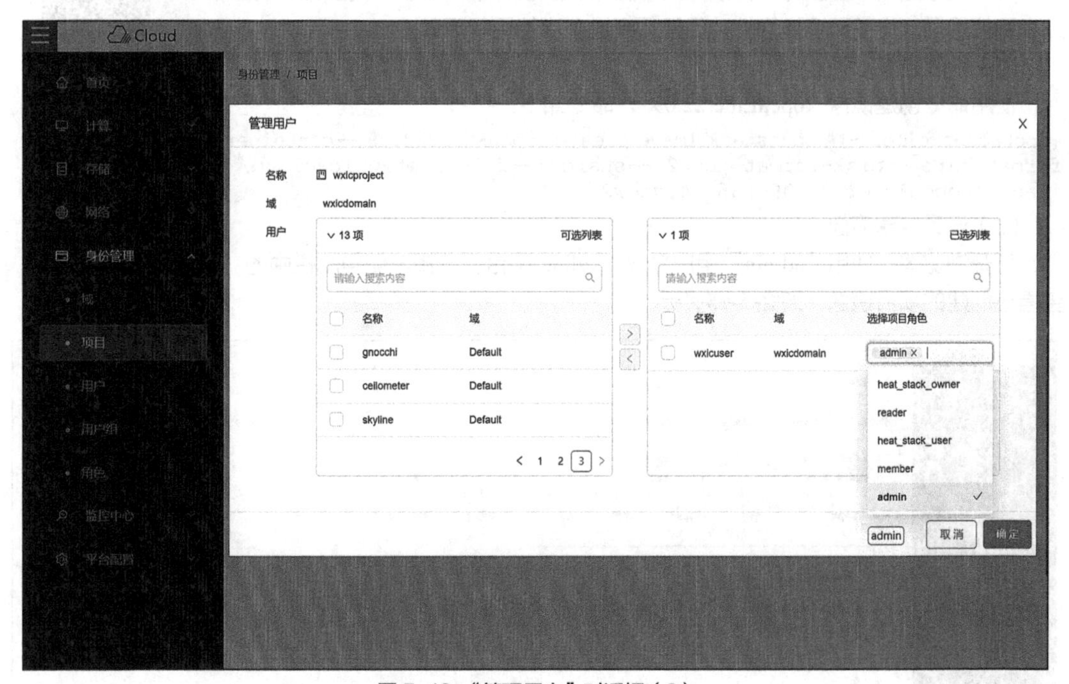

图 5-13 “管理用户”对话框（2）

上述操作完成后，从平台退出登录。使用创建的域用户“wxicuser”登录后可以看到当前域用户所在的项目、区域及配额概况，如图 5-14 所示。

图 5-14　Skyline 首页（2）

任务 5.1.2　部门资源共享

1. 共享镜像

微课 5.2　部门资源共享

共享镜像是云平台上的一种镜像共享服务，它允许用户在同一个区域或跨区域共享和使用同一份基础镜像软件，而无须经过多次的重复下载。

（1）创建镜像

使用命令创建镜像“openEuler22.09”，命令如下。

```
[root@controller ~]# openstack image create  openEuler22.09 --container-
format bare --disk-format qcow2 --shared --file /opt/wxic-cloud/
images/ openEuler-22.09-x86_64.qcow2
```

（2）设置共享镜像

打开浏览器，使用“admin”用户登录 Skyline 平台，选择左侧导航栏中的“计算→镜像”选项，查看已创建的项目镜像，如图 5-15 所示。

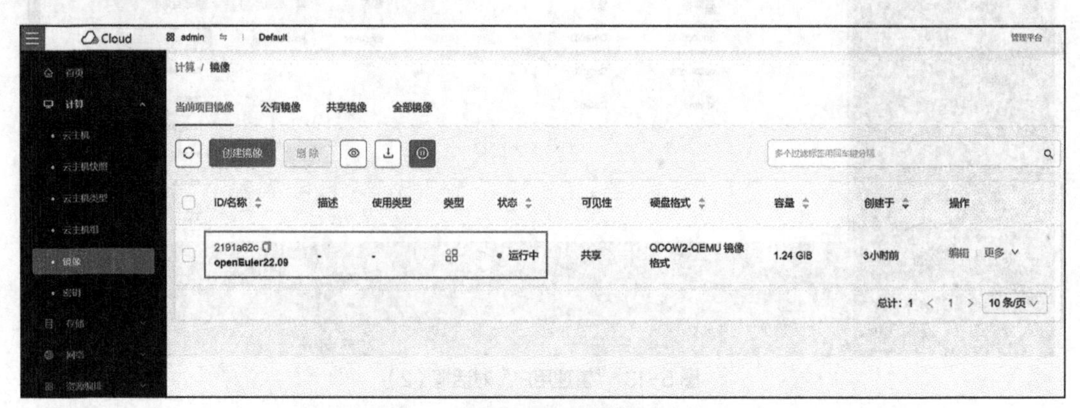

图 5-15　Skyline 镜像管理界面（1）

将镜像“openEuler 22.09”添加到“wxicproject”项目中并启用共享，命令如下。

```
[root@controller ~]# openstack image add project  openEuler22.09 wxicproject
[root@controller ~]# openstack image set openEuler22.09 --project wxicproject --accept
```

操作完成后，使用“wxicuser”用户登录 Skyline 平台，选择左侧导航栏中的“计算→镜像”选项，再选择右侧的“共享镜像”选项卡，即可看到共享镜像，如图 5-16 所示。

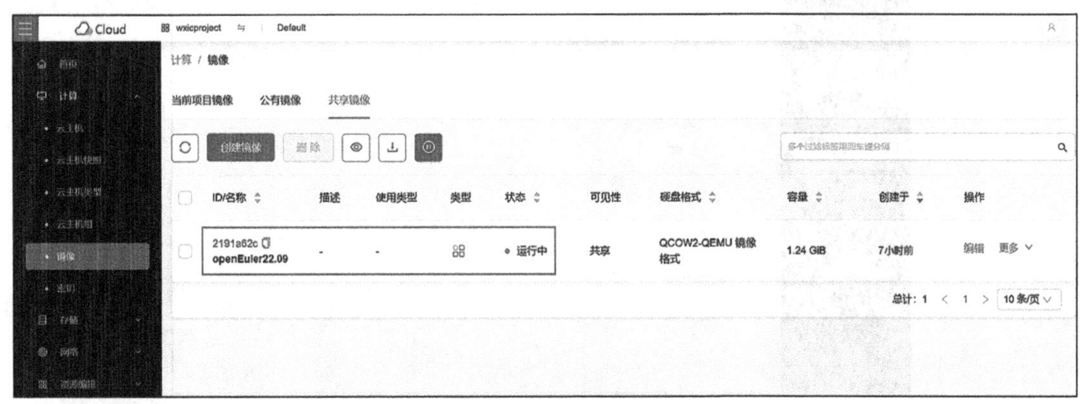

图 5-16　Skyline 镜像管理界面（2）

2. 共享网络和子网

在云平台上，共享网络和子网可以让多台云主机共享同一个网络，提高网络使用率和降低成本。

（1）共享网络

共享网络是指一个网络被多台云主机所使用，这些云主机可以互相通信，共享同一个网络空间。

创建共享网络，命令如下。

```
[root@controller ~]# openstack network create wxic-net --share --external
```

切换至其他用户查看共享网络，如图 5-17 所示。

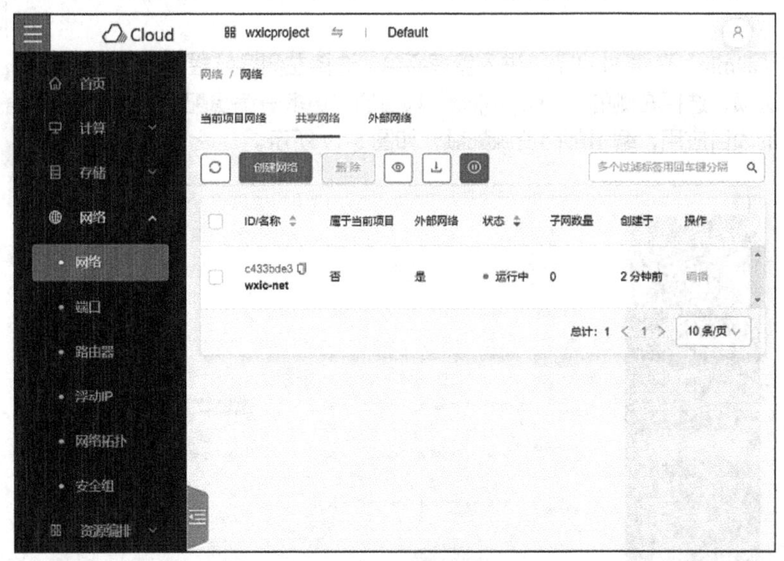

图 5-17　Skyline 网络管理界面（1）

（2）共享子网

共享子网是指一个子网被多台云主机所使用，这些云主机可以互相通信，共享同一个子网空间。每台云主机可以拥有一个独立的 IP 地址，与共享子网中的其他云主机进行通信。

在控制节点上创建“wxic-subnet”子网，相关命令如下。

```
[root@controller ~]#openstack subnet create --subnet-range 192.168.200.0/24
--gateway 192.168.200.2 --network wxic-net wxic-subnet
```

切换至其他用户查看共享子网，如图 5-18 所示。

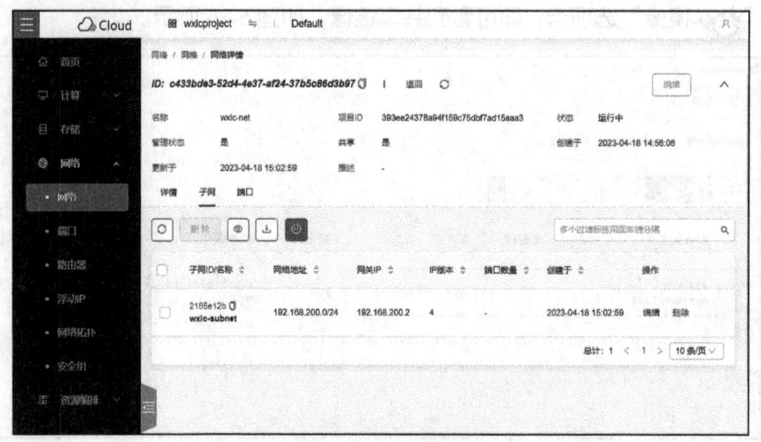

图 5-18　Skyline 网络管理界面（2）

使用共享网络和共享子网服务可以减少私有 IP 地址的浪费，降低使用成本，并且可以方便地管理多台云主机之间的通信和连接。

任务 5.1.3　企业云平台资源规划的最佳实践

1. 规划各部门可用的资源配额

企业环境下，通常由平台管理员设置和调整云平台资源的使用配额。平台管理员会通过评估部门的业务负载，来将资源分配给不同的主机、环境，如 CPU、内存、存储、带宽等。

微课 5.3　企业云平台资源规划的最佳实践

使用用户“admin”登录 Skyline，进入管理平台，选择左侧导航栏中的“身份管理→项目”选项，选择右侧的“wxicproject”项目的“更多→编辑配额”选项，可以根据部门的项目应用，编辑部门资源配额，如图 5-19 所示。

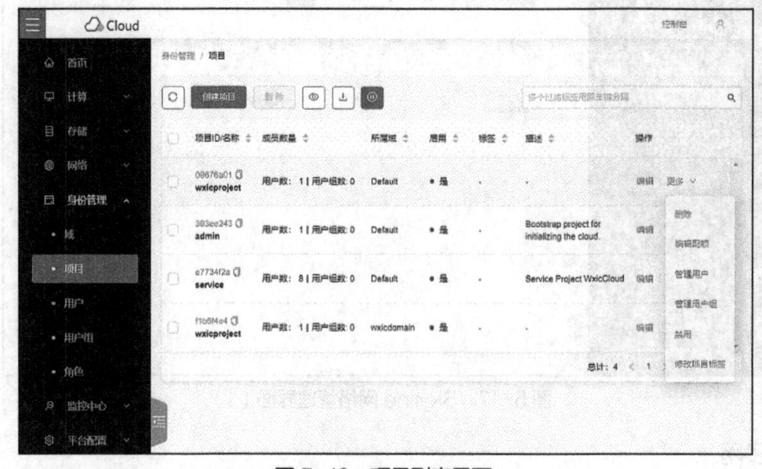

图 5-19　项目列表界面

根据实际需求修改配额数目，修改后确认即可。使用修改配额数目的项目中的用户，登录 Skyline 查看修改情况，如图 5-20 所示。

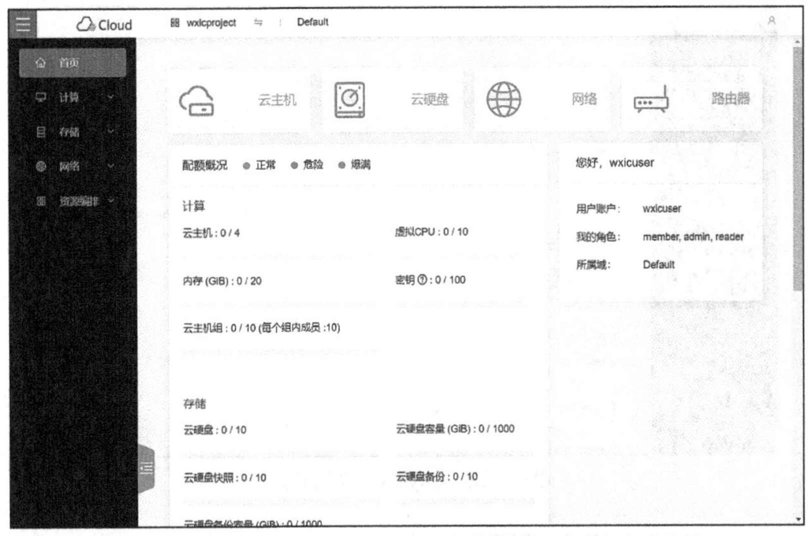

图 5-20　Skyline 首页（3）

2. 资源使用监控

云平台资源使用监控是保障云服务稳定性和性能优化的重要手段。一般而言，云平台资源使用监控工具可以提供以下功能。

（1）实时监控云资源的使用情况，如 CPU 使用率、存储空间使用率、带宽使用率等。

（2）对不同的资源使用情况进行统计分析，如对 CPU 使用量的历史数据进行统计分析，以便更好地进行资源规划和优化。

（3）报警功能，当某个资源使用量超过预设的阈值时，可以自动发送警报通知管理员，以便及时采取措施。

（4）灵活的可视化界面，以便管理员轻松地查看资源使用情况，并进行直观的分析和决策。

（5）对不同应用的资源使用情况进行详细的跟踪，以便更好地进行应用调优和性能优化。

在任务 5.1.3 中修改配额数目的基础上，使用已修改配额数目的项目中的用户“wxicuser”登录 Skyline 平台，查看云主机资源使用情况，如图 5-21 所示。

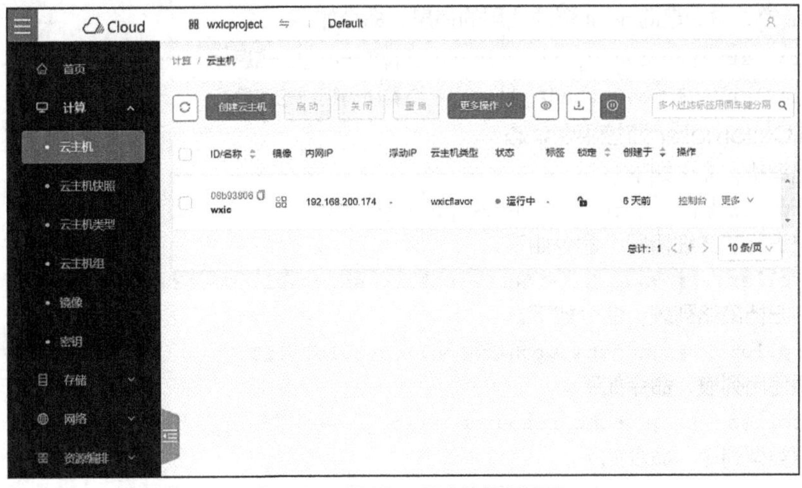

图 5-21　Skyline 云主机管理界面

返回首页，可以查看当前所属部门下的项目应用到的主机资源配额概况，如图 5-22 所示。

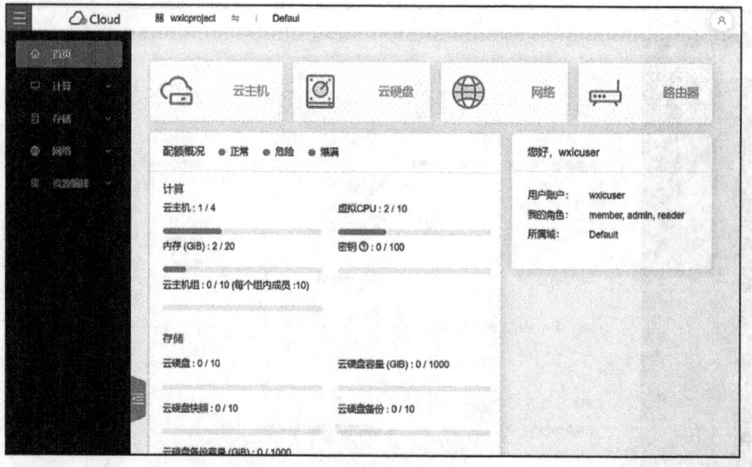

图5-22 Skyline首页（4）

任务5.2 云平台监控管理

云平台监控管理扮演着不可或缺的核心角色，其关键价值在于对系统进行实时和深度的洞察，及时发现并预防任何可能引发服务中断的潜在故障。

任务5.2.1 OpenStack平台自带监控

1. 查看平台状态及使用情况

OpenStack中的服务是分布式部署的，因此，各个服务的运行状态决定了此系统的可用性。用户可以通过OpenStack提供的接口来查看服务的运行状态，命令如下。

微课5.4
OpenStack平台自带监控

```
[root@controller ~]# nova service-list
[root@controller ~]# neutron agent-list
[root@controller ~]# cinder service-list
[root@controller ~]# heat service-list
```

使用nova相关命令查询nova资源使用的信息，命令如下。

```
[root@controller ~]# nova usage         # 查看单一租户的信息
[root@controller ~]# nova usage-list   # 查看所有租户的信息
```

2. 使用Ceilometer监控平台状态

列出归档策略，命令如下。

```
[root@controller ~]# gnocchi archive-policy list
```

查看归档策略的详细情况，命令如下。

```
[root@controller ~]# gnocchi archive-policy show ceilometer-high
```

显示当前归档策略列表，命令如下。

```
[root@controller ~]# gnocchi archive-policy-rule list
```

获得资源使用列表，命令如下。

```
[root@controller ~]# gnocchi resource list
```

获得资源类型列表，命令如下。

```
[root@controller ~]# gnocchi resource-type list
```

管理计量项，命令如下。

```
[root@controller ~]# gnocchi metric list
```

3. 使用 Aodh 警报服务监控平台状态

Aodh 支持多种警报类型，如基于阈值的警报、复合警报、基于事件的警报等，可以选择适合对应场景的警报类型，来定义相应的警报规则，具体的使用案例如下。

（1）基于阈值的警报

设置特定实例，基于 CPU 使用率限制，创建基于阈值的警报的示例，命令如下。

```
[root@controller ~]#aodh alarm create \
  --name cpu_hi \
  --type gnocchi_resources_threshold \
  --description 'instance running hot' \
  --metric cpu_util \
  --threshold 70.0 \
  --comparison-operator gt \
  --aggregation-method mean \
  --granularity 600 \
  --evaluation-periods 3 \
  --alarm-action 'log://' \
  --resource-id INSTANCE_ID \
  --resource-type instance
```

上述命令将创建一个警报，当指定实例的平均 CPU 使用率超过 70%时，将触发该警报。该警报会考虑连续 3 个时段的数据，每个时段的时间间隔为 10min。

（2）复合警报

创建一个复合警报，当任何一个指定实例的 CPU 使用率达到 80%时，将触发该警报，警报将通过 HTTP 请求通知给指定的地址，命令如下。

```
[root@controller ~]#aodh alarm create \
  --name meta \
  --type composite \
  --composite-rule '{"or": [{"threshold": 0.8, "metric": "cpu_util", \
    "type": "gnocchi_resources_threshold", "resource_id": INSTANCE_ID1, \
    "resource_type": "instance", "aggregation_method": "last"}, \
    {"threshold": 0.8, "metric": "cpu_util", \
    "type": "gnocchi_resources_threshold", "resource_id": INSTANCE_ID2, \
    "resource_type": "instance", "aggregation_method": "last"}]}' \
  --alarm-action 'http://example.org/notify'
```

（3）基于事件的警报

基于电源状态创建事件警报的示例，命令如下。

```
[root@controller ~]#aodh alarm create \
  --type event \
  --name instance_off \
  --description 'Instance powered OFF' \
  --event-type "compute.instance.power_off.*" \
  --enable True \
  --query "traits.instance_id=string::INSTANCE_ID" \
  --alarm-action 'log://' \
  --ok-action 'log://' \
  --insufficient-data-action 'log://'
```

也可以加入其他事件，例如，在实例开机但进入错误状态时，命令如下。

```
[root@controller ~]#aodh alarm create \
  --type event \
  --name instance_on_but_in_err_state \
  --description 'Instance powered ON but in error state' \
  --event-type "compute.instance.power_on.*" \
```

```
--enable True \
--query "traits.instance_id=string::INSTANCE_ID;traits.state=string::error" \
--alarm-action 'log://' \
--ok-action 'log://' \
--insufficient-data-action 'log://'
```

（4）警报检索与删除

查看所有警报，命令如下。

```
[root@controller ~]# aodh alarm list
```

使用 OpenStack 相关命令，删除 cpu_hi 警报，命令如下。

```
[root@controller ~]# openstack alarm delete cpu_hi
```

使用 OpenStack 相关命令，查看当前警报列表，命令如下。

```
[root@controller ~]# openstack alarm list
```

任务 5.2.2 安装 Zabbix 监控

1. 任务准备

目前，在 openEuler 操作系统的官方软件仓库中，openEuler 22.09 尚未针对 Zabbix 监控提供预先构建的官方软件包支持，最新的长期支持版本为 openEuler 22.03，故在此案例中使用 openEuler 22.03。

微课 5.5 安装 Zabbix 监控

登录 OpenStack 云平台，分发两台 openEuler 22.03 的云主机，云主机类型使用 4 个 vCPU/12GB 内存/60GB 硬盘。节点规划如表 5-1 所示。

表 5-1 节点规划

IP 地址	主机名	节点
192.168.100.11	zabbix-server	Zabbix Server 节点
192.168.100.12	zabbix-agent	Zabbix Agent 节点

2. 安装 Zabbix Server

云主机创建完成后，使用 SecureCRT 连接 Zabbix Server 节点，修改主机名为“zabbix-server”，命令如下。

```
[root@localhost ~]# hostnamectl set-hostname zabbix-server
```

添加 openEuler 22.03 官方提供的扩展源，命令如下。

```
[root@zabbix-server ~]#  dnf config-manager --add-repo  \
https://repo.oepkgs.net/openeuler/rpm/openEuler-22.03-LTS/extras/x86_64/
```

安装数据库服务，命令如下。

```
[root@zabbix-server ~]# dnf -y install mariadb mariadb-server --nogpgcheck
```

安装 Zabbix 服务，命令如下。

```
[root@zabbix-server ~]# dnf -y install --nogpgcheck zabbix-server-mysql
zabbix-web-mysql zabbix-nginx-conf zabbix-sql-scripts zabbix-selinux zabbix-agent
```

启动数据库服务，并设置其为开机自启动，命令如下。

```
[root@zabbix-server ~]# systemctl enable --now mariadb
```

登录数据库，创建 Zabbix 库（中文编码格式），命令如下。

```
[root@zabbix-server ~]# mysql
MariaDB [(none)]> create database zabbix character set utf8mb4 collate utf8mb4_bin;
Query OK, 1 row affected (0.000 sec)
MariaDB [(none)]> create user zabbix@localhost identified by 'zabbix';
Query OK, 0 rows affected (0.001 sec)
```

授予 Zabbix 用户访问权限，命令如下。

```
MariaDB [(none)]> grant all privileges on zabbix.* to zabbix@localhost identified by 'zabbix';
Query OK, 0 rows affected (0.001 sec)
```

设置 log_bin_trust_function_creators 的值，防止在安装或升级 MySQL 后需要解决一些函数和存储过程的语法问题，命令如下。

```
MariaDB [(none)]> set global log_bin_trust_function_creators = 1;
Query OK, 0 rows affected (0.000 sec)
```

在 Zabbix Server 节点上导入初始模式和数据，系统提示输入新创建的密码，命令如下。

```
[root@zabbix-server ~]# zcat /usr/share/doc/zabbix-sql-scripts/mysql/server.sql.gz | mysql --default-character-set=utf8mb4 -uzabbix -pzabbix zabbix
```

导入数据库后，使用数据库编辑命令，禁用 log_bin_trust_function_creators 选项，命令如下。

```
[root@zabbix-server ~]# mysql -e "set global log_bin_trust_function_creators = 0;"
```

编辑 Zabbix 配置文件/etc/zabbix/zabbix_server.conf，修改数据库密码为 zabbix，命令如下。

```
DBPassword=zabbix
```

启动 Zabbix 服务器和代理进程，使其在系统启动时自启，命令如下。

```
[root@zabbix-server ~]# zabbix_server -c /etc/zabbix/zabbix_server.conf
zabbix_server: symbol lookup error: zabbix_server: undefined symbol: usmDESPrivProtocol
# 出现这种情况时需要先升级 net-snmp，再启用 zabbix-server 服务
[root@zabbix-server ~]# dnf -y install net-snmp  net-snmp-devel
[root@zabbix-server ~]# dnf -y install net-snmp-utils
[root@zabbix-server ~]# mv /etc/zabbix/zabbix_server.conf /etc/zabbix/zabbix-server.conf
[root@zabbix-server ~]# systemctl restart zabbix-server zabbix-agent nginx php-fpm
[root@zabbix-server ~]# systemctl enable zabbix-server zabbix-agent nginx php-fpm
```

使用浏览器访问 http://192.168.100.11/zabbix，可在 Zabbix 安装界面中设置语言为简体中文（zh-CN），单击右下角的“下一步”按钮，进入下一步操作，如图 5-23 所示。

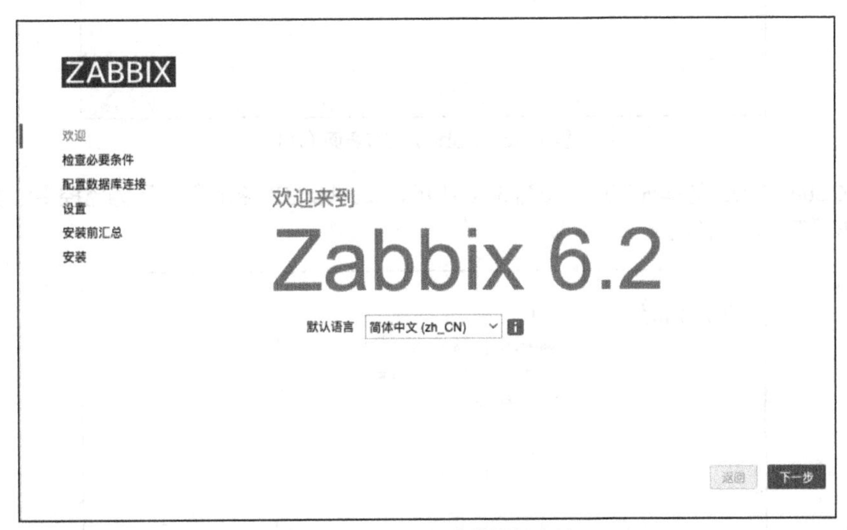

图 5-23　Zabbix 安装界面（1）

选择“检查必要条件”选项，显示 PHP 版本信息等内容，检查完成后提示 OK，单击“下一步”按钮，如图 5-24 所示。

配置数据库连接，填写连接数据库的必要信息，数据库名称为 zabbix，密码为 Zabbix，填写内容如图 5-25 所示，确认后单击“下一步”按钮。

ZABBIX

检查必要条件

欢迎
检查必要条件
配置数据库连接
设置
安装前汇总
安装

	当前值	要求的	
PHP版本	8.1.8RC1	7.4.0	OK
PHP选项"memory_limit"	128M	128M	OK
PHP选项"post_max_size"	16M	16M	OK
PHP选项"upload_max_filesize"	2M	2M	OK
PHP选项"max_execution_time"	300	300	OK
PHP选项"max_input_time"	300	300	OK
PHP支持的资料库	MySQL		OK
PHP bcmath	于		OK
PHP mbstring	于		OK
PHP选项"mbstring.func_overload"	关闭	关闭	OK

返回　下一步

图5-24　Zabbix安装界面（2）

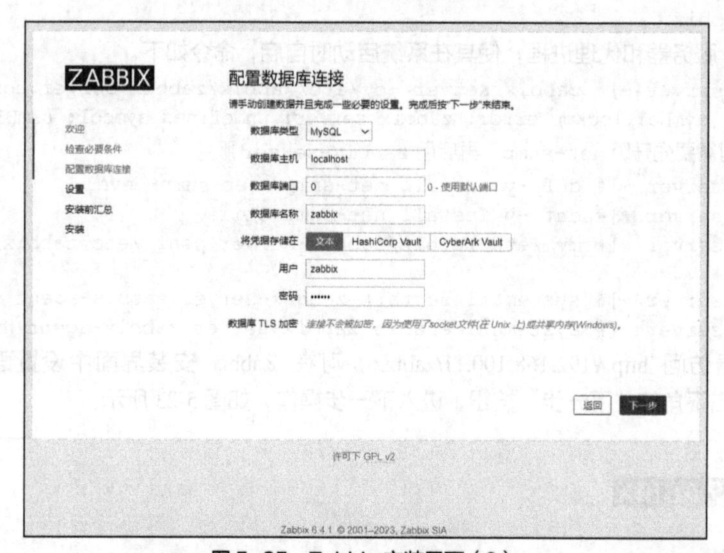

图5-25　Zabbix安装界面（3）

填写Zabbix Server的详细信息，如图5-26所示，“Zabbix主机名称”用于为监控平台设置名称，确认后单击“下一步”按钮。

ZABBIX

设置

欢迎
检查必要条件
配置数据库连接
设置
安装前汇总
安装

Zabbix主机名称　OpenStack云平台
默认时区　系统: (UTC+00:00) UTC
默认主题　蓝

返回　下一步

图5-26　Zabbix安装界面（4）

在完成安装前，平台配置概况如图 5-27 所示，检查所设置的参数，确认后单击“下一步”按钮。

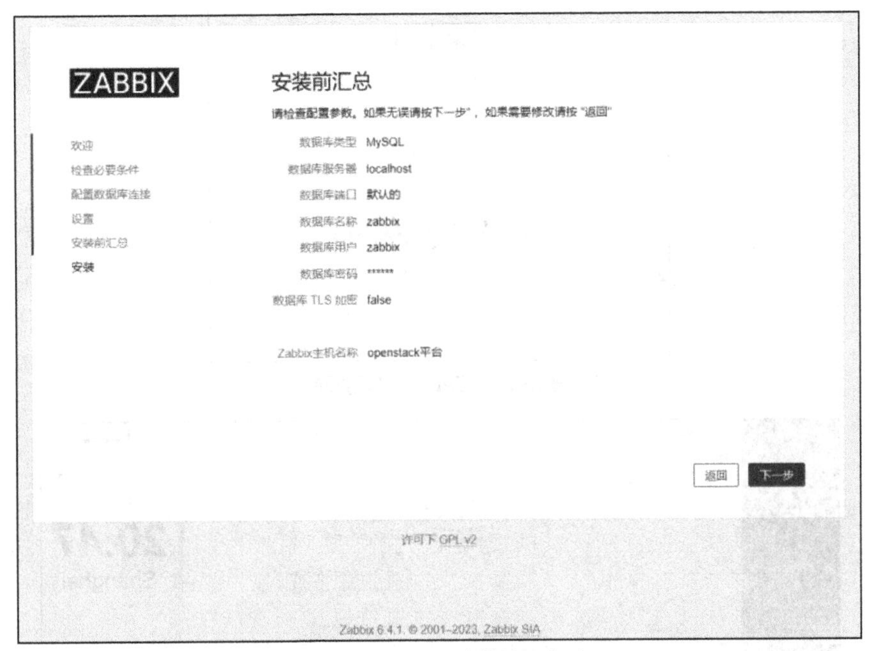

图 5-27　Zabbix 安装界面（5）

此时，Zabbix 服务已经配置完成，安装成功后如图 5-28 所示。

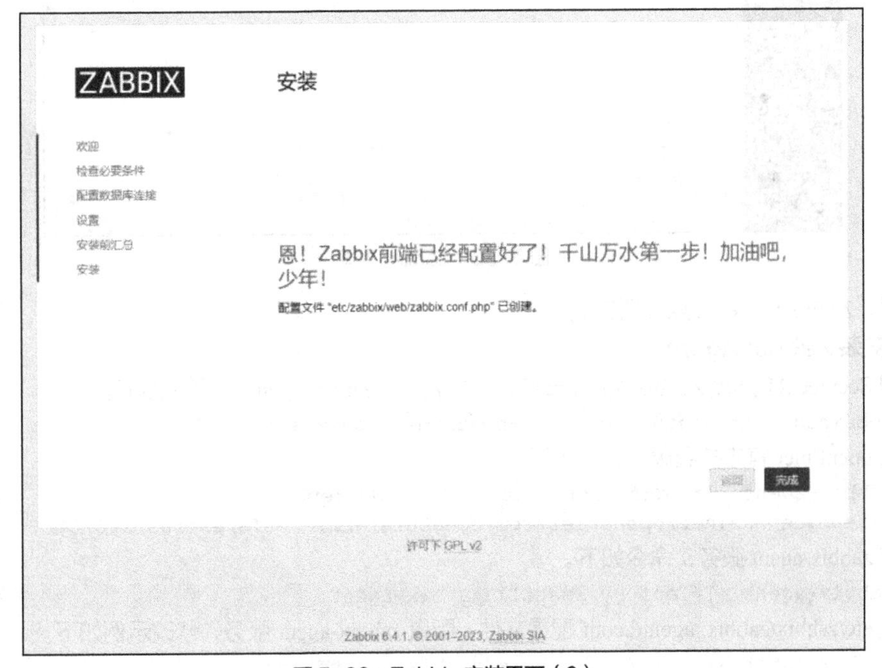

图 5-28　Zabbix 安装界面（6）

单击“完成”按钮后，进入 Zabbix 登录界面，如图 5-29 所示。

使用默认的用户名称和密码（分别为 Admin 和 zabbix）登录，进入 Zabbix 主页，如图 5-30 所示。

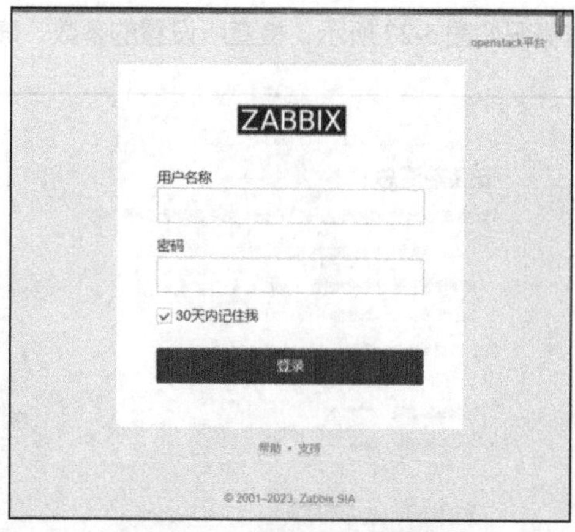

图 5-29　Zabbix 登录界面

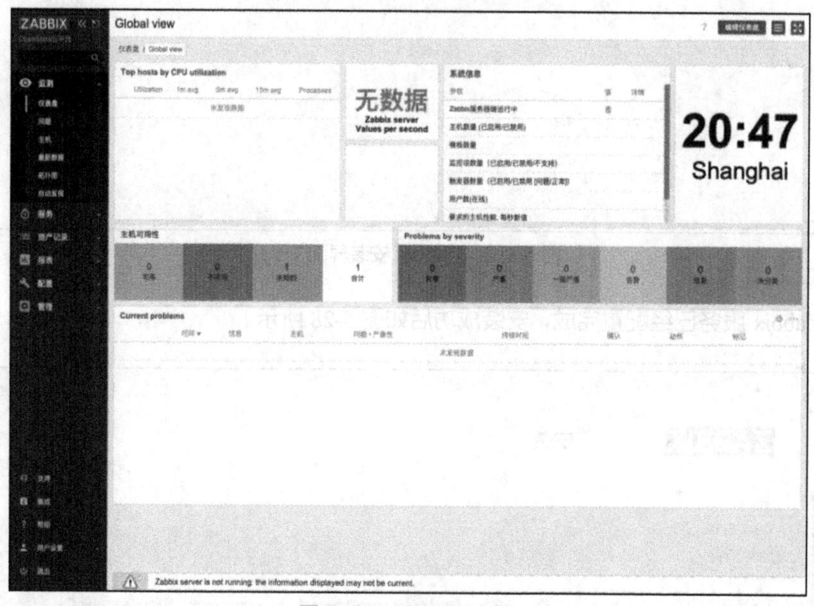

图 5-30　Zabbix 主页

此时，Zabbix Server 节点部署完成。

3. 安装 Zabbix Agent

使用 SecureCRT 连接 Zabbix Agent 节点，主机名为“zabbix-agent”，命令如下。

```
[root@localhost ~]# hostnamectl set-hostname zabbix-agent
```

添加 openEuler 官方扩展源，命令如下。

```
[root@zabbix-agent ~]#  dnf config-manager --add-repo  \
https://repo.oepkgs.net/openeuler/rpm/openEuler-22.03-LTS/extras/x86_64/
```

安装 zabbix-agent 服务，命令如下。

```
[root@zabbix-agent ~]# dnf -y install zabbix-agent
```

修改/etc/zabbix/zabbix_agentd.conf 配置文件，配置 zabbix-agent 服务，修改示例如下。

```
[root@zabbix-agent ~]# vi /etc/zabbix/zabbix_agentd.conf
Server=192.168.100.11
ServerActive=192.168.100.11
Hostname=Zabbix-agent
[root@zabbix-agent ~]# grep -n '^'[a-Z] /etc/zabbix/zabbix_agentd.conf
```

启动 zabbix-agent 服务，命令如下。

```
[root@zabbix-agent ~]# systemctl start zabbix-agent
```

回到 Web 界面，选择左侧导航栏中的“监测→主机”选项，单击右上角的“创建主机”按钮，弹出“添加主机”对话框，在该对话框中填写配置信息，将 zabbix-agent 节点添加到被监控机器中，主机群组选择“Linux servers”，填写信息如图 5-31 所示，填写完成后单击“添加”按钮。

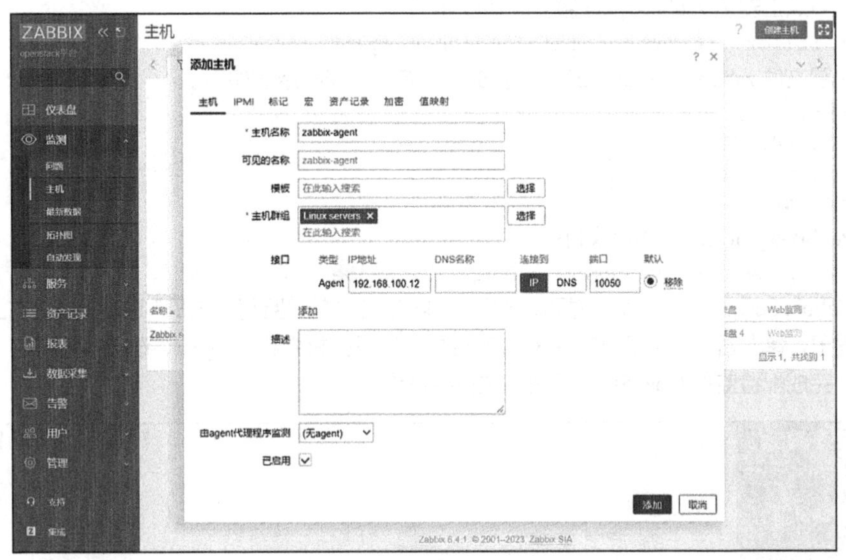

图 5-31 “添加主机”对话框

选择左侧导航栏中的“监测→主机”选项，选择“条件检索”选项，并单击标签值“cpu”来查看对 CPU 的图形化监控，如图 5-32 所示。

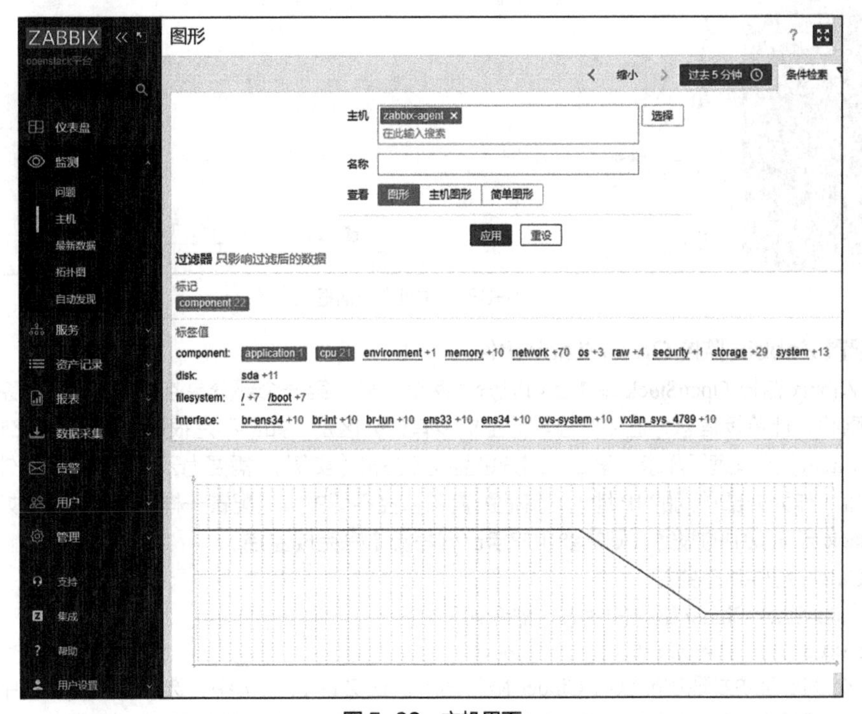

图 5-32 主机界面

4. 配置 Zabbix 监控 OpenStack 平台宿主机

监控 OpenStack 平台宿主机时，需要在控制节点上部署 zabbix-agent 服务。添加 openEuler 官方扩展源，具体操作如下。

```
[root@controller ~]#  dnf config-manager --add-repo  \
https://repo.oepkgs.net/openeuler/rpm/openEuler-22.03-LTS/extras/x86_64/
```

在控制节点上安装 zabbix-agent 服务，命令如下。

```
[root@controller ~]# dnf -y install zabbix-agent
```

修改配置文件 zabbix_agentd.conf，指定 Server 节点，修改示例如下。

```
[root@ controller ~]# vi /etc/zabbix/zabbix_agentd.conf
Server=192.168.100.11
ServerActive=192.168.100.11
Hostname=controller
```

启动 zabbix-agent 服务，命令如下。

```
[root@ controller ~]# systemctl start zabbix-agent
```

回到 Zabbix Sever 节点的 Web 界面，选择左侧导航栏中的“监测→主机”选项，单击右上角的“创建主机”按钮，弹出图 5-33 所示的“主机”对话框，填写主机名称、Agent，选择主机群组，单击“更新”按钮，完成配置监控 OpenStack 平台宿主机的操作。

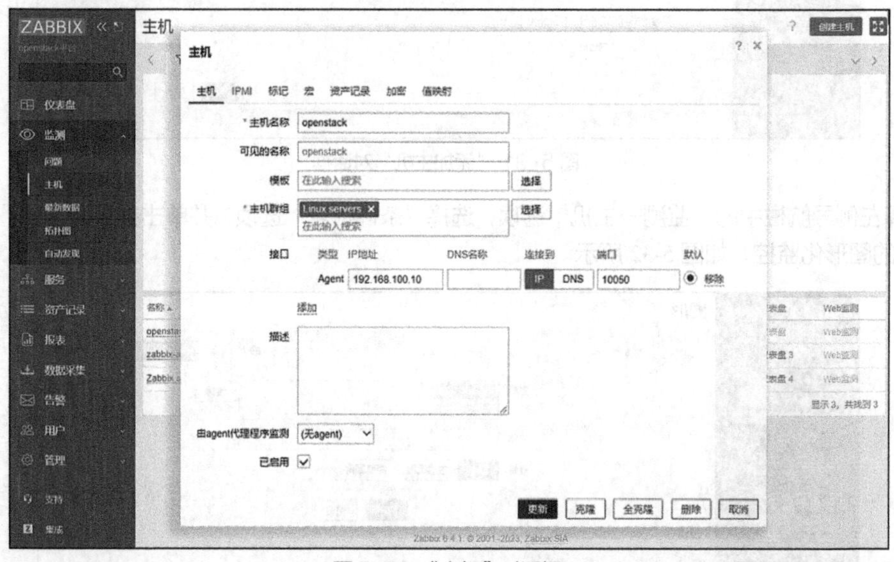

图 5-33 “主机”对话框

5. 配置 Zabbix 监控 OpenStack 组件

使用 Zabbix 监控 OpenStack 基本上可以分为两个部分：第一个部分是监控物理节点的系统信息，如控制、网络、计算等信息，这和监控其他主机没有什么区别，如果不是很严格的要求，Zabbix 自带的监控项（item）足以满足需求；第二个部分是监控云主机（实例），需要为云主机绑定一个外网 IP（浮动 IP）地址，作为外部访问的 IP 地址，其他配置和监控不同主机的配置一样。除此之外，如果想要监控 OpenStack 云环境中的服务，则需要用户自己编写脚本来完成监控。

（1）Cinder 组件状态监控

获取 OpenStack Cinder 组件的状态，命令如下。

```
[root@controller ~]# cinder service-list
```

使用 Zabbix 自动发现功能获取 Cinder 服务信息。在 Zabbix 主页中，选择左侧导航栏中的“数据采集→自动发现”选项，单击右上角“创建发现规则”按钮，相应界面如图 5-34 所示。

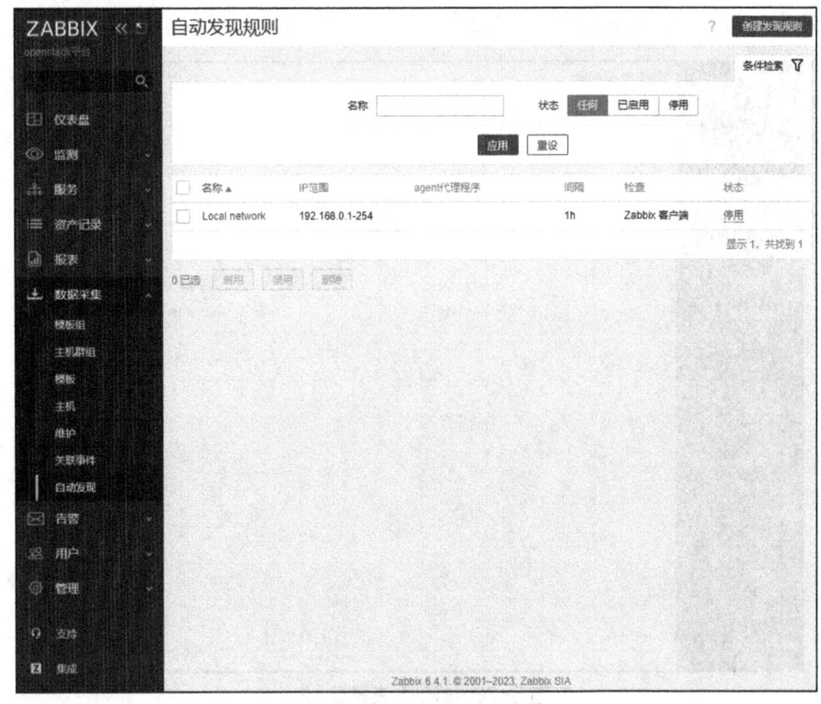

图 5-34　规则列表界面（1）

在“自动发现规则”界面中设置名称、IP 范围、更新间隔等，在“检查”区域中添加键值“openstack.cinder.discovery”，确认后单击“添加”按钮，如图 5-35 所示。

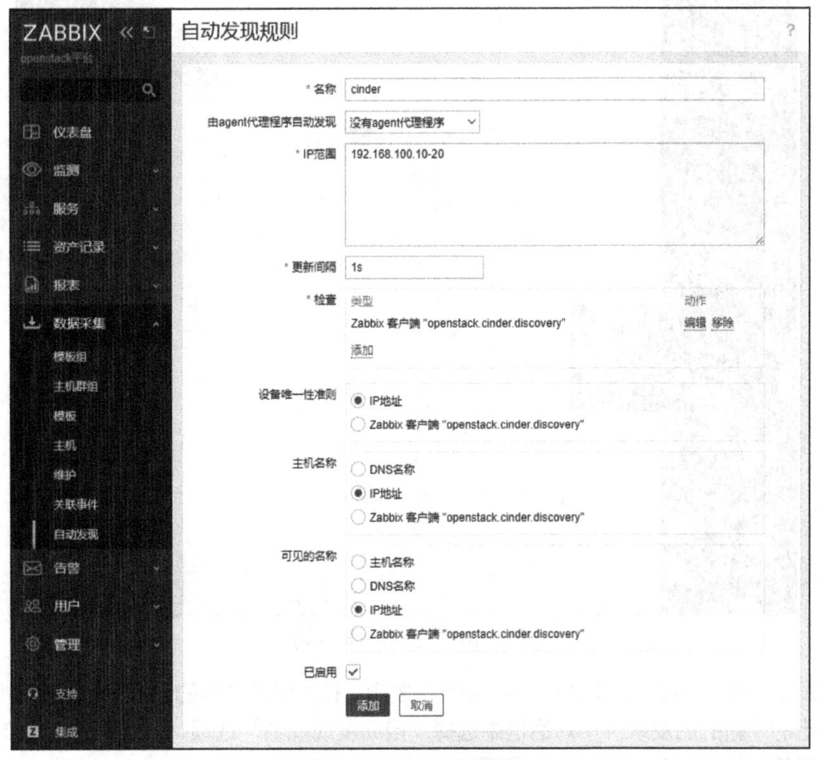

图 5-35　规则创建界面

添加完成后，提示“自动发现规则已创建”，如图 5-36 所示。

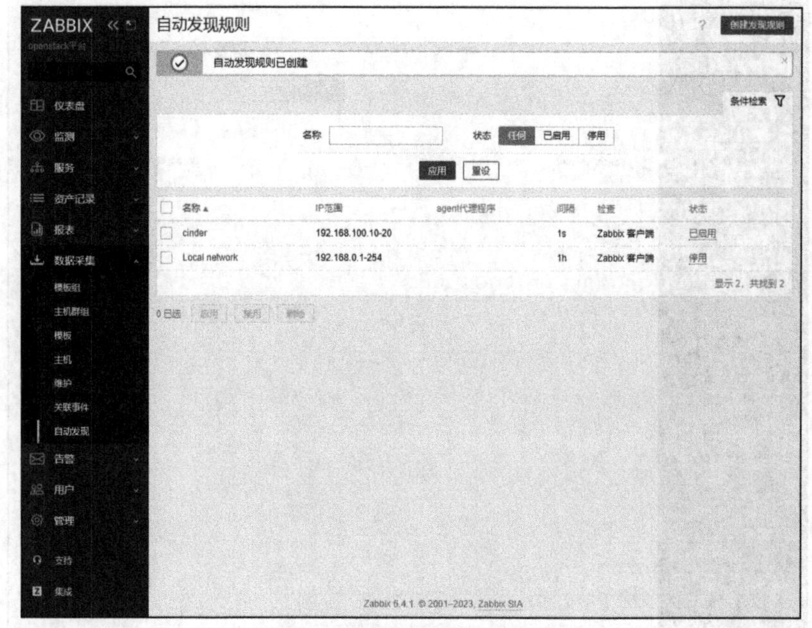

图 5-36　规则列表界面（2）

返回主页，选择左侧导航栏中的“告警→动作→发现动作”选项，如图 5-37 所示，进入“发现动作”界面。

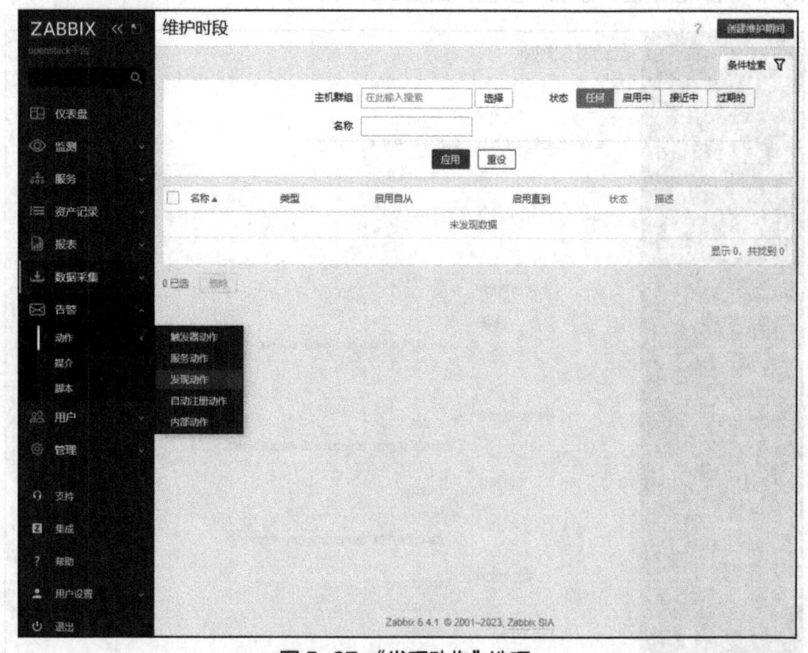

图 5-37　“发现动作”选项

单击“发现动作”界面右上角的“创建动作”按钮，在弹出的“新的动作”对话框中单击“添加”按钮，在弹出的“新的触发条件”对话框中选择“自动发现规则”选项，设置类型、操作者、自动发现规则，创建新的触发条件，如图 5-38 所示。

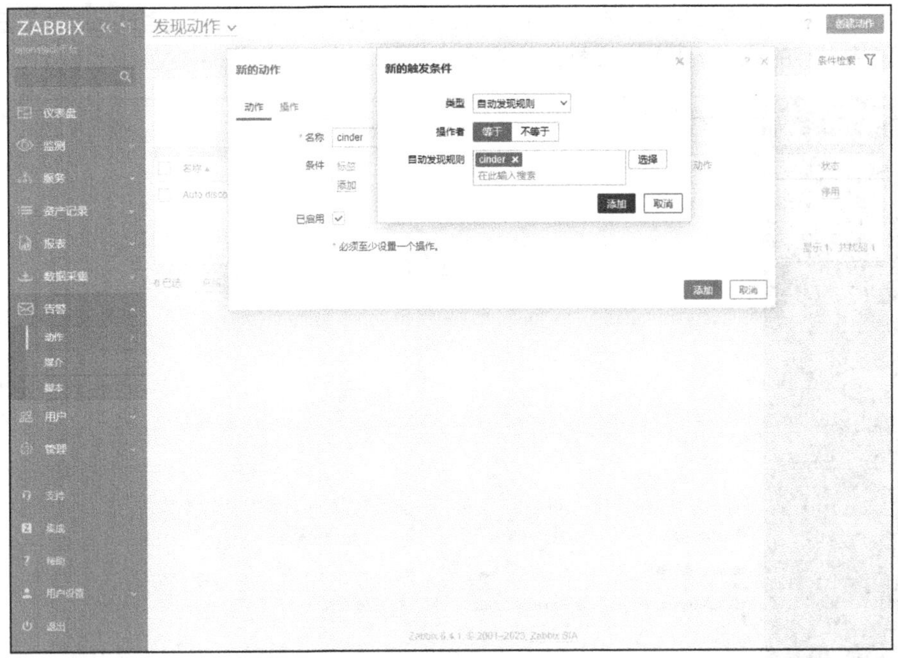

图 5-38　触发条件界面（1）

新的触发条件创建完成后，单击“添加”按钮，返回“新的动作”对话框，选择“操作”选项卡，在“操作”文本框中单击“添加”按钮，选择链接模板“Linux by Zabbix agent”，确认后单击“添加”按钮，如图 5-39 所示。

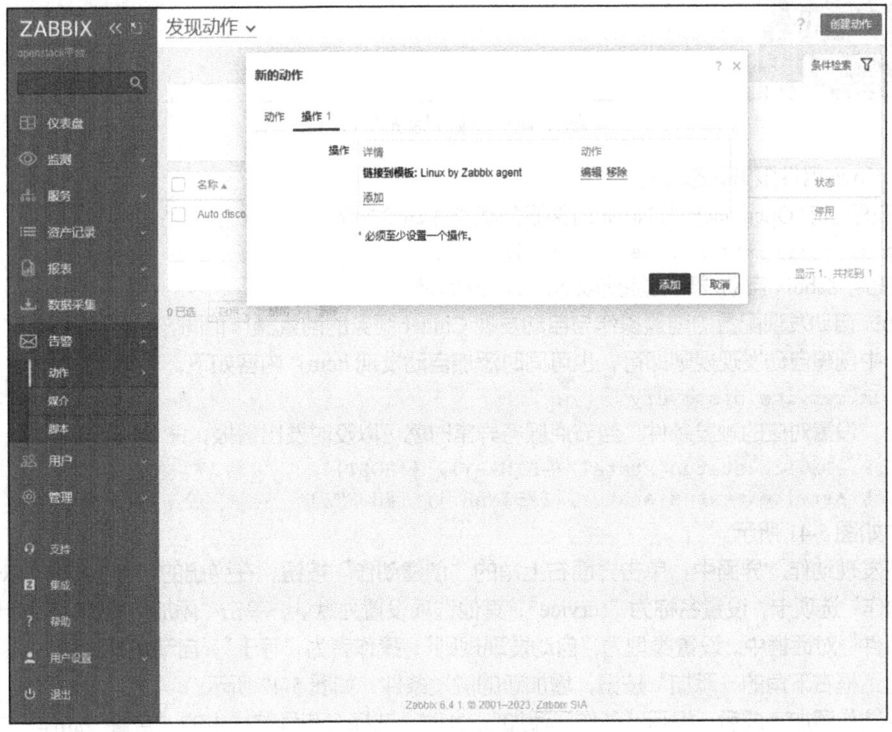

图 5-39　添加动作界面

此时，返回主页，选择左侧导航栏中的“监测→自动发现”选项，会发现通过 item 规则已经监控到宿主机控制节点的 Cinder 服务，如图 5-40 所示。

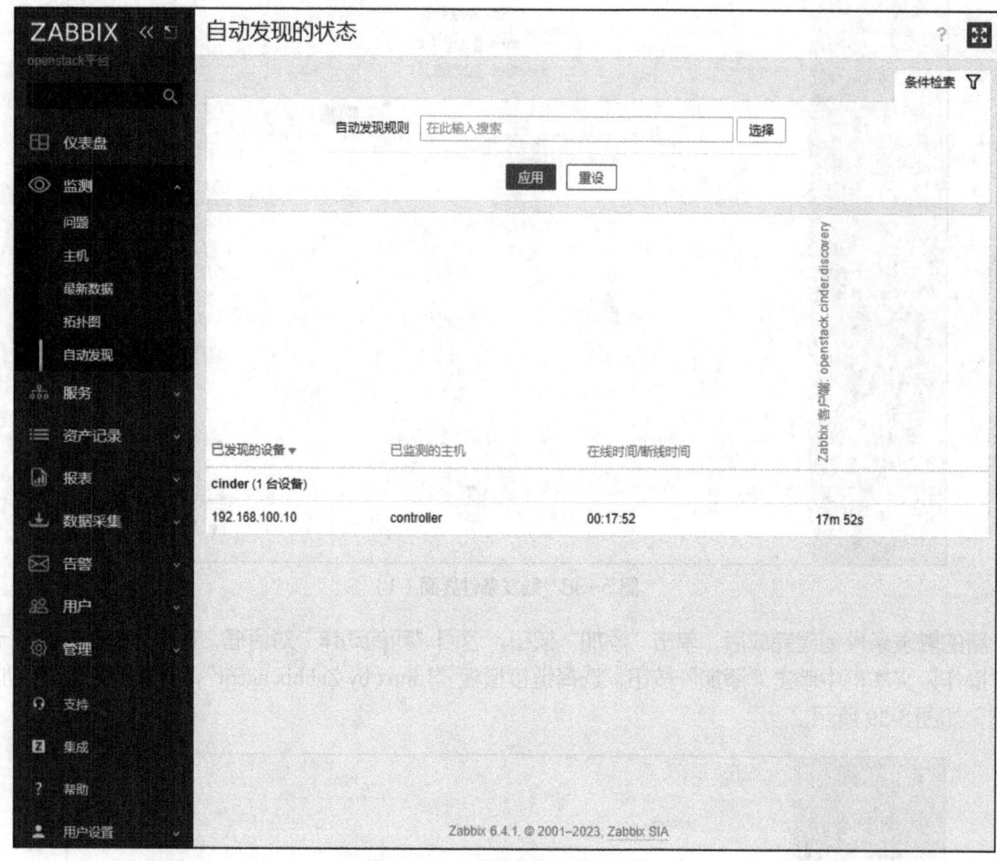

图 5-40　规则列表界面（3）

（2）Nova 组件服务状态监控

① 获取当前 OpenStack 的 Nova 组件服务状态，命令如下。

```
[root@controller ~]# nova service-list
```

② 使用 Zabbix 自动发现功能获取 Nova 服务信息。

Zabbix 自动发现配置的前置操作与自动发现 Cinder 服务的前置操作相同，在“数据采集”的“自动发现”中创建自动发现规则即可，也可同时添加自动发现 item，内容如下。

```
openstack.service.discovery
```

另外，设置对应的触发条件，当节点服务异常时就可以及时发出警报，命令如下。

```
openstack.service.status[state, {#BINARY}, {#HOST}]
openstack.service.status[status, {#BINARY}, {#HOST}]
```

具体如图 5-41 所示。

在“发现动作”界面中，单击界面右上角的“创建动作”按钮，在弹出的“新的动作”对话框中选择“动作”选项卡，设置名称为“service”，其他选项设置为默认，单击“添加”按钮。在弹出的“新的触发条件”对话框中，设置类型为“自动发现规则”，操作者为“等于”，自动发现规则为“service”，单击该对话框右下角的“添加”按钮，增加新的触发条件，如图 5-42 所示。

新的动作添加完成后，提示“动作已添加”。此时，选择左侧导航栏中的“告警→动作”选项，可以看到新增的 service 自动发现规则，如图 5-43 所示。

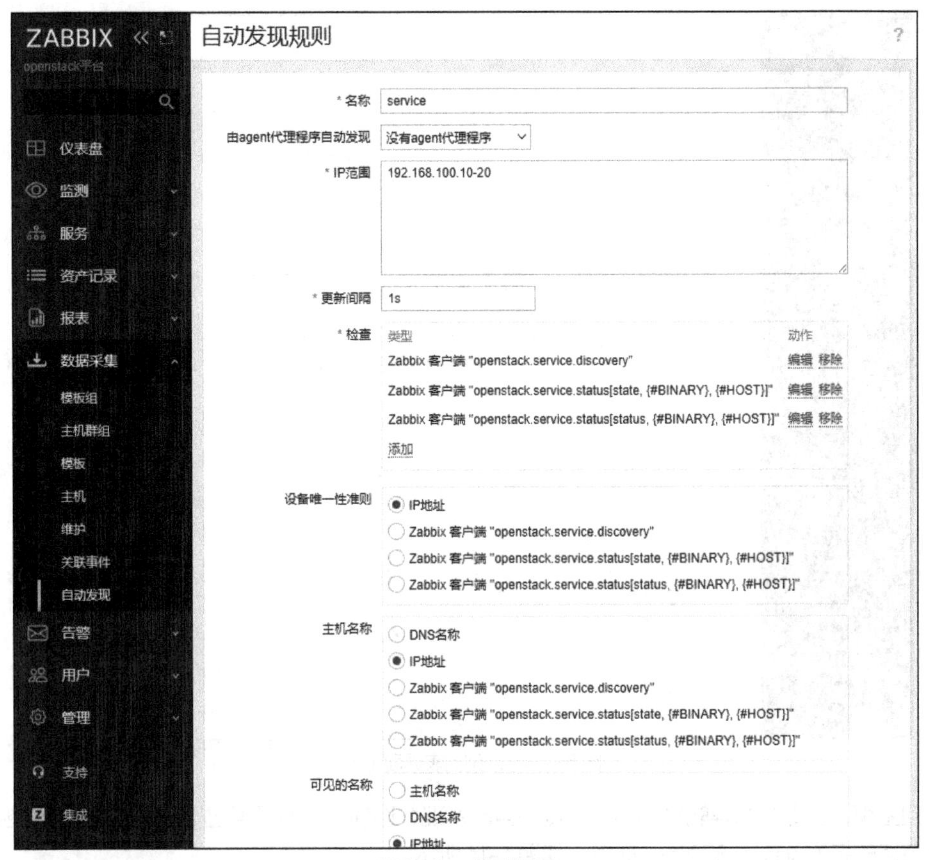

图 5-41　规则界面

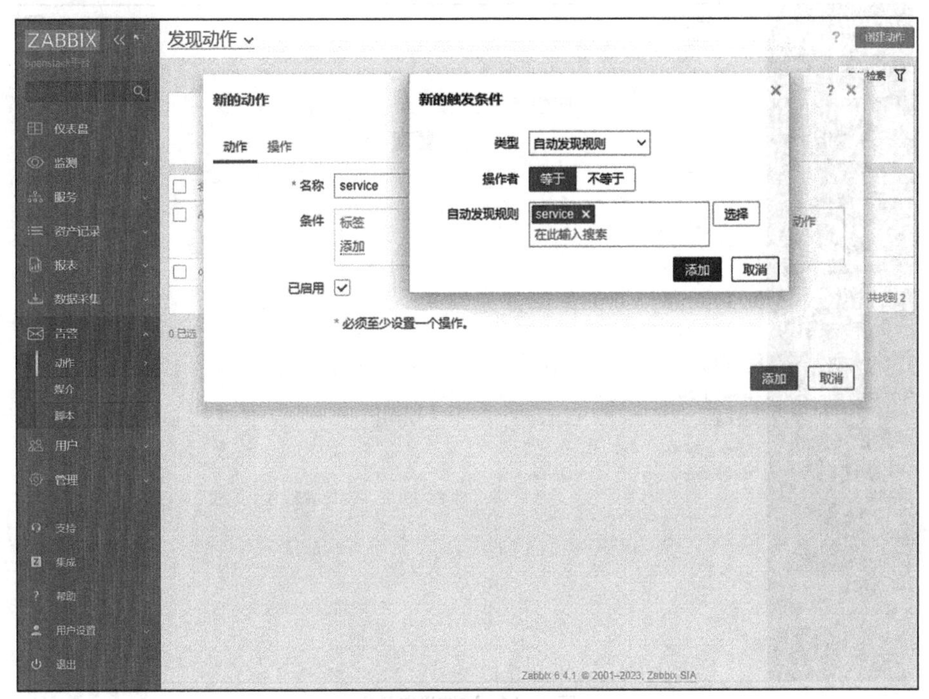

图 5-42　触发条件界面（2）

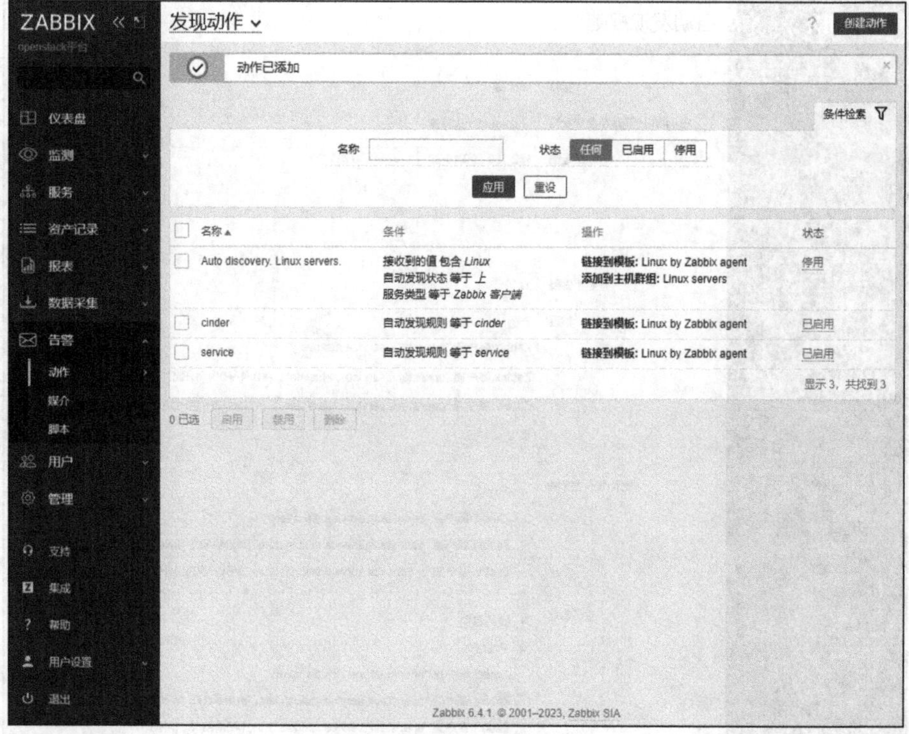

图 5-43 规则列表界面（4）

返回主页，选择左侧导航栏中的“监测→自动发现”选项，会发现通过 item 规则已经监控到宿主机控制节点的名为 service 的 Nova 服务，如图 5-44 所示。

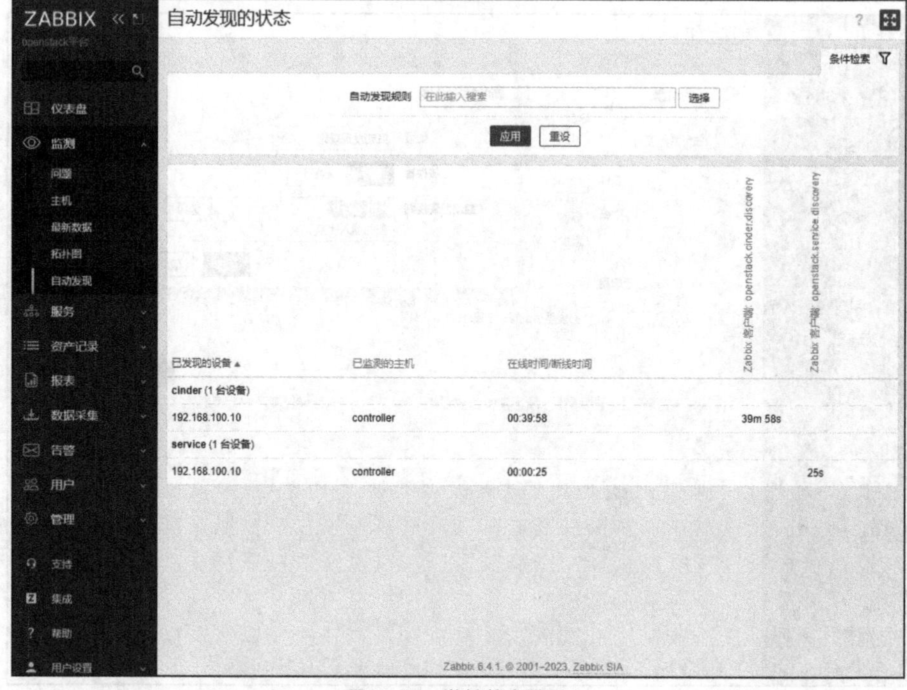

图 5-44 监控状态界面

任务 5.2.3　查看 OpenStack 宿主机状态

1. 使用 Zabbix 监控宿主机状态

在 5.2.2 小节的配置 Zabbix 监控 OpenStack 平台宿主机的基础上，通过浏览器进入 Zabbix Server 节点的 Zabbix 界面，选择左侧导航栏中的“监测→主机”选项，选择“条件检索”选项，选择指定的监控项，并使用“图形”形式进行查看，如图 5-45 所示。

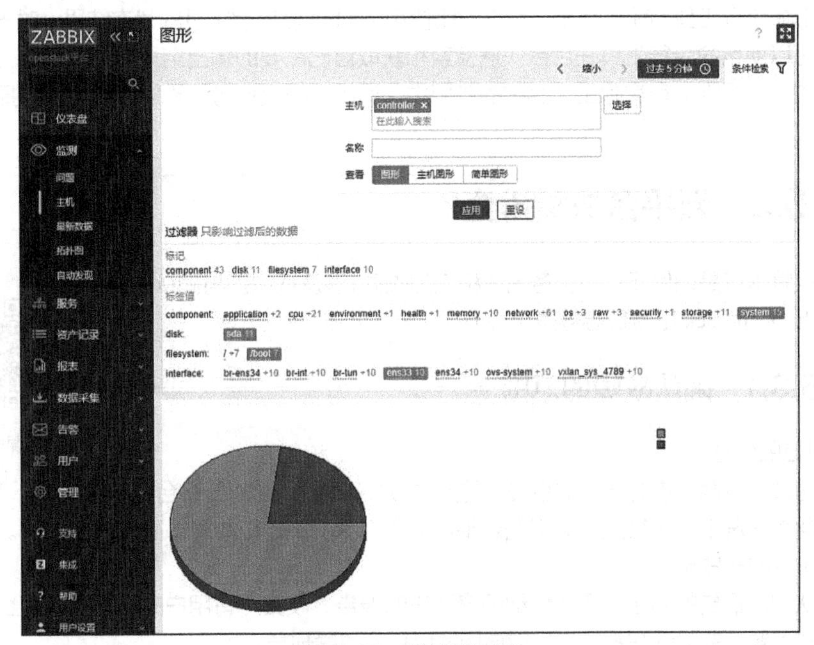

图 5-45　监控数据界面

选择“memory”“cpu”监控视图，如图 5-46 所示。

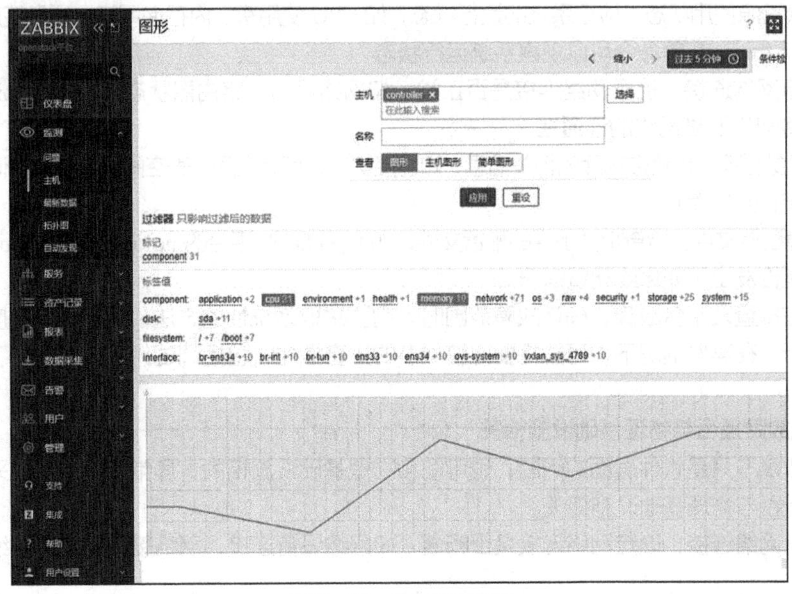

图 5-46　监控信息界面

2. 使用命令行工具查看宿主机状态

登录 OpenStack 宿主机控制节点，使用命令行工具查看宿主机磁盘使用情况，命令如下。

```
[root@controller ~]# df
```

查看宿主机内存使用情况，命令如下。

```
[root@controller ~]# free -m
```

查看宿主机 CPU 使用情况，命令如下。

```
[root@controller ~]# top
```

通过以上命令行工具的输出结果，可以对比使用 Zabbix 监控宿主机的状态结果。对于成熟的运维人员而言，可以使用命令行工具在后台快速检索和获取自己需要的输出结果，且可视化工具提供的是简洁、易用的直观界面。在实际工作实践的应用中，通常会根据业务的需求、规模来选择适合的监控数据查看方式。

任务 5.3 云平台故障排查

云平台故障排查是确保云计算服务高效稳定的关键，涉及从识别与确认故障范围到详细分析 OpenStack 服务日志等多个环节。

微课 5.6　云平台故障排查

任务 5.3.1 确认故障的范围

1. 故障范围排查

云平台的故障范围排查需要采取系统性的方法，从不同的角度对云平台进行诊断和分析。通常情况下，运维人员会根据云平台的运行状况来定位故障，确认故障范围。以下是针对云平台故障排查的一些步骤。

（1）确认用户报告的问题：需要仔细阅读用户的报告，收集所有用户的反馈，了解到底是哪些用户遇到了哪些问题，并尝试在系统上模拟该问题以复现故障现象。

（2）检查日志文件：检查相关组件的日志文件以查找任何异常。例如，在 OpenStack 平台中，可以通过查看 Nova、Neutron、Glance 等组件的日志文件来发现问题。

（3）检查资源使用状态：检查系统的运行状态，如 CPU 使用率、内存使用率、硬盘使用率等。使用类似 top、ps、free、df 等命令可以观察系统运行状态。

（4）检查网络连接：检查网络连接是否正常，如网络接口、路由器状态等。使用 ping 命令或者 traceroute 命令可以测试网络的连通性。

（5）检查数据库：检查数据库的相关信息，如连接状态、存储空间、表空间等。使用类似 mysql、psql 等命令可以查看数据库状态。

（6）检查配置文件：检查相关组件的配置文件，如 OpenStack 平台的 nova.conf 文件、neutron.conf 文件等，了解是否发生了变化或配置错误。

所以，在排查云平台故障、确认故障范围时，需要采取系统性的方法，从不同的角度对云平台进行诊断和分析。在某些情况下，可能需要比较复杂的排查操作，包括调试代码、回滚补丁等操作来解决问题。

2. 确认故障是否与物理基础设施相关

云平台故障不只是软件层面的问题，也可能和物理基础设施相关，通常情况下，会采取以下步骤来确认故障是否与物理基础设施相关。

（1）确认网络连接：确定网络连接是否畅通，如网络设备连接、网络接口连接、网络线缆连接等是否畅通。

（2）确认电源连接：确保设备的电源连接正常且设备正常工作，可以通过查看电源指示灯或者电源状态信息进行确认。

（3）确认硬件故障：如果设备存在硬件故障，如硬件组件损坏、内存故障等，那么需要检查硬件问题。

（4）确认数据中心环境：确保数据中心环境（包括温度、湿度、噪声等级、供电线路等）符合设备要求。

（5）测试备用电源和冷却系统：在一些特定条件下，如当停电或者故障发生时，备用电源和冷却系统可能非常重要，确保它们都能正常工作是很重要的。

（6）确认硬件状态：检查硬件设备是否损坏或者发生故障，应用工具（如 MemTest86、Smartmontools、lshw 等）进行检查。

物理基础设施的检查和诊断是日常维护云平台正常运行的关键举措，可及时查出故障源头，以便尽快解决故障，恢复设备的正常工作。实际实施过程中，运维人员也可以寻求专业硬件技术人员的协助和支持。

任务 5.3.2　OpenStack 服务日志分析

1. 各服务日志目录说明

OpenStack 中每个服务的日志目录位置和命名并不统一，/var/log 目录是日志文件夹的主要存储位置，根据服务名对文件夹进行进一步的划分，各服务日志文件名也因版本和配置而有所不同。列出常见的文件目录结构，命令和结果如下。

```
[root@controller ~]# ls /var/log/
anaconda        dnf.log          httpd        neutron       README     vmware-network.1.log     xferlog
audit           dnf.rpm.log      keystone     nginx         secure     vmware-network.2.log     zabbix
btmp            firewalld        lastlog      nova          skyline    vmware-network.3.log
ceilometer      glance           libvirt      openvswitch   spooler    vmware-network.log
chrony          gnocchi          maillog      placement     sssd       vmware-vgauthsvc.log.0
cinder          hawkey.log       mariadb      private       swift      vmware-vmsvc-root.log
cron            heat             memcached_exporter  prometheus  swtpm  vmware-vmtoolsd-root.log
dnf.librepo.log horizon          messages     rabbitmq      tallylog   wtmp
```

这里列出了 OpenStack 常用服务的日志目录和文件，具体介绍如下。

（1）Nova 服务主要日志：OpenStack 计算服务 Nova 的日志位于/var/log/nova，默认权限拥有者是 nova 用户。需要注意的是，并不是每台服务器上都包含所有的日志文件，如 nova-controller.log 仅在计算节点上生成，命令和结果如下。

```
[root@controller ~]# ll /var/log/nova/
总用量 4228
-rw-r--r-- 1 nova nova 4315000  4月 29 16:28 nova-controller.log
```

/var/log/nova/nova-controller.log：记录虚拟机实例在启动和运行中产生的日志。

（2）Neutron 服务主要日志：网络服务 Neutron 的日志默认存放在/var/log/neutron 目录中，命令和结果如下。

```
[root@controller ~]# ll /var/log/neutron/
总用量 9696
-rw-r--r-- 1 neutron neutron  12226  4月 29 15:50 dhcp-agent.log
-rw-r--r-- 1 neutron neutron   2099  4月 29 15:39 l3-agent.log
-rw-r--r-- 1 neutron neutron   5469  4月 18 09:06 metadata-agent.log
```

```
-rw-r--r-- 1 neutron neutron 4510328  4月 29 16:29 openvswitch-agent.log
-rw-r--r-- 1 root    root          0  4月 18 09:06 privsep-helper.log
-rw-r--r-- 1 neutron neutron 4648100  4月 29 16:29 server.log
```

① /var/log/neutron/server.log：记录与Neutron API服务相关的日志。

② /var/log/neutron/dhcp-agent.log：记录关于dhcp-agent的日志。

③ /var/log/neutron/l3-agent.log：记录关于Neutron L3代理相关活动的日志。

④ /var/log/neutron/metadata-agent.log：记录通过Neutron代理给Nova元数据服务的相关日志。

⑤ /var/log/neutron/openvswitch-agent.log：记录与Open vSwitch相关操作的日志项，在具体实现OpenStack网络时，如果使用了不同的插件，则会有相应的日志文件名。

⑥ /var/log/neutron/privsep-helper.log：记录关于特权分离助手相关活动的日志。

（3）Cinder服务主要日志：块存储Cinder产生的日志默认存放在/var/log/cinder目录中，命令和结果如下。

```
[root@controller ~]# ll /var/log/cinder/
总用量 11404
-rw-r--r-- 1 cinder cinder 10994772  4月 29 16:29 api.log
-rw-r--r-- 1 cinder cinder      608  4月 18 09:14 cinder-api.log
-rw-r--r-- 1 cinder cinder     3516  4月 25 10: 27 schedulerS.log
```

① /var/log/cinder/api.log：记录Cinder服务中API服务器的日志文件。

② /var/log/cinder/cinder-api.log：记录关于cinder-api服务的日志。

③ /var/log/cinder/scheduler.log：记录关于Cinder调度服务的日志。

（4）Keystone服务主要日志：身份认证Keystone服务的日志默认存放在/var/log/keystone目录中，命令和结果如下。

```
[root@controller ~]# ll /var/log/keystone/
总用量 640
-rw-r--r-- 1 root keystone 652829  4月 29 16:13 keystone.log
```

/var/log/keystone/keystone.log：记录Keystone服务的启动信息、关闭信息、用户身份认证和授权过程中产生的错误信息。

（5）Glance服务主要日志：镜像服务Glance的日志默认存放在/var/log/glance目录中，命令和结果如下。

```
[root@controller ~]# ll /var/log/glance/
总用量 516
-rw-r--r-- 1 glance glance 396554  4月 29 16:27 glance-api.log
```

/var/log/glance/glance-api.log：记录与Glance API相关的日志。

2. 如何有效查看相关服务的日志

在OpenStack中，每个组件服务都有各自的日志，查看和分析这些日志有助于用户发现及诊断各种问题。以下是一些有效查看OpenStack相关服务日志的方法。

（1）使用命令行工具查看日志。在OpenStack节点上安装的组件服务可以通过命令行工具进行访问和管理，因此可以使用该工具查看日志，命令如下。

```
[root@controller ~]# tail -f /var/log/*.log
```

（2）使用OpenStack Dashboard查看日志。控制节点上安装了OpenStack Dashboard组件，可以通过它访问Horizon Web界面，从而可以查看组件服务日志。

（3）使用ELK和Prometheus开源日志管理平台进行日志的查看。这些开源平台提供了完整的日志解决方案，包括可视化、警报、搜索、过滤器等。

无论使用哪种方法，总是应该能够根据需要轻松地搜索、分析和查看OpenStack服务产生的各种日志信息，从而保持OpenStack环境的最佳状态。

任务 5.3.3 常见故障及解决方案

1. OpenStack 服务故障排查

OpenStack 云平台中的服务故障排查是一项需要耐心和技巧的任务。通常情况下，可以尝试以下故障排查的基本步骤。

第一，记录错误信息。OpenStack 服务往往会输出一些错误信息和日志，这些信息能很好地提示服务出现的故障。针对异常故障，首先应该在服务日志中查找相关信息。

第二，检查服务状态。可以在控制节点上使用命令行工具或面板查看服务的状态。对于故障服务，可以使用 systemctl status 命令来查看其详细的状态信息。

第三，检查网络连接。所有 OpenStack 服务都涉及网络通信，有时网络问题是产生错误的原因。可以使用 ping、nslookup、tracerout 等命令来检查相应 IP 地址或主机名称在网络上的可到达性。

第四，检查配置信息。在 OpenStack 的配置文件中可能存在错误或不一致的配置项，导致服务无法正常运行。

第五，再次缩小故障范围。如果在前 4 个步骤中没有找到故障的根本原因，那么应该再次缩小故障范围。可以使用 strace、tcpdump、sysdig、gdb 和其他调试工具来进一步逐层排查故障。

以下提供了 OpenStack 关键服务的故障排查方法与流程。

（1）Keystone 服务故障排查

① 查看端口 5000 是否处于开放状态，命令和结果如下。

```
[root@controller ~]# netstat -ntlp |grep 5000
tcp6      0        0 :::5000        :::*     LISTEN         60190/httpd
```

② 确保数据库服务处于 active 状态，命令和结果如下。

```
[root@controller ~]# service mariadb status
Active: active (running)
```

③ 查看 Keystone 的 endpoint 端点是否存在，命令和结果如下。

```
[root@controller ~]# openstack endpoint list |grep keystone
| 732b68762f7d40139
  f9f7f436dcf8ca7    | RegionOne | keystone | identity | True | admin    | http://contro
                                                                            ller:5000/v3/ |
| cd510665cea64461a
  eb162d1958b9ae9    | RegionOne | keystone | identity | True | public   | http://contro
                                                                            ller:5000/v3/ |
| dcdd9b274eec4a918
  87d0ff8a1ce27d1    | RegionOne   keystone | identity   True | internal | http://contro
                                                                            ller:5000/v3/ |
```

（2）Glance 服务故障排查

① 查看数据库服务是否处于 active 状态，命令和结果如下。

```
[root@controller ~]# service mariadb status
Active: active (running)
```

② 查看 Glance 服务的 endpoint 是否存在，命令和结果如下。

```
[root@controller ~]# openstack endpoint list |grep glance
| 2daab67774e84f86
  98413fce6b9721a6  | RegionOne   | glance   | image   | True  | internal | http://cont
                                                                          roller:9292 |
| 60da8a5ca61d46ac
  98f68a75329eea24  | RegionOne   | glance   | image   | True  | admin    | http://cont
                                                                          roller:9292 |
| af496de941b34440
  b79ee007b27d2768  | RegionOne     glance   | image     True  | public   | http://cont
                                                                          roller:9292 |
```

③ 若 Glance 后端存储对接了其他存储，查看 Glance 后端存储目录的用户和用户组是否为 glance，则命令和结果如下。

```
[root@controller ~]# ll /var/lib/glance
总用量 0
drwxr-x--- 2 glance glance 94 4月 29 15:47 images
```

```
[root@controller ~]# ll /var/lib/glance/images
总用量 927936
-rw-r----- 1 glance glance 21233664 4月 29 15:47 163f479d-f31b-4be3-867c-eec831c6c8b3
-rw-r----- 1 glance glance 928972800 4月 23 13:00 63bf7f04-8fdf-4724-b94c-00526cf7f2bb
```

（3）Nova服务故障排查

① 查看数据库服务是否处于active状态，命令和结果如下。

```
[root@controller ~]# service mariadb status
Active: active (running)
```

② 查看Nova服务的endpoint是否存在，命令和结果如下。

```
[root@controller ~]# openstack endpoint list |grep nova
| 28c37de3ddea4d34
  8ea6baa2c729f998 | RegionOne | nova | compute | True | admin    | http://controll
                                                                     er:8774/v2.1 |
| 62938d4be9ce4293
  8cd363e4dfd89335 | RegionOne | nova | compute | True | public   | http://controll
                                                                     er:8774/v2.1 |
| d149ee6bda63407e
  a0399c12f7554b23 | RegionOne | nova | compute | True | internal | http://controll
                                                                     er:8774/v2.1 |
```

查看Nova服务的状态，命令和结果如下。

```
[root@controller ~]# nova service-list
nova CLI is deprecated and will be a removed in a future release
+----------+--------+--------+--------+--------+-----+--------+--------+------+
|id        |Binary  | Host   |Zone    | Status |State|Updated_|Disabled|Forced|
|          |        |        |        |        |     |at      |Reason  |down  |
+----------+--------+--------+--------+--------+-----+--------+--------+------+
| 0721267a-| nova-  |control |internal|enabled |up   |2023-04-|-       |False |
| a9e4-4d82-|conduct|ler     |        |        |     |29T08:  |        |      |
| 863e-e18f5|or     |        |        |        |     |13:42.  |        |      |
| 3750bfa  |        |        |        |        |     |000000  |        |      |
| b167c2bf-|nova-sc |control |internal|enabled |up   |2023-04-|-       |False |
| 82b2-48db-|heduler|ler     |        |        |     |29T08:  |        |      |
| a3a9-552f7|       |        |        |        |     |13:39.  |        |      |
| c8f3abd  |        |        |        |        |     |000000  |        |      |
| b78cfc1d-2|nova-  |compute |nova    |enabled |up   |2023-04-|-       |False |
| 0b5-403b-8|compute|        |        |        |     |29T08:  |        |      |
| 247-8cc50b|       |        |        |        |     |13:45.  |        |      |
| 2142a7   |        |        |        |        |     |000000  |        |      |
+----------+--------+--------+--------+--------+-----+--------+--------+------+
```

③ 若Nova服务后端修改了存储位置，查看Nova实例存储目录的权限是否发生了变化，则命令和结果如下。

```
[root@controller ~]# ll /var/lib/nova/
总用量 0
drwxr-xr-x 2 nova nova 6  1月 26 23:19 buckets
drwxr-xr-x 2 nova nova 6  1月 26 23:19 instances
drwxr-xr-x 2 nova nova 6  1月 26 23:19 keys
drwxr-xr-x 2 nova nova 6  1月 26 23:19 networks
drwxr-xr-x 2 nova nova 6  1月 26 23:19 tmp
```

（4）Neutron服务故障排查

① 查看数据库服务是否处于active状态，命令和结果如下。

```
[root@controller ~]# service mariadb status
Active: active (running)
```

② 查看Neutron服务的endpoint是否存在，命令和结果如下。

```
[root@controller ~]# openstack endpoint list |grep neutron
```

```
| 28c37de3ddea4d34 | RegionOne | neutron | network | True | internal | http://cont |
| 8ea6baa2c729f998 |           |         |         |      |          | roller:9696 |
| 62938d4be9ce4293 | RegionOne | neutron | network | True | public   | http://cont |
| 8cd363e4dfd89335 |           |         |         |      |          | roller:9696 |
| d149ee6bda63407e | RegionOne   neutron | network   True | admin    | http://cont |
| a0399c12f7554b23 |           |         |         |      |          | roller:9696 |
```

③ 查看 Open vSwitch 服务是否处于 active 状态，命令和结果如下。

```
[root@controller ~]# neutron agent-list
neutron CLI is deprecated and will be removed in the Z cycle. Use openstack CLI instead.
+---------------+---------+-----------+-------------+-----+-------+-----------+
|               |         |           |             |     | admin_|           |
| id            | agent_  | host      |availability_| alive| state_| binary   |
|               | type    |           | zone        |     | up    |           |
+---------------+---------+-----------+-------------+-----+-------+-----------+
| 32ffae82-943c-| DHCP    | controller| nova        | :-) | True  | neutron-  |
| 46de-9581-33ee| agent   |           |             |     |       | dhcp-agent|
| b15fc691      |         |           |             |     |       |           |
| 54bc4f28-87eb-| Open    |           |             |     |       | neutron-  |
| 4134-98ce-7ffb| vSwitch | compute   |             | :-) | True  | openvswitch|
| b046ca44      | agent   |           |             |     |       | -agent    |
| cfe9c306-8e4c-| Metadat |           |             |     |       | neutron-  |
| 48f0-89d9-409f| a agent | controller|             | :-) | True  | metadata- |
| 098a2ca1      |         |           |             |     |       | agent     |
| db18ac26-2640-|         |           |             |     |       | neutron-  |
| 4c5a-8a26-5812| L3 agent| controller| nova        | :-) | True  | l3-agent  |
| ad29b0fa      |         |           |             |     |       |           |
| ee14474c-3aa6-| Open    |           |             |     |       | neutron-  |
| 4dbd-98dc-4d0b| vSwitch | controller|             | :-) | True  | openvswitch|
| 87eabb90      | agent   |           |             |     |       | -agent    |
+---------------+---------+-----------+-------------+-----+-------+-----------+
```

④ 查看 Neutron 各服务是否处于 active 状态，命令和结果如下。

```
[root@controller ~]# systemctl status neutron*
neutron-openvswitch-agent
Active: active (running)
neutron-dhcp-agent
Active: active (running)
neutron-l3-agent
Active: active (running)
neutron-metadata-agent
Active: active (running)
neutron-server
Active: active (running)
```

（5）Cinder 服务故障排查

① 查看数据库服务是否处于 active 状态，命令如下。

```
[root@controller ~]# service mariadb status
```

② 查看 Cinder 服务的 endpoint 是否存在，命令和结果如下。

```
[root@controller ~]# openstack endpoint list |grep cinder
| 084d088d48104408 | RegionOne | cinderv3 | volumev3 | True | admin  | http://controller: |
| aacca31dd32bc750 |           |          |          |      |        | 8776/v3/%          |
|                  |           |          |          |      |        | (project_id)s      |
| 26e1ba67e2c14c13 | RegionOne | cinderv3 | volumev3 | True | public | http://controller: |
| 9aa1268d60e4e875 |           |          |          |      |        | 8776/v3/%          |
|                  |           |          |          |      |        | (project_id)s      |
```

```
| e3dbbb701d4f4473 b8e663ca1128f1d8 | RegionOne  cinderv3 | volumev3  True | internal | http://controller:8776/v3/%(project_id)s |
```

③ 查看 cinder-volumes 卷组是否处于可用状态或者是否还有剩余空间，命令如下。

```
[root@compute ~]#  vgdisplay
```

④ 查看 Cinder 各服务是否处于 active 状态，命令如下。

```
[root@controller ~]# systemctl status openstack-cinder*
```

（6）Swift 服务故障排查

① 查看数据库服务是否处于 active 状态，命令如下。

```
[root@controller ~]# service mariadb status
```

② 查看 Swift 服务的 endpoint 是否存在，命令和结果如下。

```
[root@controller ~]# openstack endpoint list |grep swift
| 4e0c94fe19a44140 97613ebc231e6c3e | RegionOne | swift | object-store | True | public   | http://controller:8080/v1/AUTH_%(project_id)s |
| 5eb8ecd38aca4eb5 8212d55d9b562bd6 | RegionOne | swift | object-store | True | internal | http://controller:8080/v1/AUTH_%(project_id)s |
| d215d07651814b82 af929d788bab20e4 | RegionOne   swift | object-store   True | admin    | http://controller:8080/v1 |
```

查看/srv/node/是否已满，命令和结果如下。

```
[root@compute ~]# du -h /srv/node/
0        /srv/node/sdb1/objects
0        /srv/node/sdb1
0        /srv/node/sdb2/objects
0        /srv/node/sdb2
0        /srv/node/sdb3/objects
0        /srv/node/sdb3
0        /srv/node
```

总之，在 OpenStack 环境中排查故障需要耐心和技巧，需要有丰富的经验并熟练使用相关工具，需要不断重复以上步骤，以最小化故障范围并找到故障发生的根本原因。

2. 宿主机故障排查

OpenStack 宿主机故障会给云平台的正常运转带来很大威胁。宿主机的故障排查需要优先查看宿主机的进程状态，通过使用 top 或者 ps 命令可以查看宿主机上所有正在运行的进程。

如果发现宿主机正在运行任何意外的进程，如病毒或者恶意软件，则需要对宿主机进行一次完整的检测，可以通过查看/var/log/wtmp 来发现异常信息。

/var/log/wtmp 是一个二进制文件，记录了每个用户的登录次数和持续时间等信息。该日志文件永久记录了每个用户的登录、注销及系统的启动、停止的事件。因此，随着系统正常运行时间的增加，该文件的大小会越来越大，其增加的速度取决于系统用户登录的次数。该日志文件可以用来查看用户的登录记录，但是需要使用 last 命令访问该文件以获得这些信息，命令如下。

```
[root@controller ~]# last -f /var/log/wtmp
```

last 命令的输出结果会以逆序从后向前显示用户的登录记录，last 也能根据用户、终端 tty 或时间显示相应的记录，命令和结果如下。

```
[root@controller ~]# last root -f /var/log/wtmp
[root@controller ~]# last -t 2024-01-01 08:00:00 -f /var/log/wtmp
```

3. 网络故障排查

OpenStack 网络故障是一项常见的问题。通常会使用以下方式排查 OpenStack 网络故障。

（1）确认网络拓扑：检查 OpenStack 网络拓扑，包括网络配置、VLAN 的创建和配置。确保网络拓扑中的物理网络、网络配置和网络连接工作正常，如图 5-47 所示。

图 5-47　网络拓扑

（2）检查虚拟网络配置：检查虚拟交换机、租户路由器和所有相关虚拟网络设备。确保网络设备可以成功启动，并启用相应的服务。

（3）检查网络资源：检查 OpenStack 网络资源，确保没有冲突或资源耗尽的问题。确保计算节点的正常状态，以及计算节点与其他节点之间的正常通信，命令如下。

```
[root@controller ~]# neutron agent-list
```

（4）检查网络服务：检查 OpenStack 网络服务是否正常，测试和检查所有网络服务是否正常工作，命令如下。

```
[root@controller ~]# systemctl status neutron-*
```

（5）检查虚拟机网络：检查虚拟机的网络连接，确保供应商网络类型、网络地址及网关 IP 地址配置正确。

（6）检查虚拟路由器：检查路由器的接口，确保虚拟路由器正常工作。

（7）检查日志和错误：检查所有错误信息，找到任何有用的日志，以及/var/log 中的任何有用信息。

网络故障往往是频繁出现和难以解决的问题，需要仔细地分析和检查各种网络组件及服务，以找到故障的根本原因。

4. 存储故障排查

OpenStack 平台存储故障排查通常很棘手，因为存储层是 OpenStack 的核心组成部分之一，其出现故障可能会导致整个系统带宽不足、易出现瓶颈和错误。以下是一些可能有用的建议，以帮助用户排查 OpenStack 存储故障。

（1）检查磁盘空间：检查 OpenStack 存储的基本磁盘空间和网络文件系统等，如 NFS，确保磁盘正确格式化且空间充足，命令如下。

```
[root@compute ~]# df -Th
```

（2）检查存储状态：检查存储节点的状态，如 Ceph 集群、逻辑卷管理（Logical Volume Manager，LVM）和各类存储后端的状态。确保存储后端能够成功启动，并启用相应的存储服务，命令如下。

```
[root@compute ~]# systemctl status openstack-nova*
```

（3）检查硬件状态：检查存储系统的硬件状态，确保磁盘等硬件设备正常工作。

（4）检查存储资源：检查存储资源，确保没有冲突或资源耗尽的问题。如果有默认配额，则要调整配额以满足实际存储需求。

在 OpenStack 中新建的云主机都存放在计算节点的/var/lib/nova/instances 目录下，当存储空间不够用时，可以将云主机转移到新的存储位置。查看云主机存放目录的命令如下。

```
[root@compute ~]# ll /var/lib/nova/instances/
```

（5）检查卷状态：根据卷的类型和创建方法，验证目标设备的正确性和状态，尽可能排除卷与其预期功能不符的情况，命令和结果如下。

```
[root@compute ~]# lsblk
NAME                                                 MAJ:MIN RM SIZE  RO TYPE MOUNTPOINTS
sda                                                  8:0     0  80G   0  disk
├sda1                                                8:1     0  1G    0  part / boot
└sda2                                                8:2     0  79G   0  part
└rl-root                                             253:0   0  79G   0  lym  /
sdb                                                  8:16    0  40G   0  disk
├sdb1                                                8:17    0  10G   0  part /srv/node/
                                                                              sdb1
├sdb2                                                8:18    0  10G   0  part /srv/node/
                                                                              sdb2
 └sdb3                                               8:19    0  10G   0  part /srv/node/
                                                                              sdb3
sdc                                                  8:32    0  80G   0  disk
├sdc1                                                8:33    0  20G   0  part
| ├cinder--volumes-cinder--volumes—pool_tmeta        253:1   0  20G   0  lym
|| └cinder--volumes-cinder--volumes--pool            253:3   0  19G   0  lym
| | └cinder--volumes-cinder--volumes—pool_tdata      253:2   0  19G   0  lym
|└cinder--volumes-cinder--volumes--pool              253:3   0  19G   0  lym
├sdc2                                                8:34    0  20G   0  part
└sdc3                                                8:35    0  30G   0  part
sr0                                                  11:0    1  1024M 0  rom
```

（6）检查日志和错误：检查所有与存储相关的错误信息并找到有用的日志信息，如 Cinder、Glance 的日志信息。

项目小结

本项目首先在读者了解云平台日常管理策略的基础上，进行了具体业务策略管理的介绍，然后对云平台常见的平台监控和日志分析工具进一步进行了对比，以便读者了解目前云平台下常用的警报系统及日志分析工具，从而在企业业务场景下，能够根据需求选定不同的开源工具，实现对业务的监控、警报和分析。

在云平台监控管理知识的基础上，本项目介绍了及时发现和获取云平台出现的故障的方法。此时，掌握云平台故障排查思路和流程就显得至关重要，包括确认故障的范围、故障关联性，在定位故障的同时，本项目介绍了如何进行云平台日志的查看与分析，并根据分析结果进行故障排查。

拓展知识

Zabbix 监控系统模板应用

生产环境下，每一台主机的监控项都有很多，一个一个地添加这些监控项费时费力，何况可能不止有一台主机。用户可以把一个对应服务的监控项添加到一个模板中，这样能方便以后的操作。同时，Zabbix 的模板功能支持自定义，在配置警报的过程中，可以通过钉钉或微信等即时通信软件进行警报，以便有效、快速解决问题。

知识巩固

1. 单选

（1）Zabbix 监控宿主机状态需要在宿主机上安装的组件是（　　）。

A. Zabbix Server　B. Zabbix Agent　C. Zabbix Proxy　D. Zabbix Web 界面

（2）使用云平台进行部署的应用突然出现故障，无法正常访问，下列不属于故障排查流程的是（　　）。

A. 查看日志文件　B. 重启应用

C. 删除整个系统并重新构建　D. 检查依赖库是否完整

（3）当云平台出现故障时，首先需要完成的是（　　）。

A. 分析事件日志　B. 打电话寻求技术支持

C. 定位故障范围　D. 重启系统

（4）OpenStack 平台自带的监控系统是（　　）。

A. Ceilometer　B. Zabbix　C. Nagios　D. Prometheus

（5）需要对云平台进行监控管理的原因是（　　）。

A. 验证云平台的高可用性　B. 提高云平台的性能

C. 避免故障和数据丢失　D. 以上所有答案都是

2. 填空

（1）云计算平台中需要对各种资源进行________管理和控制，以确保其使用的公平性和效率。

（2）在云计算平台中，不同应用程序所需的资源量可能不同，因此需要进行对资源的________规划和管理。

（3）云计算平台需要进行________监控，以便及时检查系统健康状态和处理异常。

（4）________是一款基于开源 Elasticsearch 和 Logstash 的企业级日志管理工具，用于收集、存储、搜索和分析日志数据。

（5）________是一款免费的、开源的、跨平台的监控和警报工具，通过 HTTP、TCP、UDP 等方式获取指定的监控数据。

3. 简答

（1）为什么需要对云平台进行资源规划？

（2）如何衡量云平台管理策略的成功与否？

（3）云平台监控和日志分析工具可通过哪些方式提供报告和警报？

拓展任务

容器化部署 Zabbix 监控系统

Zabbix 是一款功能强大且应用广泛的企业级监控解决方案，Zabbix 的传统部署方式在面临复杂环境和大规模应用时，可能会面临组件众多、依赖关系复杂及升级维护困难等问题。因此，采用容器化部署 Zabbix，不仅可以简化部署流程，缩短上线时间，还能有效提升系统的可移植性、可扩展性和可靠性。

微课 5.7 容器化部署 Zabbix 监控系统

任务步骤

1. 规划 Zabbix-Server 节点

使用 openEuler 22.09 操作系统容器化部署 Zabbix 监控系统，Zabbix-Server 节点规划如表 5-2 所示。

表 5-2 Zabbix-Server 节点规划

IP	主机名	节点
172.128.11.176	zabbix-server	Zabbix-Server

2. 配置 Zabbix-Server 节点基础环境

修改主机名，命令如下。

```
[root@localhost ~]# hostnamectl set-hostname zabbix-server
[root@localhost ~]# exec bash
```

关闭防火墙和 SELinux，命令如下。

```
[root@zabbix-server ~]# systemctl disable --now firewalld
[root@zabbix-server ~]# setenforce 0
[root@zabbix-server ~]# sed -i \
's/^SELINUX=.*/SELINUX=permissive/g' /etc/selinux/config
```

修改 Yum 源镜像加速地址为华为云镜像站，命令如下。

```
[root@zabbix-server ~]# sed -i \
's|http://repo.openeuler.org/|https://mirrors.huaweicloud.com/openeuler/|g' \
 /etc/yum.repos.d/openEuler.repo
```

3. 安装 Docker 环境

安装 Docker 服务，命令如下。

```
[root@zabbix-server ~]# dnf -y install docker-engine
```

设置 Docker 服务开机自启动并立即启动，命令如下。

```
[root@zabbix-server ~]# systemctl enable --now docker
```

4. 容器化部署 Zabbix 监控系统

创建自定义 Docker 网络“zabbix-net”，设置其子网和有效 IP 地址范围，命令如下。

```
[root@zabbix-server ~]# docker network create --subnet 172.20.0.0/16 \
--ip-range 172.20.240.0/20 zabbix-net
6417a2c31db9cb223608ff434f2deddfdb1984ac3f8e2449ed4c9b5259339e40
```

创建 MySQL 容器实例，设置环境变量来初始化数据库，命令如下。

```
[root@zabbix-server ~]# docker run --name mysql-server --restart always
--privileged=true  -t \
```

```
-e MYSQL_DATABASE="zabbix" \
-e MYSQL_USER="zabbix" \
-e MYSQL_PASSWORD="zabbix_pwd" \
-e MYSQL_ROOT_PASSWORD="root_pwd" \
--network=zabbix-net \
-d mysql:8.1
53b8d7809f6451d79c6e449052dfe939755f691a3d97735b7b66f500730c93af
```

创建 Zabbix-Server 容器实例，设置环境变量配置数据库连接，命令如下。

```
[root@zabbix-server ~]# docker run --name zabbix-server-mysql -t \
-e DB_SERVER_HOST="mysql-server" \
-e MYSQL_DATABASE="zabbix" \
-e MYSQL_USER="zabbix" \
-e MYSQL_PASSWORD="zabbix_pwd" \
-e MYSQL_ROOT_PASSWORD="root_pwd" \
-e ZBX_JAVAGATEWAY="zabbix-java-gateway" \
--network=zabbix-net \
-p 10051:10051 \
--restart unless-stopped \
-d zabbix/zabbix-server-mysql:latest
```

创建 Zabbix-Web 容器实例，连接到前面创建的 MySQL 和 Zabbix-Server 容器实例，并设置环境变量配置后端连接。同时，将容器的 8080 端口映射到主机的 80 端口，命令如下。

```
[root@zabbix-server ~]# docker run --name zabbix-web-nginx-mysql -t \
-e ZBX_SERVER_HOST="zabbix-server-mysql" \
-e DB_SERVER_HOST="mysql-server" \
-e MYSQL_DATABASE="zabbix" \
-e MYSQL_USER="zabbix" \
-e MYSQL_PASSWORD="zabbix_pwd" \
-e MYSQL_ROOT_PASSWORD="root_pwd" \
--network=zabbix-net \
-p 80:8080 \
--restart unless-stopped \
-d zabbix/zabbix-web-nginx-mysql:latest
c50a7bc69a4d1e9cba5dcb64761f82bb8ee1e6cec51a282fe4747db797633661
```

5. 访问登录页面

Zabbix 监控系统部署完成后，使用浏览器访问地址 http://172.128.11.176 便可以看到 Zabbix 登录页面，如图 5-48 所示，使用用户名 Admin、密码 zabbix 登录，可以进入 Zaabbix 主页。

图 5-48　Zabbix 登录页面（容器化部署）

项目6 云基础架构平台应用

学习目标

【知识目标】

① 学习云基础设施及架构的工作方式。
② 学习云应用集群架构系统。

【技能目标】

① 掌握云应用系统部署方式。
② 掌握云应用系统的迁移方法。
③ 掌握高可用架构的部署与验证方法。

【素养目标】

① 养成细致入微、严谨务实的工作态度。
② 培养持续学习和自我提升的能力。
③ 锻炼独立思考和自主决策的能力。

项目概述

此前，小张通过了解云平台的管理策略，使用开源工具对 OpenStack 组件及公司线上业务进行实时监控，保障了公司业务的正常运行。同时，故障的应急处理为公司挽回了客户，减少了损失。接下来，公司要对线上产品做进一步迭代更新，需要小张根据评估开发量进行线上云应用系统部署。小张需要提供云系统的部署方案，对云应用系统进行迁移、备份及通过高可用架构部署等灾备措施来保证线上业务的稳定性。

知识准备

6.1 云基础架构平台

云基础架构平台是连接底层基础设施与上层服务架构的“桥梁”，它借助虚拟化、分布式计算等技术构建弹性、安全的资源环境。

6.1.1 云基础设施

云计算是通过互联网提供计算资源的一种模式，它允许用户通过网络访问和使用存储在数据中心的大量计算资源，如服务器、存储设备、数据库、网络设备和软件应用等，云计算由云基础设施虚拟化平台提供支持。

1. 云基础设施概述

云基础设施是指由云服务供应商构建、提供和管理的基于云计算的技术设施，包括云服务器、云存储、云数据库和云网络等。它们被部署在数据中心，具有相当优秀的扩展性、灵活性和自动化能力，可以满足不同规模、不同类型和不同业务场景的需求。云基础设施的使用将大大简化企业和组织的 IT 管理及维护工作，降低 IT 成本，提高 IT 资源利用率，同时能够为用户提供更加高效、安全、可靠和智能的云服务。

2. 云基础设施运作机制

云基础设施的高效运作是通过将资源与物理硬件分离并通过虚拟化技术将它们放置在云上进行交付来实现的。云基础设施的运作主要依赖以下几个机制。

（1）虚拟化技术的应用。云基础设施使用虚拟化技术，将物理资源如服务器、存储、网络等转化为虚拟资源，并由虚拟化管理软件来管理和分配这些资源。这种方式能够提高资源利用率和灵活性，降低物理资源的成本和维护费用。

（2）自动化技术的应用。云基础设施通过自动化技术来实现自动化部署、自动化伸缩、自动化备份、自动化监控等操作。这种方式可以大大减少人工干预，提高操作效率，降低故障率，并保证服务的高可用性。

（3）高可用架构设计。云基础设施采用了高可用架构，通过多种技术手段，包括数据冗余、负载均衡、故障转移、自动备份等技术手段来保证服务的可用性。这种方式可以减少业务中断时间，提高业务的可靠性。

（4）提供弹性扩容能力。云基础设施具有弹性扩容能力，能够快速地分配和回收资源，并根据业务负载自动进行资源的伸缩，满足业务的变化需求。

（5）采用按需计费方式。云基础设施采用按需计费的方式，根据用户的实际使用情况来收取费用，用户可以根据自己的需求和财务情况，灵活地控制成本。

3. 构建云基础设施的要求

构建云基础设施是组织迁移到云的第一步。对于实施云战略的组织，需要遵循以下稳健的云基础设施构建要求来确定可行性。

（1）具有软件兼容性。用户选择的云基础设施和虚拟组件需要支持用户已用于业务应用程序的软件平台。由于选择云基础设施服务是一项长期决策，因此需要确保它可以针对用户的业务进行扩展。

（2）具有网络兼容性。需要为服务器和存储的虚拟化配置网络组件和基础设施。组织还应确保其网络组件和基础设施在采用云时易于迁移。网络兼容性意味着用户拥有所需数量的服务器和网络设备，可以通过云支持用户的数据和应用程序。

（3）云基础设施资源可见。对于被迁移到云基础设施的资源，能够可见十分重要。组织应该让所有迁移应用程序的利益相关者了解云基础设施的变化。由于业务应用程序在迁移到云的过程中会经历大量的平台和云基础设施变化，因此，需要向使用相关数据和应用程序的每个用户清楚地传达确切的过渡组件。

（4）具有系统集成和自动化能力。云基础设施的硬件和软件组件的集成及自动化是采用云解决方案时较重要的一步。网络、存储和服务器等元素需要自动化，以按需支持业务应用程序。

6.1.2 云基础架构

云计算由互联网上的多个软件平台、数据库、网络设备和服务器支持，它在云基础架构的帮助下建立在云架构上。

云架构是指规划和使用云基础设施资源的蓝图或框架，用户能够在计算环境中协同使用各种技术。云架构被视为云计算的后端功能平台，它定义了如何组织、管理和利用云资源来为用户提供服务和应用。

1. 云基础架构类型

云基础架构为多种类型的云平台提供服务。但是在不同的部署方式下，不同类型的云基础架构存在一些基本差异，如表6-1所示。

表6-1 云基础架构类型

云基础架构类型	优点	缺点
私有云基础架构由单个组织使用和管理。私有云使用的云基础设施由其内部IT团队开发和维护，架构更安全	私有云基础架构为用户提供对云平台的更多控制和更高的灵活性	从长远来看，私有云基础架构成本高昂
公有云基础架构使用第三方云服务供应商的服务，并利用多租户环境以更低的成本获得数据存储和计算能力。多租户环境是单个云平台被多个租户或用户划分和使用的	公有云基础架构的开销低于其他云基础架构的开销，并提供无限的可扩展性	采用公有云基础架构的缺点是在公共服务器中存在数据安全风险
混合云基础架构是公有云基础架构和私有云基础架构的组合。它通过私有云平台实现安全的数据存储，并降低公有云计算资源的成本	混合云基础架构可以确保用户对业务应用程序的控制和具备较高的灵活性，同时提供具有成本效益的解决方案	实施混合云基础架构需要密集的规划和控制开销

2. 云基础架构组件

构建云计算模型以支持组织业务时，云基础架构承载了一系列关键组件来部署和运行业务应用。构成云基础架构并用于部署业务应用的基本组成部分如下。

（1）硬件

云基础架构中的硬件主要指构成数据中心的物理基础设施，包括服务器集群、存储设备、网络交换机，以及其他关键硬件设施。这些硬件资源被合理布局在不同地理位置的数据中心中，以确保云计算环境的高效稳定运行。借助虚拟化技术，这些硬件资源被抽象、整合到资源池中，从而提供弹性的、可扩展的、高可用的云服务。

（2）虚拟化层

虚拟化层是云平台中的核心组成部分之一，它模拟了完整的计算机系统，能够独立运行操作系统和应用程序。通过虚拟化技术，物理硬件的计算能力被抽象出来，并按需分配给云端的虚拟资源实例，用户可以通过简单的界面轻松访问和管理虚拟资源。

（3）云存储系统

云存储系统是云基础架构的关键组成部分，负责用户数据的持久化存储与管理。云存储系统提供了数据复制、备份、恢复、压缩/解压、访问控制等多种功能服务，使组织能够在任何地点通过互联网安全地存储和访问其数据。

（4）网络服务

云平台的网络服务涵盖了路由、交换、负载均衡等网络策略，它们共同构建起支撑云计算服务的网络基础架构。其中，防火墙等安全措施是网络服务不可或缺的一部分，负责保护数据传输过程中的安全性。

综上所述，以上各个组成部分共同构成了云基础架构，它们通过紧密协作提供高性能、高可靠性和高度弹性的云服务。

3. 使用云基础架构的好处

使用云基础架构有很多好处，包括以下几个方面。

（1）灵活性

云基础架构可以根据需求快速进行扩容和缩减，使企业根据业务需求调整资源规模，从而提高资源的利用率。

（2）可靠性

云基础架构通过多地域、多可用区的部署，确保服务能够持续提供，避免单节点故障。

（3）安全性

云基础架构提供了严格的安全控制和保护措施，包括访问控制、数据加密、漏洞修复等，保证了企业数据和信息的安全性。

（4）成本控制

云基础架构采取了按需付费的方式，企业可以根据实际使用情况付费，避免了大量投资设备的压力和高成本。

（5）易用性

云基础架构提供了标准化的接口和管理工具，使企业人员可以方便地管理和使用云服务，提高生产力。

6.2 云应用集群架构系统

云应用集群架构系统是一种构建高可用、高性能和可扩展应用的解决方案，通过整合多节点资源，借助负载均衡技术（如 Keepalived 与 HAProxy 等组件），实现对大规模并发请求的有效处理及故障快速恢复，确保服务的连续性和稳定性。

6.2.1 Keepalived 架构

1. Keepalived 简介

（1）Keepalived 概述

Keepalived 是使用 C 语言编写的开源软件，可以根据节点的健康状态自动切换 IP 地址，以实现高可用性。它提供了一种自动切换技术，可以将多台服务器组成一台虚拟服务器，这台虚拟服务器提供一个指定的 IP 地址，当其中的一台服务器失效或出现故障时，将自动切换到其他服务器上，以维护系统的高可用性。同时，Keepalived 提供了多种监控方式来检测节点的健康状态，如监控 ICMP、TCP、HTTP 等方式，以及通过脚本的方式实现自定义健康检查。除此之外，Keepalived 将配置信息存储在本地文件中，使用 Linux 虚拟服务器（Linux Virtual Server，LVS）、HAProxy 等负载均衡软件，可以实现无盘的状态同步。

Keepalived 主要通过虚拟路由冗余协议（Virtual Router Redundancy Protocol，VRRP）实现高可用性，VRRP 的出现就是为了解决路由器的单节点故障问题，它能保证当个别节点宕机时，整个网络可以不间断地运行。所以，Keepalived 既具有配置管理 LVS 的功能，可对 LVS 下面的节点进行健康检查，又可以实现系统网络服务的高可用性。

（2）VRRP

在现实的网络环境中，主机之间的通信都是通过配置静态路由完成的，而主机之间的路由器一旦出现故障，通信就会失败。因此，在这种通信模式中，路由器就成了一个单点瓶颈，为了解决这个瓶颈而引入了 VRRP。

VRRP 是一种主备模式的协议，通过 VRRP 可以在网络发生故障时透明地进行设备切换且不影响主机间的数据通信，其中涉及两个概念：物理路由器和虚拟路由器。VRRP 可以将两台或者多台物理路由器虚拟成一台虚拟路由器，这台虚拟路由器通过虚拟 IP 地址（一个或多个 IP 地址）对外提供服务，而在虚拟路由器内部，多台物理路由器协同工作，同一时间只有一台物理路由器对外提供服务，这台物理路由器被称为主路由器（即 master 角色）。它拥有对外提供的虚拟 IP 地址，提供各种网络功能，如地址解析协议（Address Resolution Protocol，ARP）请求、数据转发等，而其他物理路由器不拥有对外提供的虚拟 IP 地址，也不提供对外网络功能，仅仅接收主路由器的 VRRP 状态通告信息，这些路由器被统称为备份路由器（即 backup 角色）。当主路由器失效时，处于 backup 角色的备份路由器将重新进行选举，产生一个新的主路由器继续对外服务，整个切换过程对于用户来说完全透明。

在一台虚拟路由器中，只有主路由器会一直发送 VRRP 报文信息，备份路由器只接收主路由器发送过来的报文信息，用来监控主路由器运行状态。因此，不会发生主路由器抢占的现象，除非它的优先级更高，而当主路由器不可用时，备份路由器就无法收到主路由器发送过来的报文信息，于是就认定主路由器出现了故障，接着多台备份路由器会进行选举，优先级最高的备份路由器将成为新的主路由器，这种选举并进行角色切换的过程非常快，因此保证了服务的持续可用性。

2. Keepalived 体系结构

Keepalived 是一款高度模块化的软件，结构简单，但扩展性很强，官方给出的 Keepalived 的体系结构如图 6-1 所示。

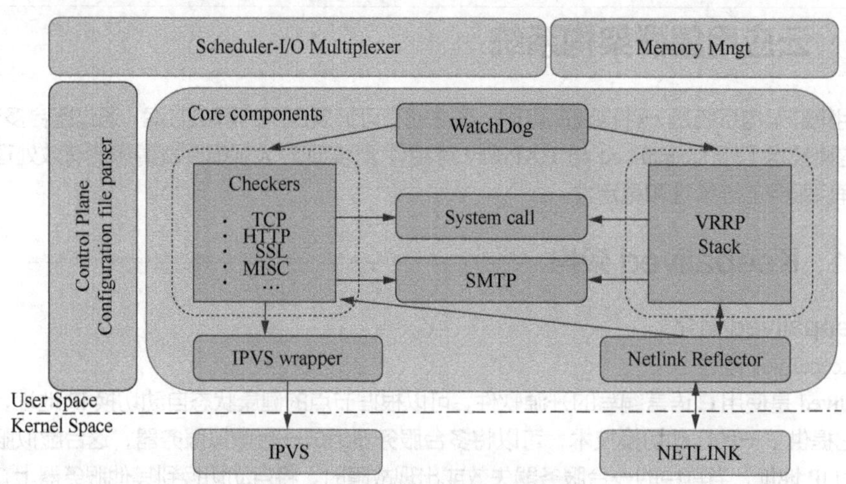

图 6-1　Keepalived 的体系结构

该体系结构 IP 地址虚拟服务器中，处于内核的是 IP 虚拟服务器（IP Virtual Server，IPVS）和 NETLINK，其中 NETLINK 提供高级路由及其他相关的网络功能，如果在负载均衡器上启用 iptables/netfilter，则会直接影响它的性能。对图 6-1 中部分模块的功能介绍如下。

① VRRP Stack：LVS 中负责直接路由（Direct Routing，DR）地址调度，负责负载均衡器之间的失败切换（FailOver）。

② Checkers：负责检查调度器后端的 Real Server（真实服务器）或者 Upstream Server（上游服务器）的健康状况。

③ WatchDog：负责监控 Checkers 和 VRRP Stack 的状况。

④ IPVS wrapper：用来发送设定的规则到内核 IPVS 中。

⑤ Netlink Reflector：用来设定 VRRP 的虚拟 IP（Virtual IP，VIP）地址。

⑥ System call：系统调用。

Keepalived 正常运行时，会启动 3 个进程，分别为 core（核心进程）、check 和 vrrp。对这 3 个进程的介绍如下。

① core：为 Keepalived 的核心进程，负责主进程的启动、维护及全局配置文件的加载和解析。

② check：负责健康检查，包括常见的各种检查方式。

③ vrrp：用来实现 VRRP。

3. LVS 简介

（1）LVS 概述

LVS 目前已经被集成到 Linux 内核模块中。LVS 主要用于负载均衡，如 Web 客户端想要访问后端服务，Web 请求会先经过 LVS 调度器，LVS 调度器根据预设的算法决定如何将请求分发送给后端的所有服务器。

（2）LVS 的基本原理

LVS 的核心功能就是实现负载均衡，负载均衡方案有以下 4 种。

① 基于域名系统（Domain Name System，DNS）的域名轮流解析方案。

② 基于客户端的调度访问方案。

③ 基于应用层系统的调度方案。

④ 基于 IP 地址的调度方案。

其中效率最高的是基于 IP 地址的调度方案。该方案其实就是将请求转发给对应的 IP 地址+端口号，LVS 集群系统的 IP 负载均衡技术是通过其核心模块 IPVS 来实现的。

LVS 的基本原理如图 6-2 所示。

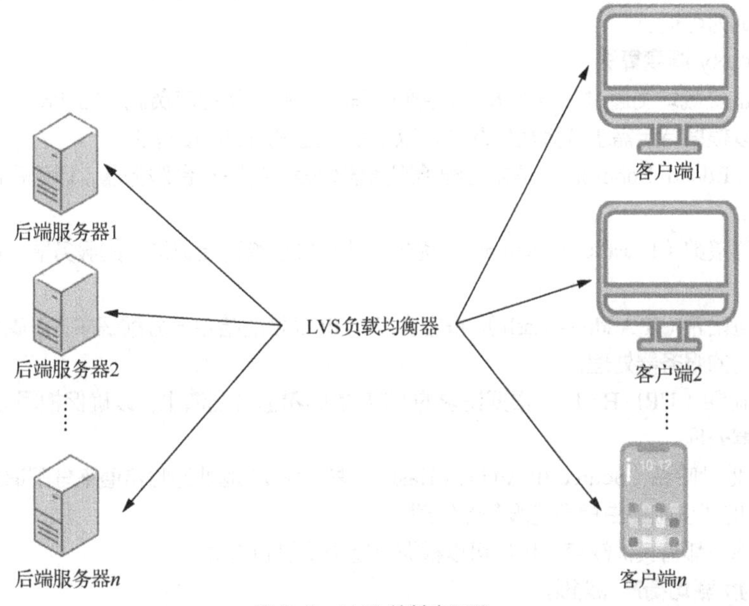

图 6-2 LVS 的基本原理

LVS 负载均衡器会虚拟化一个 VIP，对于客户端来说，它事先只知道这个 VIP。客户端将请求发送给 VIP，LVS 负载均衡器会将请求转发给后端服务器中的一个，这些服务器被称为 Real Server。转发的规则是通过设置 LVS 负载均衡器的负载均衡算法来实现的，如随机分配、按照权值分配等。

后端服务器提供的功能是一致的，不论请求转发到哪台后端服务器，最终得到的结果是一致的，所以对于客户端来说，它并不关心有多少台后端服务器在提供服务，它只关心访问的 VIP 是多少。后端服务器处理完请求后，根据 LVS 负载均衡器的不同模式，会选择不同的方式将数据返回给客户端。

6.2.2 HAProxy 架构

1. HAProxy 简介

HAProxy 是一款具备高可用性、负载均衡功能，以及基于 TCP（网络的第 4 层）和 HTTP（网络的第 7 层）应用的代理软件，支持虚拟主机，并为多种场景提供解决方案。HAProxy 的特性如下。

（1）HAProxy 适用于负载大的 Web 站点

负载大的 Web 站点通常需要会话保持或 7 层处理，HAProxy 完全可以支持数以万计的并发连接。其运行模式使得它可以简单且安全地整合到用户当前的架构中，同时可以保护用户的 Web 服务器不被暴露到公网中。

（2）HAProxy 支持连接拒绝

保持连接开放所带来的维护成本相对较低，因此，在网站面对恶意攻击蠕虫（Attack Bots）时，HAProxy 可通过支持连接拒绝来有效地遏制蠕虫的破坏性。这一特性已被用于帮助易于遭受小型 DDoS 攻击的网站免受损害，这是目前许多其他负载均衡器尚未具备的独特优势。

（3）HAProxy 支持全透明代理（已具备硬件防火墙的典型特性）

可以使用客户端的 IP 地址或者任何其他地址来连接后端服务器。这个特性仅在 Linux 2.4/2.6 内核中安装好 cttproxy 补丁后才可以使用。这个特性也使得为某些特殊服务器处理部分流量的同时不修改服务器的地址成为可能。

2. HAProxy 调度算法

HAProxy 的负载均衡通过分发请求到多台服务器来平衡应用程序负载。HAProxy 提供了几种调度算法，用户可以根据特定需求选择最佳算法。以下是一些常用的调度算法。

（1）轮询（Round Robin）。将请求分配给服务器列表中的下一台服务器，以确保所有服务器都得到平等的机会。

（2）最少连接数（Least Connections）。优先将请求分配给连接数最少的服务器，以便更好地平衡负载。

（3）IP 地址哈希（IP Address Hash）。根据客户端 IP 地址的哈希值分配请求，以确保同一客户端的请求始终由相同的服务器处理。

（4）URL 哈希（URL Hash）。根据请求的 URL 的哈希值分配请求，以确保相同 URL 的请求始终由相同的服务器处理。

（5）源 IP 地址哈希（Source IP Address Hash）。基于源 IP 地址的哈希值来分配请求，以确保来自同一源 IP 地址的请求始终由相同的服务器处理。

这些调度算法都有其优缺点，用户可以根据特定情况进行选择。

3. 常见的负载均衡产品对比

市面上常见的负载均衡产品主要有两种：硬件产品和软件产品。对这两种负载均衡产品的介绍如下。

（1）硬件产品。硬件产品（如 F5 和 Array 等商用负载均衡器）由专业维护团队进行维护，花销较大，规模较小的架构并不需要使用这种负载均衡产品。

（2）软件产品。软件产品（如 Nginx、HAProxy、LVS 等开源的负载均衡软件）成本较低。

在构建经济高效且广泛受认可的软件架构时，常见的一种方案是 Web 前端采用 Nginx/HAProxy+Keepalived，Nginx/HAProxy 负责 Web 前端的负载均衡，Keepalived 负责提供高可用功能，后端采用 MySQL 一主多从和读写分离的方式，并采用 LVS+Keepalived 的架构。

常见的负载均衡产品对比如表 6-2 所示。

表 6-2 常见的负载均衡产品对比

负载均衡产品	优点	缺点
Nginx	1. Nginx 工作在网络的第 7 层之上，可以针对 HTTP 服务实现一些分流的策略，如针对域名、目录结构进行分流。 2. Nginx 对网络稳定性的依赖非常弱，理论上只要能 ping 通就能实现负载均衡功能。 3. Nginx 安装和配置比较简单，对其进行测试比较方便，它基本上能把错误以日志形式输出。 4. Nginx 可以通过端口检测到服务器内部的故障。 5. Nginx 不仅仅是一款优秀的负载均衡器/反向代理软件，还是功能强大的 Web 应用服务器。 6. Nginx 可作为中层反向代理使用	1. Nginx 仅能支持 HTTP、HTTPS 和 E-mail 协议，这会导致其适用范围较小。 2. 针对后端服务器的健康检测，Nginx 只支持通过端口来检测，不支持通过 URL 来检测
HAProxy	1. HAProxy 支持虚拟主机。 2. HAProxy 的优点能够弥补 Nginx 的一些缺点，如支持 Session 的直接保持、Cookie 的引导；同时支持通过获取指定的 URL 来检测后端服务器的状态。 3. HAProxy 与 LVS 类似，本身只是一款负载均衡软件；单纯从效率上来讲，HAProxy 比 Nginx 有更快的负载均衡速度，在并发处理上也是优于 Nginx 的。 4. HAProxy 支持 TCP 的负载均衡转发，可以对 MySQL 进行负载均衡	1. 不支持邮局协议（Post Office Protocol，POP）/简单邮件传送协议（Simple Mail Transfer Protocol，STMP）。 2. 不支持 SPDY（Speedy）协议。 3. 不支持 HTTP cache 功能。 4. 重载配置时需要重启进程，虽然也是软重启，但 Nginx 的重载更为平滑和友好。 5. 对多进程模式的支持不够好
LVS	1. 抗负载能力强，工作在网络的第 4 层之上，仅用作分发，没有流量的产生，这个特点也决定了它在负载均衡软件中是性能较强的，对内存和 CPU 资源的消耗比较低。 2. 配置性比较低，因为没有太多可配置的内容，所以并不需要太多接触，大大降低了人为出错的概率。 3. 工作稳定，这是因为其本身的抗负载能力很强，自身有完整的双机热备方案，如 LVS+Keepalived，但在项目实施中用得最多的还是 LVS/DR+Keepalived。其中，DR 指 Disater Recovery，表示灾难恢复 4. 应用范围比较广，因为 LVS 工作在第 4 层之上，所以它几乎可以对所有应用做负载均衡，包括 HTTP、数据库、在线聊天室等	1. 软件本身不支持正则表达式处理，不能实现动静分离。 2. 如果网站应用比较庞大，则 LVS/DR+Keepalived 实施起来比较复杂，特别是在 Windows 操作系统中，实施、配置及维护过程会比较复杂，相对而言，Nginx/HAProxy+Keepalived 就简单多了

综上所述，Nginx、HAProxy 和 LVS 各有优缺点，没有好坏之分，要选择使用哪个作为负载均衡器，要以实际的应用环境来决定。

项目实施

任务 6.1 云应用系统部署

微课 6.1 云应用系统部署

在云应用系统中，部署 LNMP（Linux、Nginx、MySQL 和 PHP）架构下的 WordPress 应用，旨在利用云计算资源的弹性和可管理性，实现灵活配置与优化资源利用。本任务将介绍如何部署 LNMP 架构下的 WordPress 应用，主要包括基础环境准备、安装 LNMP 环境和部署 WordPress 等内容。

任务 6.1.1 基础环境准备

1. 创建 2 个 vCPU/4GB 内存/20GB 硬盘规格的云主机

使用部署好的 OpenStack 云平台，导入 openEuler22.09 镜像，创建网络、子网，修改安全组，创建规格为 2 个 vCPU/4GB 内存/20GB 硬盘的云主机。

（1）导入镜像

使用 openstack image 命令，导入 openEuler-22.09.x86_64.qcow2 镜像，并将其命名为 openEuler22.09，命令和结果如下。

```
[root@controller ~]# openstack image create openEuler22.09 --disk-format qcow2
--container-format bare --progress --file /opt/wxic-cloud/images/openEuler-22.09.x86_
64.qcow2
    [=======================>] 100%
```

（2）创建网络和子网

① 使用 openstack network 命令，创建名为 wxic-net 的共享网络，命令如下。

```
[root@controller ~]# openstack network create wxic-net --share --external
```

② 使用 openstack subnet 命令，创建子网，配置网关，命令如下。

```
[root@controller ~]# openstack subnet create --subnet-range 192.168.200.0/24
--gateway subnet
```

③ 管理安全组规则。登录 Dashboard，选择左侧导航栏中的“网络→安全组”选项，单击右上方的“添加规则”按钮，修改安全组为三进三出（即添加 3 个入口访问规则和 3 个出口访问规则）。添加完成后，查看安全组规则详情，如图 6-3 所示。

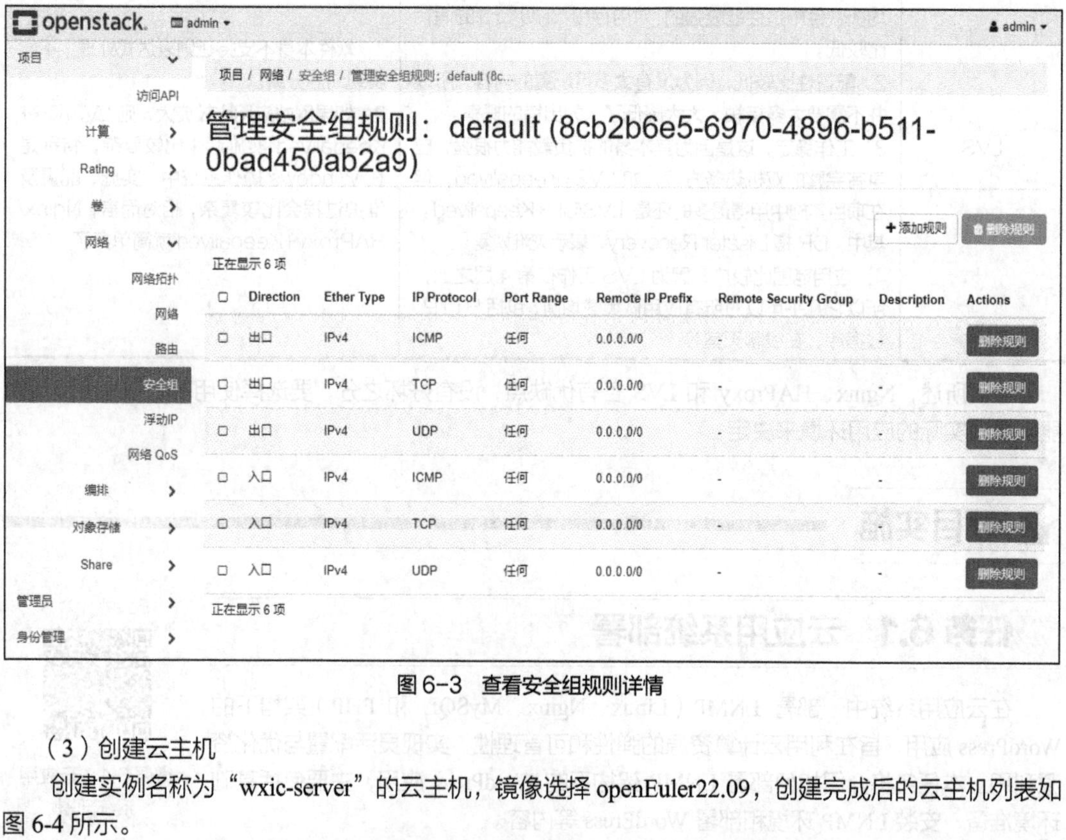

图 6-3 查看安全组规则详情

（3）创建云主机

创建实例名称为“wxic-server”的云主机，镜像选择 openEuler22.09，创建完成后的云主机列表如图 6-4 所示。

图 6-4　创建完成的云主机列表

2. 使用 SecureCRT 连接云主机

使用 SecureCRT 工具，输入云主机 IP（Hostname）与用户名（Username）信息，其余选项保持默认，如图 6-5 所示，单击“Connect”按钮快速连接云主机。

在弹出的窗口中输入 root 用户的登录密码，即可连接，连接成功界面如图 6-6 所示。

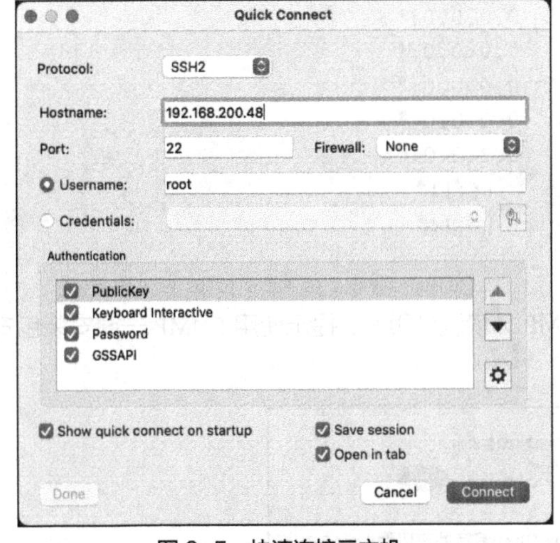

图 6-5　快速连接云主机

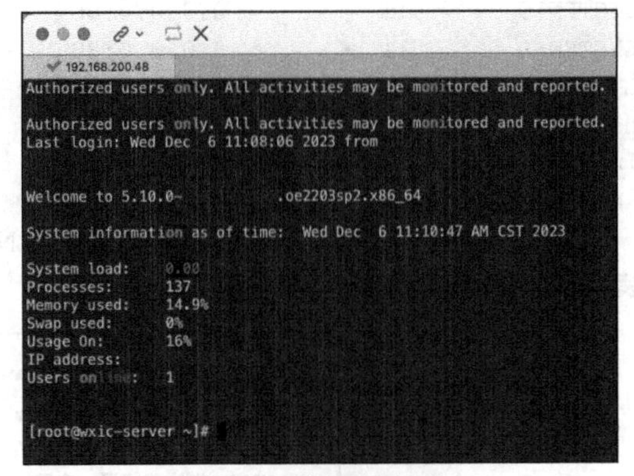

图 6-6　连接成功界面

任务 6.1.2　安装 LNMP 环境

LNMP 是目前主流的 Web 网站服务器架构，是指 Linux 操作系统下 Nginx+MySQL+PHP 的网站服务器架构，它有多种部署和实现方法。此处采用官方网站上提供的 LNMP 一键安装包，使用该安装包将自动部署好 LNMP 环境。

访问 https://soft.vpser.net/lnmp/，在打开的页面中选择 lnmp2.0-full.tar.gz 包进行下载，相关命令如下。

```
[root@wxic-server ~]# wget https://soft.lnmp.com/lnmp/lnmp2.0-full.tar.gz
[root@wxic-server ~]# tar -zxvf lnmp2.0-full.tar.gz
```

执行install.sh脚本，部署LNMP环境，命令如下。

```
[root@wxic-server ~]# cd lnmp2.0-full
[root@wxic-server lnmp2.0-full]# ./install.sh
```

此处选择安装数据库的版本，直接按Enter键表示选择安装默认版本，命令如下。

```
......
Enter your choice (1, 2, 3, 4, 5, 6, 7, 8, 9, 10 or 0):
```

输入密码000000后默认全部按Enter键即可，命令和结果如下。

```
......
Do you want to enable or disable the InnoDB Storage Engine?
Default enable,Enter your choice [Y/n]:
......
Enter your choice (1, 2, 3, 4, 5, 6, 7, 8, 9, 10, 11, 12, 13):
......
Enter your choice (1, 2 or 3):
......
Press any key to install...or Press Ctrl+c to cancel
```

等待执行结束后，终端显示nginx、php-fpm等服务处于运行状态并显示最终运行结果，结果如下。

```
......
nginx (pid 264857) is running...
php-fpm is runing!
 SUCCESS! MySQL running (265440)
State  Recv-Q Send-Q Local Address:Port Peer Address:PortProcess
LISTEN 0      128          0.0.0.0:22        0.0.0.0:*
LISTEN 0      50           0.0.0.0:3306      0.0.0.0:*
LISTEN 0      4096         0.0.0.0:111       0.0.0.0:*
LISTEN 0      511          0.0.0.0:80        0.0.0.0:*
LISTEN 0      511          0.0.0.0:80        0.0.0.0:*
LISTEN 0      128            [::]:22           [::]:*
LISTEN 0      4096           [::]:111          [::]:*
Install lnmp takes 59 minutes.
Install lnmp V2.0 completed! enjoy it.
```

通过浏览器访问http://192.168.200.48，进入LNMP环境测试页面，提示使用LNMP一键安装包安装成功，如图6-7所示。

图6-7 LNMP环境测试页面

任务 6.1.3 部署 WordPress

1. 官方网站下载源代码包

WordPress 是一款流行的、免费的、开源的内容管理系统（Content Management System，CMS）。它拥有直观的用户界面和强大的功能，可以让用户轻松地创建和管理网站、博客等。WordPress 拥有丰富的主题和插件库，可用于增强网站的功能和扩展样式。它支持自定义域名和自托管，因此用户可以将其网站部署在自己的服务器上，并完全按照自己的喜好定制网站。WordPress 已经成为许多企业网站、新闻网站和个人博客的首选平台之一，使用 WordPress，用户可以轻松地创建、发布和管理内容，与访客互动。

用户可以通过访问 WordPress 官方网站来下载最新可用的 WordPress 版本。

2. 解压并进行相关配置

通过官方网站下载 WordPress 压缩包后，上传 wordpress-6.2-zh_CN.tar.gz 包并解压，命令如下。

```
[root@wxic-server ~]# tar -zxvf wordpress-6.2-zh_CN.tar.gz
```

创建数据库。登录数据库，创建 WordPress 数据库并赋予其远程登录权限，命令和结果如下。

```
[root@wxic-server ~]# mysql -uroot -p000000
mysql> create database wordpress;
Query OK, 1 row affected (0.00 sec)
mysql> grant all privileges on *.* to root@localhost identified by '000000' with grant
option;
Query OK, 0 rows affected (0.00 sec)
mysql> grant all privileges on *.* to root@'%' identified by '000000' with grant option;
Query OK, 0 rows affected (0.00 sec)
```

删除/home/wwwroot/default 目录下的 index.html 文件，命令如下。

```
[root@wxic-server ~]# rm -rf /home/wwwroot/default/index.html
```

切换到/root/wordpress 目录，将该目录下的所有文件复制到/home/wwwroot/default 目录下，并赋予所有用户读写的权限，命令如下。

```
[root@wxic-server ~]# cp -rvf wordpress/* /home/wwwroot/default/
[root@wxic-server ~]# chmod -R 775 /home/wwwroot/default/
```

在/home/wwwroot/default 目录下可以看到一个 wp-config-sample.php 文件，其为 WordPress 默认的配置文件，将其复制一份并将复制生成的文件重命名为 wp-config.php，命令如下。

```
[root@wxic-server ~]# cd /home/wwwroot/default/
[root@wxic-server default]# cp wp-config-sample.php wp-config.php
```

编辑配置文件，修改配置内容，命令如下。

```
[root@wxic-server default]# vi wp-config.php
// ** Database settings - You can get this info from your web host ** //
/** The name of the database for WordPress */
define( 'DB_NAME', 'wordpress' );

/** Database username */
define( 'DB_USER', 'root' );

/** Database password */
define( 'DB_PASSWORD', '000000' );

/** Database hostname */
define( 'DB_HOST', 'localhost' );
```

```
/** Database charset to use in creating database tables. */
define( 'DB_CHARSET', 'utf8' );

/** The database collate type. Don't change this if in doubt. */
define( 'DB_COLLATE', '' );
```

3. 查看WordPress界面

WordPress配置文件修改完成后，在浏览器地址栏中输入IP地址（192.168.200.48），进入WordPress安装界面，在其中配置好站点信息，单击左下方的“安装WordPress”按钮，如图6-8所示。

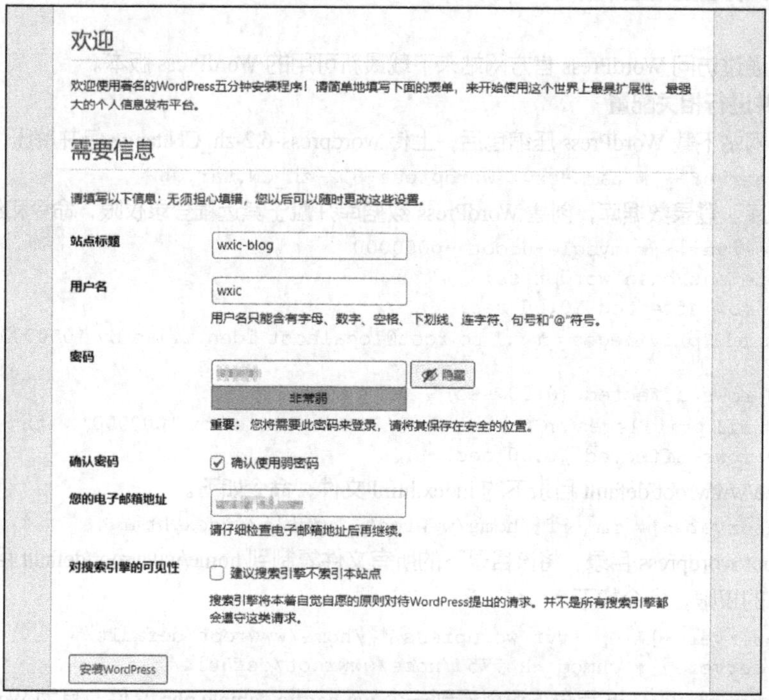

图6-8 WordPress安装界面

安装完成后，使用设置的用户名和密码登录WordPress，其登录界面如图6-9所示。

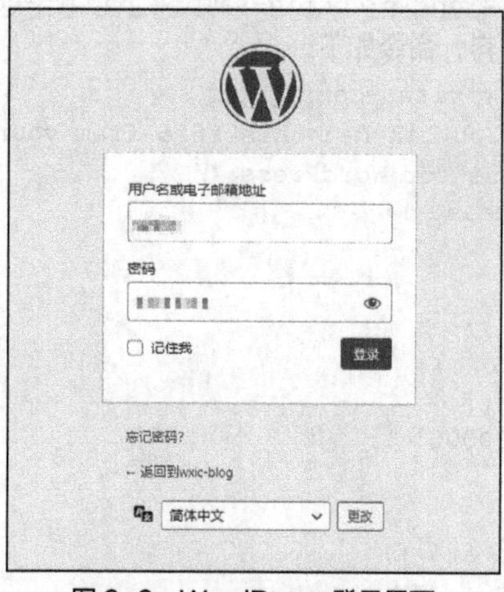

图6-9 WordPress登录界面

登录成功后，进入 WordPress 应用的后台仪表盘界面，如图 6-10 所示。

图 6-10　WordPress 应用的后台仪表盘界面

至此，WordPress 应用部署完成！

任务 6.2　云应用系统迁移

云应用系统迁移的核心在于保障数据与配置的无缝迁移。本任务将介绍云主机快照技术在迁移过程中的应用及相应的迁移策略，以实现从原环境到目标环境的高效过渡，保障业务的连续性。

微课 6.2　云应用系统迁移

任务 6.2.1　云主机快照

1. 快照功能

OpenStack 提供的快照功能适用于云服务器实例、云磁盘和云卷。OpenStack 的快照功能可以让用户在任何时间保存云服务器实例、云磁盘和云卷的一个完整的状态副本，以及在需要时恢复整个状态。快照功能是基于写时复制（Copy-On-Write）技术实现的，所以快照不会占用过多的空间，也不影响计算性能。

当需要撤回某些操作、防止数据丢失，或者测试新的应用程序时，OpenStack 快照功能能够方便地创建虚拟机的副本，并允许修改和试验，同时保证原始数据不会被破坏。因此，快照功能可以让系统管理员轻松、简单地恢复虚拟机状态，提高运行效率和降低成本。

2. 在 Dashboard 界面中创建快照

登录 OpenStack Dashboard 界面，选择左侧导航栏中的“项目→计算→实例”选项，选择云主机右侧下拉列表中的“创建快照”选项，进入创建快照界面，输入快照名称，输入完成后，单击“创建快照”按钮即可。

确认创建已完成后，使用 openstack image list 命令查看通过 Dashboard 界面创建的快照，命令如下。

```
[root @ controller ~]# openstack image list
```

3. 使用命令创建快照

（1）使用 openstack server image 命令，为 wxic-server 节点创建名为 wxic-wordpress1 的快照，命令

如下。

```
[root@controller ~]# openstack server image create wxic-server --name wxic-wordpress1
```

（2）创建完成后，使用openstack image list命令查看镜像列表，命令如下。

```
[root@controller ~]# openstack image list
```

4. 压缩快照

（1）使用openstack image命令，获取当前主机的快照列表，查看快照ID，命令和结果如下。

```
[root@controller ~]# openstack image list
+--------------------------------------+------------------+---------+
| ID                                   | Name             | Status  |
+--------------------------------------+------------------+---------+
| 375af26b-bea5-469b-943c-cbdf864cc511 | cirros           | active  |
| f984d422-df85-4d04-a8c0-d6b449b53bc3 | openEuler22.09   | active  |
| f2595d3c-0fd7-4965-8018-fd5add4a0825 | wxic-wordpress   | active  |
| 199f41aa-de9e-4eef-aaaf-b66a331a000b | wxic-wordpress1  | active  |
+--------------------------------------+------------------+---------+
```

（2）使用glance命令，将生成的wxic-wordpress快照导出到当前目录下，命令如下。

```
[root@controller ~]#glance image-download --file wxic-wordpress.qcow2
f2595d3c-0fd7-4965-8018-fd5add4a0825
```

（3）使用qemu-img命令，对导出的wxic-wordpress快照进行压缩，压缩后的快照文件名为wxic-wordpress1，命令如下。

```
[root@controller ~]# qemu-img convert -c -O qcow2 wxic-wordpress.qcow2
wxic-wordpress1.qcow2
```

（4）对wxic-wordpress快照文件压缩完成后，使用du命令对比压缩前后的文件大小，命令如下。

```
[root@controller ~]# du -sh wxic-wordpress.qcow2
[root@controller ~]# du -sh wxic-wordpress1.qcow2
```

（5）通过结果可以看出，压缩后的快照所占用的空间明显小于压缩前的，对快照大小的压缩操作成功，命令和结果如下。

```
[root@ controller ~]# qemu-img convert -c -O qcow2 wxic-wordpress.qcow2
wxic-wordpress1.qcow2
[root@ controller ~]# ll
总用量 2478032
-rw-------. 1 root root       1305  4月 20  14:15 anaconda-ks.cfg
-rw-r-r--. 1 root root   912392192  5月  9  17:53 wxic-wordpress1.qcow2
-rw-r-r--. 1 root root  1625096192  5月  9  17:50 wxic-wordpress.qcow2
[root@ controller ~]# du -sh wxic-wordpress.qcow2
1.6G    wxic- wordpress.qcow2
[root@ controller ~]# du -sh wxic-wordpress1.qcow2
871M    wxic- wordpress1.qcow2
```

任务6.2.2 云应用迁移

1. 对任务6.1中部署的WordPress云主机创建快照

（1）使用Skyline管理平台，选择左侧导航栏中的“计算→云主机”选项，选择云主机“wxic-wordpress”，对任务6.1中部署的WordPress云主机创建快照。设置快照名称后，单击“确定”按钮，如图6-11所示。

（2）返回主页，选择左侧导航栏中的“计算→云主机快照”选项，可查看到名为wxic-wordpress的快照已创建成功，如图6-12所示。

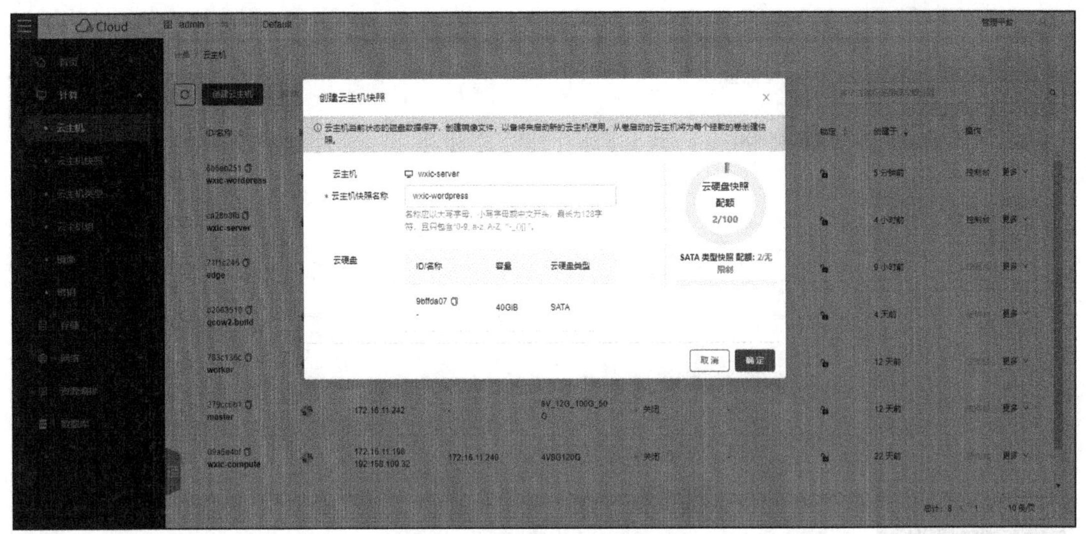

图 6-11　云主机快照创建界面

图 6-12　云主机快照界面

2. 使用快照启动新的云主机

（1）使用 Skyline 管理平台，选择左侧导航栏中的“计算→云主机”选项，在创建云主机界面中进行基础配置，在“全部架构”选项中设置云主机规格为“ecs.c7.large.2”，启动源为“云主机快照”，并选择云主机快照为“wxic-wordpress”，如图 6-13 所示。

云主机快照类型配置完成后，单击“下一步：网络配置”按钮，设置网络类型为“public-wxicnet”，设置虚拟网卡为“public-wxicnet”“自动分配地址”，如图 6-14 所示。

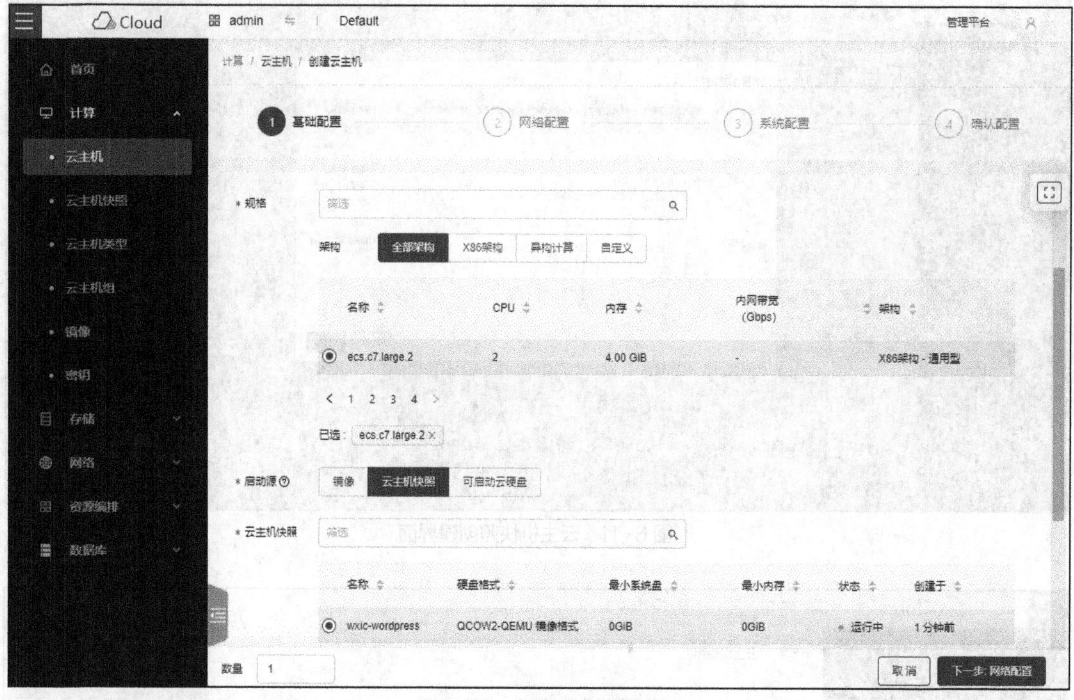

图 6-13 创建云主机界面

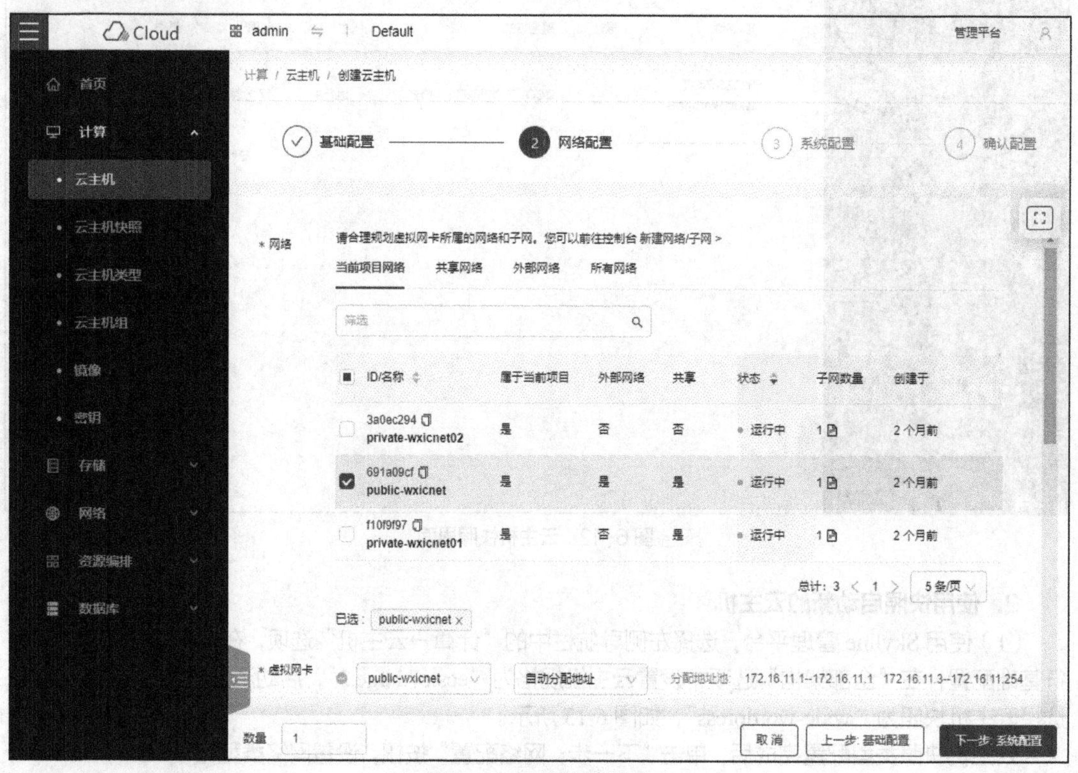

图 6-14 网络配置界面

设置安全组类型为“wxic-Security”，如图 6-15 所示。

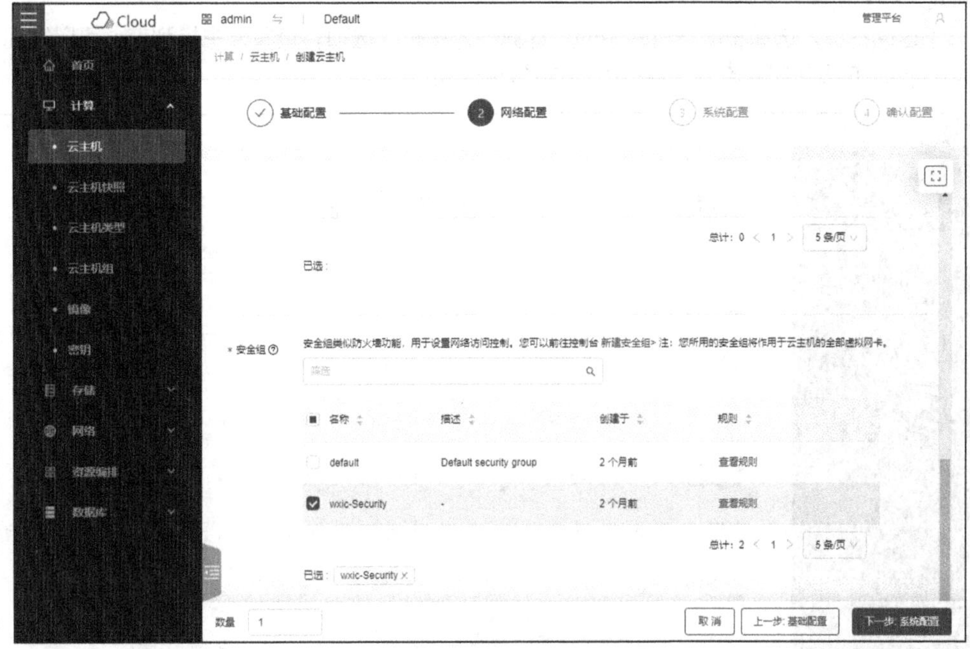

图 6-15　设置安全组类型界面

网络及安全组配置完成后，单击“下一步：系统配置”按钮，填写基本信息及登录密码，如图 6-16 所示。

图 6-16　填写基本信息及登录密码

（2）单击“下一步：确认配置”按钮，确认启动源为云主机快照，网络及系统配置确认无误后，单击“确定”按钮创建云主机。

（3）返回主页，选择左侧导航栏中的“计算→云主机”选项，查看云主机列表，可以发现名为wxic-wordpress的云主机状态为“运行中”，如图6-17所示。

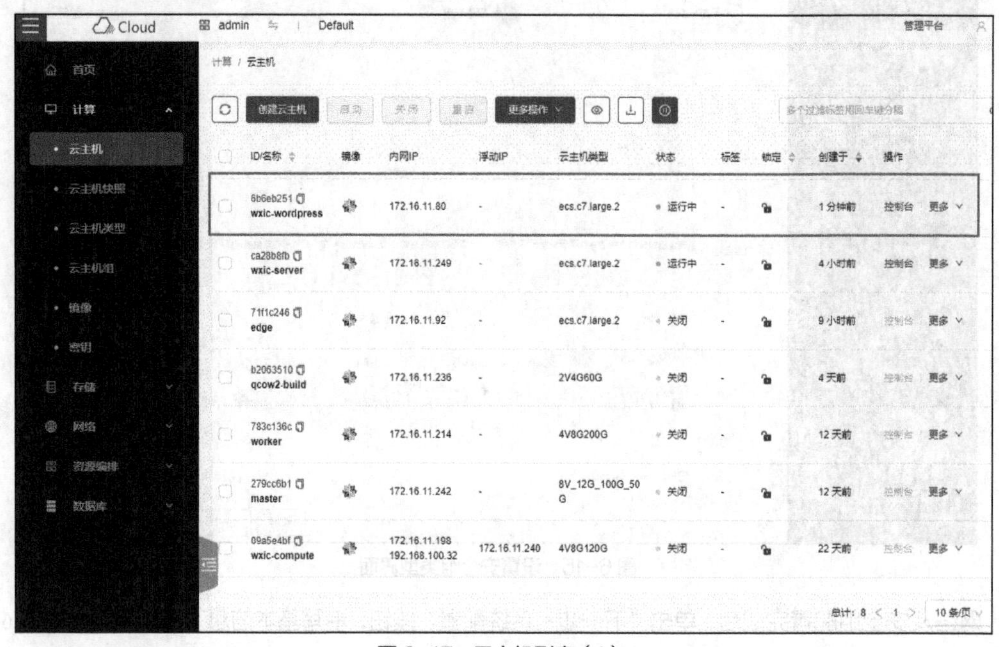

图6-17 云主机列表（1）

3. 验证服务状态是否正常可用

通过快照的方式启动基于WordPress的云主机。启动后，同样需要验证快照环境下的服务状态是否正常可用。

在浏览器上以http://IP地址的方式进入WordPress博客界面，验证云主机内服务状态是否正常可用，如图6-18所示。

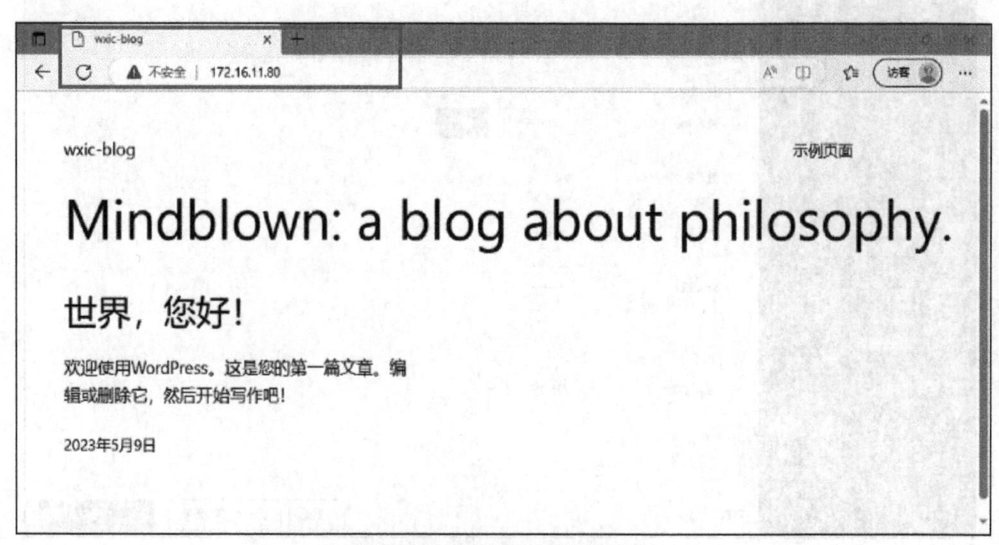

图6-18 验证云主机内服务状态

任务 6.3 高可用架构应用

本任务运用高可用架构，集成了 MySQL 主从复制机制、WordPress 博客应用，以及 HAProxy 负载均衡器。该架构可通过数据库复制保障故障切换，利用 WordPress 内容管理及 HAProxy 均衡负载，来保证架构的稳定性和可扩展性。接下来将介绍部署高可用架构应用的具体方法。

微课 6.3 高可用架构应用

任务 6.3.1 基础环境准备

1. 创建 3 台云主机

（1）节点规划

参考 6.1.1 小节的步骤，创建 3 台云主机，云主机类型为 4 个 vCPU/12GB 内存/60GB 硬盘。各节点主机名和 IP 地址规划如表 6-3 所示。

表 6-3 各节点主机名和 IP 地址规划

节点	主机名	IP 地址
数据库主节点	primary	172.16.11.17
数据库从节点 01	replica01	172.16.11.93
数据库从节点 02	replica02	172.16.11.153

注意 数据库主节点在实际部署过程中具有多重作用，主要用于在 LNMP 环境下部署 WordPress 服务及 HAProxy 负载均衡器的节点，数据库从节点 01 和数据库从节点 02 主要用于运行数据库集群的从节点。

（2）查看云主机

云主机创建完成后，使用 Skyline 管理平台查看云主机详情，如图 6-19 所示。

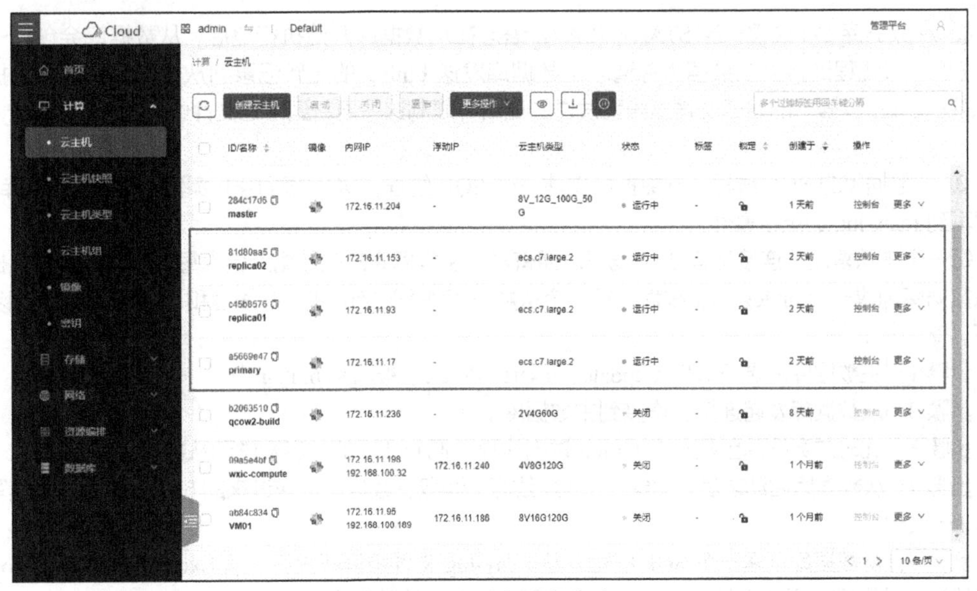

图 6-19 云主机列表（2）

2. 升级、更新系统软件包

使用命令升级、更新系统软件包，命令如下。

```
[root@primary ~]# dnf -y update && dnf -y upgrade
```

注意  数据库主节点配置完成后，对数据库从节点进行同样的操作。

任务 6.3.2 部署主从数据库

1. 主从数据库的作用和原理

（1）主从复制的含义如下。主从复制用来建立一个和主数据库完全一样的数据库，该数据库称为从数据库；主数据库一般是准实时的业务数据库。

（2）主从复制的作用如下。

① 做数据的热备：从数据库作为后备数据库，主数据库服务器出现故障后，可切换到从数据库继续工作，避免数据丢失。

② 架构的扩展：目前业务量越来越大，I/O 访问频率越来越高，单机无法满足需求，架构扩展可通过多库存储来降低磁盘 I/O 访问的频率，提高单机的 I/O 性能。

③ 读写分离，使数据库能支撑更大的并发数：该作用在报表中尤其重要。由于部分报表结构查询语言（Structure Query Language，SQL）语句的执行速度非常慢，导致锁表，从而影响前台服务。如果前台使用主节点，报表使用从节点，那么报表 SQL 语句将不会造成前台锁，保证了前台运行速度。

（3）主从复制的原理如下。数据库中有一个 bin-log 文件，记录了所有 SQL 语句。主从复制的原理就是把主数据库的 bin-log 文件的 SQL 语句复制到从数据库中，在从数据库的 relay-log 文件中再执行一次 SQL 语句。下面的主从配置就是围绕这个原理进行的，具体需要通过以下 3 个线程来操作。

① binlog 输出线程：每当从数据库连接到主数据库的时候，主数据库都会创建一个线程，并发送 binlog 内容到从数据库中。在从数据库中，当开始发送的时候，从数据库就会创建两个线程进行处理。

② 从数据库 I/O 线程：当 START SLAVE 语句在从数据库开始执行之后，从数据库会创建一个 I/O 线程，该线程连接到主数据库并请求主数据库发送 binlog 的更新记录到从数据库中。从数据库 I/O 线程读取主数据库的 binlog 输出线程发送的更新并复制这些更新到本地文件中，其中包括 relay-log 文件。

③ 从数据库的 SQL 线程：从数据库创建一个 SQL 线程，这个线程用于读取并执行从数据库 I/O 线程写到 relay-log 的更新事件。

每一个主从复制的连接都有 3 个线程。拥有多个从数据库的主数据库会为每一个连接到主数据库的从数据库创建一个 binlog 输出线程，每一个从数据库都有它自己的 I/O 线程和 SQL 线程，详细步骤如下。

步骤 1：主数据库的更新事件（update、insert、delete）被写到 binlog 中。

步骤 2：从数据库发起连接，连接到主数据库。

步骤 3：此时主数据库创建一个 binlog 输出线程，把 binlog 的内容发送到从数据库中。

步骤 4：从数据库启动之后，创建一个 I/O 线程，读取主数据库传送过来的 binlog 内容并将其写入 relay-log 文件。

步骤 5：从数据库创建一个 SQL 线程，从 relay-log 文件中读取内容，从 Exec_Master_Log_Pos 位置开始执行读取到的更新事件，并将更新内容写入从节点的数据库中。

2. 数据库主节点的安装与配置

（1）配置/etc/hosts 文件，命令和结果如下。

```
[root@primary ~]# vi /etc/hosts
172.16.11.17 primary
172.16.11.93 replica01
172.16.11.153 replica02
```

（2）安装数据库 MariaDB，命令如下。

```
[root@primary ~]# dnf -y install mariadb mariadb-server
```

（3）启动数据库并设置开机自启动，命令如下。

```
[root@primary ~]# systemctl enable --now mariadb
```

（4）初始化数据库，并设置 MariaDB 数据库的 root 用户的密码，命令和结果如下。

```
[root@primary ~]# mysql_secure_installation
......
Enter current password for root (enter for none): #此处直接按 Enter 键即可
OK, successfully used password, moving on...
......
Switch to unix_socket authentication [Y/n] n        #无须切换身份认证
......
Change the root password? [Y/n] y   #修改密码
New password:
Re-enter new password:
Password updated successfully!
Reloading privilege tables..
 ... Success!
......
Remove anonymous users? [Y/n] y
......
Disallow root login remotely? [Y/n] n
......
Remove test database and access to it? [Y/n] y
......
Reload privilege tables now? [Y/n] y
 ......
```

（5）修改 primary 数据库主节点的数据库配置文件，在配置文件/etc/my.cnf.d/mariadb-server.cnf 的[mysqld]中增添内容，命令如下。

```
[root@primary ~]# vi /etc/my.cnf.d/mariadb-server.cnf
......
[mysqld]
......
log_bin = mysql-bin                        #记录操作日志
binlog_ignore_db = mysql                   #不同步 MySQL 系统数据库
server_id = 17                    #数据库集群中的每个节点 ID 都要不同，一般使用 IP 地址的最
#后一段的数字作为 ID，如 172.30.11.17 的 server_id 写为 17
......
```

（6）编辑完配置文件后，重启数据库服务，并进入数据库，命令和结果如下。

```
[root@primary ~]# systemctl restart mariadb
[root@primary ~]# mysql -uroot -p000000
```

```
MariaDB [(none)]>
```

（7）开放数据库主节点的数据库权限。在 primary 数据库主节点上，授权在任何客户端机器上都可以 root 用户登录到数据库，在主节点上创建一个 user 用户连接节点 wxic-mysql2 与 wxic-mysql3，并为其赋予从节点同步主节点数据库的权限，命令和结果如下。

```
MariaDB [(none)]> grant all privileges on *.* to root@'%' identified by '000000';
Query OK, 0 rows affected (0.003 sec)
MariaDB [(none)]> grant replication slave on *.* to 'user'@'replica01' identified by
'000000';
Query OK, 0 rows affected (0.002 sec)
MariaDB [(none)]> grant replication slave on *.* to 'user'@'replica02' identified by
'000000';
Query OK, 0 rows affected (0.002 sec)
```

3. 数据库从节点的安装与配置

（1）在数据库的两个从节点上配置 hosts，均添加相应内容，命令和添加的内容如下。

```
[root@replica01 ~]# vi /etc/hosts
172.16.11.17 primary
172.16.11.93 replica01
172.16.11.153 replica02
[root@replica02 ~]# vi /etc/hosts
172.16.11.17 primary
172.16.11.93 replica01
172.16.11.153 replica02
```

（2）在数据库的两个从节点上安装 MariaDB 数据，启动数据库并设置开机自启动，命令如下。

```
[root@replica01 ~]# dnf -y install mariadb mariadb-server
[root@replica02 ~]# dnf -y install mariadb mariadb-server
[root@replica01 ~]# systemctl enable --now mariadb
[root@replica02 ~]# systemctl enable --now mariadb
```

（3）在数据库的两个从节点上初始化数据库，命令和结果如下。

```
[root@ replica01 ~]# mysql_secure_installation
NOTE: RUNNING ALL PARTS OF THIS SCRIPT IS RECOMMENDED FOR ALL MariaDB
  SERVERS IN PRODUCTION USE!  PLEASE READ EACH STEP CAREFULLY!

In order to log into MariaDB to secure it, we'll need the current
password for the root user. If you've just installed MariaDB, and
haven't set the root password yet, you should just press enter here.

Enter current password for root (enter for none):
OK, successfully used password, moving on...

Setting the root password or using the unix_socket ensures that nobody
can log into the MariaDB root user without the proper authorisation.

You already have your root account protected, so you can safely answer 'n'.

Switch to unix_socket authentication [Y/n] n    #无须切换身份认证
 ... skipping.

You already have your root account protected, so you can safely answer 'n'.

Change the root password? [Y/n] y               #修改密码
New password:
Re-enter new password:
```

```
Password updated successfully!
Reloading privilege tables..
 ... Success!


By default, a MariaDB installation has an anonymous user, allowing anyone
to log into MariaDB without having to have a user account created for
them.  This is intended only for testing, and to make the installation
go a bit smoother.  You should remove them before moving into a
production environment.

Remove anonymous users? [Y/n] y
 ... Success!

Normally, root should only be allowed to connect from 'localhost'.  This
ensures that someone cannot guess at the root password from the network.

Disallow root login remotely? [Y/n] n
 ... skipping.

By default, MariaDB comes with a database named 'test' that anyone can
access.  This is also intended only for testing, and should be removed
before moving into a production environment.

Remove test database and access to it? [Y/n] y
 - Dropping test database...
 ... Success!
 - Removing privileges on test database...
 ... Success!

Reloading the privilege tables will ensure that all changes made so far
will take effect immediately.

Reload privilege tables now? [Y/n] y
 ... Success!

Cleaning up...

All done!  If you've completed all of the above steps, your MariaDB
installation should now be secure.

Thanks for using MariaDB!
```

（4）修改配置文件。修改数据库从节点 replica01 与数据库从节点 replica02 的数据库配置文件，在配置文件/etc/my.cnf.d/mariadb-server.cnf 中的[mysqld]中增添内容，命令和增添的内容如下。

```
[root@replica01 ~]# vi /etc/my.cnf.d/mariadb-server.cnf
......
[mysqld]
......
log_bin = mysql-bin
binlog_ignore_db = mysql
server_id = 93
......
[root@replica01 ~]# systemctl restart mariadb
```

```
[root@replica02 ~]# vi /etc/my.cnf.d/mariadb-server.cnf
……
[mysqld]
……
log_bin = mysql-bin
binlog_ignore_db = mysql
server_id = 153
……
[root@replica02 ~]# systemctl restart mariadb
```

（5）修改完配置文件后，重启数据库服务，分别在数据库从节点 replica01 与 replica02 上登录 MariaDB 数据库，配置数据库从节点连接数据库主节点的连接信息。其中，参数 master_host 用于指定数据库主节点名为 primary，参数 master_user 用于指定用户为 user。在数据库从节点 replica01 上执行命令，命令和结果如下。

```
[root@ replica01 ~]# mysql -uroot -p000000
MariaDB [(none)]> change master to master_host='primary',master_user='user',
master_password='000000';
Query OK, 0 rows affected (0.024 sec)
MariaDB [(none)]> start slave;
Query OK, 0 rows affected (0.001 sec)
MariaDB [(none)]> show slave status\G;
      Slave_IO_Running: Yes
      Slave_SQL_Running: Yes
```

在数据库从节点 replica02 上执行命令，命令和结果如下。

```
[root@ replica02 ~]# mysql -uroot -p000000
MariaDB [(none)]> change master to master_host='primary',master_user=
'user',master_password='000000';
Query OK, 0 rows affected (0.029 sec)

MariaDB [(none)]> start slave;
Query OK, 0 rows affected (0.001 sec)

MariaDB [(none)]> show slave status\G;
      Slave_IO_Running: Yes
      Slave_SQL_Running: Yes
```

配置完主从数据库之间的连接信息之后，启用从节点服务。使用 show slave status\G;命令查看数据库从节点服务状态，如果 Slave_IO_Running 和 Slave_SQL_Running 的状态都为 Yes，则数据点从节点服务启用成功。

任务 6.3.3 部署负载均衡器

1. HAProxy 节点 1 安装与配置

（1）安装 HAProxy 服务

在数据库主节点上安装 HAProxy 服务，命令如下。

```
[root@primary ~]# dnf -y install haproxy
```

（2）配置 HAProxy

修改数据库主节点的 HAProxy 配置文件/etc/haproxy/haproxy.cfg，命令如下。

注意

将 haproxy.cfg 配置文件原来的内容删除，并替换为以下内容，注意修改 listen status 和 listen mariadb 字段处的 IP 地址为实际 IP 地址。

```
[root@primary ~]# cat /etc/haproxy/haproxy.cfg
global
  log        127.0.0.1 local2
  chroot      /var/lib/haproxy
  maxconn     4000
  user        haproxy
  group       haproxy
  daemon
  stats socket /var/lib/haproxy/stats
defaults
  mode                    http
  log                     global
  option                  redispatch
  retries                 3
  timeout http-request    10s
  timeout queue           1m
  timeout connect         10s
  timeout client          1m
  timeout server          1m
  timeout check           10s
  maxconn                 4000
listen status
   bind                   172.16.11.17:9000
   mode                   http
   stats                  enable
   stats   uri            /stats
   stats   auth           admin:admin
   stats   admin          if TRUE
listen mariadb
   balance                roundrobin
   mode                   tcp
   option                 tcplog
   option                 tcpka
   bind                   172.16.11.17:3307
server primary   172.16.11.17:3306  check weight 1
server replica01 172.16.11.93:3306  check weight 1
server replica02 172.16.11.153:3306 check weight 1
```

在 HAProxy 配置文件中，要关注的地方主要为 listen mariadb 字段下的各个配置，相关配置的解释如下。

① balance roundrobin：新连接定向到循环顺序列表中的下一个目标，并通过服务器的权值进行修改。关于 balance 的模式还有 source、leastconn、static-rr 等，关于其他的模式，读者可以自行了解。

② mode tcp：定义路由的连接类型。Galera Cluster 使用 TCP 连接。

③ option tcplog：启用记录有关 TCP 连接的日志信息。

④ option tcpka：启用 Keepalived 功能以维护 TCP 连接。

⑤ bind 172.16.11.17:3307：HAProxy 服务监听的 IP 地址及端口，因为使用的是 primary 节点，所以不能监听 3306 端口，监听该端口时会出现冲突。

⑥ server primary 172.16.11.17:3306 check weight 1：定义 HAProxy 在路由连接中使用的节点。check 代表接受检查，weight 代表权值。

配置文件修改完成后，使用命令检查 HAProxy 的配置文件是否有问题，命令和结果如下。

```
[root@primary ~]# haproxy -f /etc/haproxy/haproxy.cfg -c
Configuration file is valid
```

2. HAProxy负载均衡主从数据库的实现

（1）启动HAProxy

在任务6.2配置HAProxy的基础上，启用HAProxy服务，命令如下。

```
[root@primary ~]# systemctl start haproxy
```

使用命令查看3307端口和9000端口，发现它们已启动，说明HAProxy服务运行正常，命令和结果如下。

```
[root@primary ~]# netstat -ntlp
Active Internet connections (only servers)
Proto RecV-Q   Send-Q   Local Address        Foreign    State    PID/Program name
                                             Address
tcp    0       0        172.16.11.17:3307    0.0.0.0:*  LISTEN   27713/haproxy
tcp    0       0        0.0.0.0:22           0.0.0.0:*  LISTEN   6871/sshd: /usr/sbi
tcp    0       0        172.16.11.17:9000    0.0.0.0:*  LISTEN   27713/haproxy
tcp    0       0        0.0.0.0:111          0.0.0.0:*  LISTEN   1/systemd
tcp6   0       0        : : :3306            : : : *    LISTEN   27387/mariadbd
tcp6   0       0        : : :22              : : : *    LISTEN   6871/sshd: /usr/sbi
tcp6   0       0        : : :111             : : : *    LISTEN   1/systemd
```

使用浏览器访问http://172.16.11.17:9000/stats，查看HAProxy服务的状态（登录用户名和密码均为admin），如图6-20所示。

HAProxy version 2.4.17-9f97155, released 2022/05/13

Statistics Report for pid 28011

\> General process information

pid = 28011 (process #1, nbproc = 1, nbthread = 2)
uptime = 0d 0h00m39s
system limits: memmax = unlimited; ulimit-n = 8033
maxsock = 8033; maxconn = 4000; maxpipes = 0
current conns = 1; current pipes = 0/0; conn rate = 1/sec; bit rate = 0.000 kbps
Running tasks: 0/15; idle = 100 %

active UP | backup UP
active UP, going down | backup UP, going down
active DOWN, going up | backup DOWN, going up
active or backup DOWN | not checked
active or backup DOWN for maintenance (MAINT)
active or backup SOFT STOPPED for maintenance
Note: "NOLB"/"DRAIN" = UP with load-balancing disabled.

Display option:
- Scope :
- Hide 'DOWN' servers
- Refresh now
- CSV export
- JSON export (schema)

External resources:
- Primary site
- Updates (v2.4)
- Online manual

status

	Queue			Session rate			Sessions						Bytes		Denied		Errors			Warnings		Server								
	Cur	Max	Limit	Cur	Max	Limit	Cur	Max	Limit	Total	LbTot	Last	In	Out	Req	Resp	Req	Conn	Resp	Retr	Redis	Status	LastChk	Wght	Act	Bck	Chk	Dwn	Dwntme	Thrtle
Frontend				1	1	-	1	1	4 000	2			497	20 343	0	0	0					OPEN								
Backend	0	0		0	0		0	0	400	0	0	0s	497	20 343	0	0		0	0	0	0	39s UP		0/0	0	0		0		

mariadb

	Queue			Session rate			Sessions						Bytes		Denied		Errors			Warnings		Server								
	Cur	Max	Limit	Cur	Max	Limit	Cur	Max	Limit	Total	LbTot	Last	In	Out	Req	Resp	Req	Conn	Resp	Retr	Redis	Status	LastChk	Wght	Act	Bck	Chk	Dwn	Dwntme	Thrtle
Frontend				0	0	-	0	0	4 000	0			0	0	0	0	0					OPEN								
primary	0	0	-	0	0		0	0	-	0	0	?	0	0		0		0	0	0	0	39s UP	L4OK in 0ms	1/1	Y	-	0	0	0s	-
replica01	0	0	-	0	0		0	0	-	0	0	?	0	0		0		0	0	0	0	39s UP	L4OK in 0ms	1/1	Y	-	0	0	0s	-
replica02	0	0	-	0	0		0	0	-	0	0	?	0	0		0		0	0	0	0	39s UP	L4OK in 0ms	1/1	Y	-	0	0	0s	-
Backend	0	0		0	0		0	0	400	0	0	?	0	0	0	0		0	0	0	0	39s UP		3/3	3	0		0	0s	

Choose the action to perform on the checked servers : Apply

图6-20 HAProxy首页（1）

（2）实现负载均衡主从数据库

① 在数据库主节点上，测试数据库从节点的负载均衡，输出数据库主节点和从节点的server_id，命令如下。

```
[root@primary ~]# for i in $(seq 1 10); do mysql -uroot -p000000 -h172.16.11.17 -P3306
-e 'select @@server_id;'; done | egrep '[0-9]'
17
17
17
17
17
17
17
17
17
17
```

```
[root@primary ~]# for i in $(seq 1 10); do mysql -uroot -p000000 -h172.16.11.93 -P3306
-e 'select @@server_id;'; done | egrep '[0-9]'
93
93
93
93
93
93
93
93
93
93
[root@primary ~]# for i in $(seq 1 10); do mysql -uroot -p000000 -h172.16.11.153 -P3306
-e 'select @@server_id;'; done | egrep '[0-9]'
153
153
153
153
153
153
153
153
153
```

② 切换至数据库从节点 replica01 并关闭数据库服务后，返回 HAProxy 首页，查看数据库从节点 replica01 的运行状态，如图 6-21 所示。

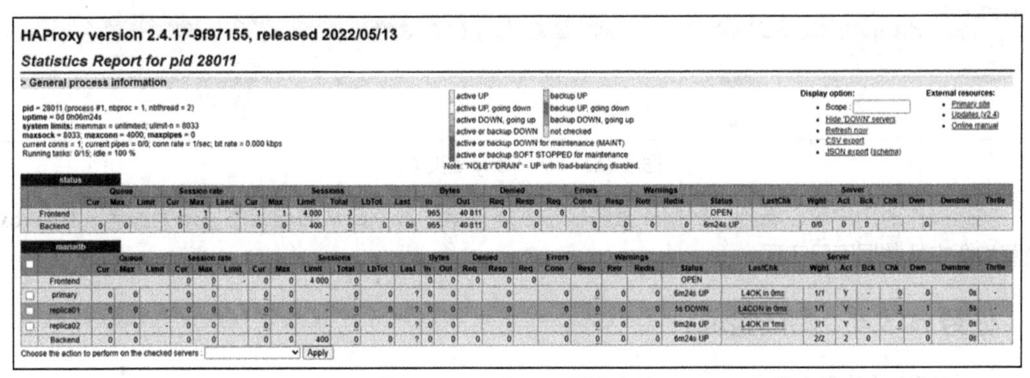

图 6-21 HAProxy 首页（2）

实际操作时用户可以看到，被关闭数据库服务的数据库从节点状态显示为红色。

③ 在数据库主节点上，再次测试数据库从节点的负载均衡，输出数据库主节点和从节点的 server_id，命令如下。

```
[root@primary ~]# for i in $(seq 1 10); do mysql -uroot -p000000 -h172.16.11.17 -P3306
-e 'select @@server_id;'; done | egrep '[0-9]'
17
17
17
17
17
17
17
17
17
```

```
17
[root@primary ~]# for i in $(seq 1 10); do mysql -uroot -p000000 -h172.16.11.153 -P3306
-e 'select @@server_id;'; done | egrep '[0-9]'
153
153
153
153
153
153
153
153
153
```

此时可以发现，除了被关闭数据库服务的数据库从节点无法使用负载均衡之外，其他节点依旧可以使用负载均衡。

④ 切换至数据库从节点 replica01 并重新启用数据库服务，待其恢复为正常状态后，返回 HAProxy 首页，查看数据库从节点 replica01 的运行状态，如图 6-22 所示。

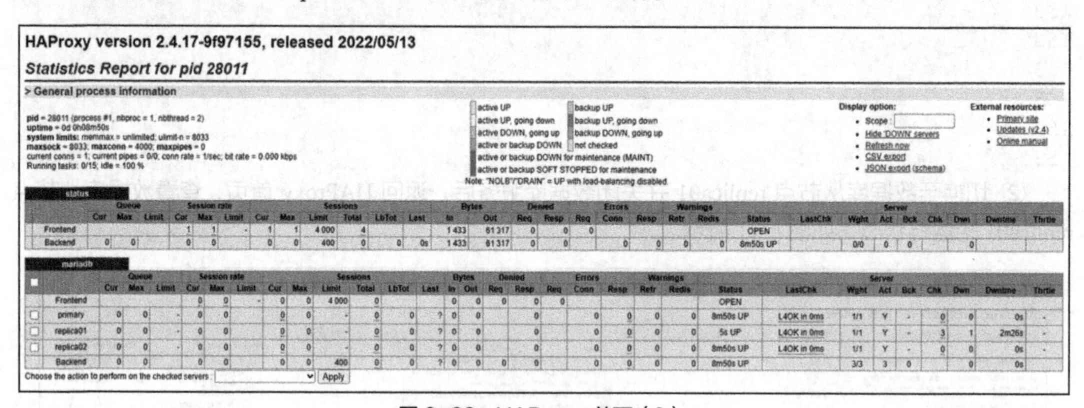

图 6-22 HAProxy 首页（3）

实际操作时用户可以看到，数据库服务状态恢复正常的数据库从节点状态显示为绿色，HAProxy 负载均衡主从数据库已实现。

任务 6.3.4 测试验证

1. 验证主从数据库同步

（1）在数据库主节点上创建数据库

在数据库主节点 primary 上创建数据库 wxic，并在数据库 wxic 中创建表 company，插入表数据，创建完成后，查看 company 表的数据，命令和结果如下。

```
[root@primary ~]# mysql -uroot -p000000
MariaDB [(none)]> create database wxic;
Query OK, 1 row affected (0.001 sec)
MariaDB [(none)]> use wxic;
Database changed
MariaDB [wxic]> create table company(id int not null primary key,name
varchar(50),addr varchar(255));
Query OK, 0 rows affected (0.017 sec)
MariaDB [wxic]> insert into company values(1,"alibaba","china");
Query OK, 1 row affected (0.003 sec)
MariaDB [wxic]> select * from company;
+----+---------+-------+
```

```
| id | name    | addr  |
+----+---------+-------+
|  1 | alibaba | china |
+----+---------+-------+
1 row in set (0.000 sec)

MariaDB [wxic]>
```

（2）在数据库从节点上验证复制功能

登录数据库从节点 replica01，进入数据库，查看数据库列表，切换至 wxic 数据库，查询表及其内容，验证从数据库的复制功能，命令和结果如下。

```
[root@replica01 ~]# mysql -uroot -p000000
MariaDB [(none)]> show databases;
+--------------------+
| Database           |
+--------------------+
| information_schema |
| mysql              |
| performance_schema |
| wxic               |
+--------------------+
4 rows in set (0.000 sec)
MariaDB [(none)]> use wxic;
Database changed
MariaDB [wxic]> show tables;
+--------------------+
| Tables_in_wxic     |
+--------------------+
| company            |
+--------------------+
1 row in set (0.000 sec)
MariaDB [wxic]> select * from company;
+----+---------+-------+
| id | name    | addr  |
+----+---------+-------+
| 1  | alibaba | china |
+----+---------+-------+
1 row in set (0.000 sec)

MariaDB [wxic]>
```

登录数据库从节点 replica02，进入数据库，查看数据库列表，切换至 wxic 数据库，查询表及其内容，验证从数据库的复制功能，命令和结果如下。

```
[root@replica02 ~]# mysql -uroot -p000000
MariaDB [(none)]> show databases;
+--------------------+
| Database           |
+--------------------+
| information_schema |
| mysql              |
| performance_schema |
| wxic               |
+--------------------+
4 rows in set (0.001 sec)
```

```
MariaDB [(none)]> use wxic;
Database changed
MariaDB [wxic]> show tables;
+--------------------+
| Tables_in_wxic     |
+--------------------+
| company            |
+--------------------+
1 row in set (0.000 sec)
MariaDB [wxic]> select * from company;
+----+---------+-------+
| id | name    | addr  |
+----+---------+-------+
| 1  | alibaba | china |
+----+---------+-------+
1 row in set (0.001 sec)
MariaDB [wxic]>
```

综上，当数据库主节点的数据库内容发生变更时，数据库从节点同步更新，验证主从数据库同步成功。

2. 验证负载均衡器

根据 HAProxy 实现负载均衡的原理，当用户在访问数据库时，默认访问的地址为 172.16.11.17:3307。HAProxy 在接收到用户请求后，会根据配置的调度算法，把请求分散到不同的数据库节点，同时多个数据库节点的信息会同步。

（1）访问负载均衡数据库

访问负载均衡数据库，命令如下。

```
[root@primary ~]# mysql -h 172.16.11.17 -P 3307 -uroot -p000000
```

（2）负载均衡器功能测试

使用 3307 端口能够访问数据库服务，但是不确定 HAProxy 功能是否正常，因为不确定后台访问的数据库到底是哪一个，此时可以通过以下方式来进行设置，即可以为每个节点的 MySQL/MariaDB 设置唯一的 server_id，设定 3 个数据库节点拥有不同的 server_id，命令如下。

```
[root@primary ~]# mysql -h 172.16.11.17 -uroot -p000000 -e "SET GLOBAL server_id=17"
[root@primary ~]# mysql -h 172.16.11.93 -uroot -p000000 -e "SET GLOBAL server_id=93"
[root@primary ~]# mysql -h 172.16.11.153 -uroot -p000000 -e "SET GLOBAL server_id=153"
```

现在 3 个数据库节点都拥有自己的专属 server_id，当通过 HAProxy 服务器访问数据库节点的时候，能确定 HAProxy 服务是否正常运行，命令和结果如下。

```
[root@primary ~]# mysql -h 172.16.11.17 -P 3307 -uroot -p000000 -e "show variables like 'server_id'"
+------------------+---------+
| Variable_name    | Value   |
+------------------+---------+
|  server_id       | 153     |
+------------------+---------+
 [root@primary ~]# mysql -h 172.16.11.17 -P 3307 -uroot -p000000 -e "show variables like 'server_id'"
+------------------+---------+
| Variable_name    | Value   |
+------------------+---------+
|  server_id       | 93      |
+------------------+---------+
[root@primary ~]# mysql -h 172.16.11.17 -P 3307 -uroot -p000000 -e "show variables like 'server_id'"
```

```
+-------------------+---------+
| Variable_name     | Value   |
+-------------------+---------+
|  server_id        |  17     |
+-------------------+---------+
```

可以看到每次访问的数据库节点都发生了变化，验证 HAProxy 负载均衡器成功。

3. 验证 WordPress 功能是否正常

（1）查看数据库中的表

登录数据库主节点 primary，进入数据库，查看数据库列表，切换至 wordpress 数据库，查看 wordpress 数据库中的表内容，命令和结果如下。

```
[root@primary ~]# mysql -uroot -p000000
MariaDB [(none)]> show databases;
+----------------------------+
| Database                   |
+----------------------------+
| information_schema         |
| mysql                      |
| performance_schema         |
| wordpress                  |
+----------------------------+
4 rows in set (0.000 sec)
MariaDB [(none)]> use wordpress;
Database changed
MariaDB [wordpress]> show tables;
+----------------------------+
| Tables_in_wordpress        |
+----------------------------+
| wp_commentmeta             |
| wp_comments                |
| wp_links                   |
| wp_options                 |
| wp_postmeta                |
| wp_posts                   |
+----------------------------+
```

登录数据库从节点 replica01，进入数据库，查看数据库列表，切换至 wordpress 数据库，查看 wordpress 数据库中的表，命令和结果如下。

```
[root@replica01 ~]# mysql -uroot -p000000
MariaDB [(none)]> show databases;
+----------------------------+
| Database                   |
+----------------------------+
| information_schema         |
| mysql                      |
| performance_schema         |
| wordpress                  |
+----------------------------+
4 rows in set (0.001 sec)
MariaDB [(none)]> use wordpress;
Database changed
MariaDB [wordpress]> show tables;
+----------------------------+
| Tables_in_wordpress        |
+----------------------------+
```

```
|  wp_commentmeta            |
|  wp_comments               |
|  wp_links                  |
|  wp_options                |
|  wp_postmeta               |
|  wp_posts                  |
+----------------------------+
```

登录数据库从节点 replica02，进入数据库，查看数据库列表，切换至 wordpress 数据库，查看 wordpress 数据库中的表，命令和结果如下。

```
[root@replica02 ~]# mysql -uroot -p000000
MariaDB [(none)]> show databases;
+----------------------------+
|  Database                  |
+----------------------------+
|  information_schema        |
|  mysql                     |
|  performance_schema        |
|  wordpress                 |
+----------------------------+
4 rows in set (0.001 sec)

MariaDB [(none)]> use wordpress;
Database changed
MariaDB [wordpress]> show tables;
+----------------------------+
|  Tables_in_wordpress       |
+----------------------------+
|  wp_commentmeta            |
|  wp_comments               |
|  wp_links                  |
|  wp_options                |
|  wp_postmeta               |
|  wp_posts                  |
+----------------------------+
```

综上，可以发现 WordPress 服务后台数据库中的表内容进行了同步。

（2）访问 WordPress 博客界面

通过浏览器访问 http://172.16.11.17，可以看到 WordPress 博客界面能被正常访问，功能使用正常，如图 6-23 所示。

图 6-23　WordPress 博客界面

项目小结

本项目首先介绍了基于云平台的常见 Web 网站架构应用，由传统架构出发，随着业务扩展，在请求量不断增加的同时，业务需求推动着技术架构的提升，包括升级和部署集群架构及分布式缓存架构。此后，本项目对高可用集群负载均衡架构的具体应用做了进一步的讲解，介绍了目前主流的架构部署方式和架构选择，从而使读者能够在实际业务场景下灵活根据业务的承载量、使用量来架设不同的架构并部署企业应用。在此基础上，本项目介绍了快照的应用和云应用的迁移。最后，本项目通过具体的项目实施，帮助读者理解并实践企业主流负载均衡系统的应用。

拓展知识

LVS 的 IP 地址负载均衡技术

负载均衡技术有很多实现方案，有基于 DNS 域名轮流解析方案、基于客户端调度访问方案、基于应用层系统的调度方案，以及基于 IP 地址的调度方案。在这些负载均衡方案中，执行效率最高的是基于 IP 地址的调度方案。

LVS 的 IP 地址负载均衡技术是通过 IPVS 模块来实现的，IPVS 是 LVS 集群系统的核心软件，LVS 集群的整个执行流程如下：访问的请求首先经过 VIP 到达负载调度器，然后由负载调度器从 Real Server 列表中选取一个 Real Server 节点响应用户的请求。当用户的请求到达负载调度器后，调度器如何将请求发送到提供服务的 Real Server 节点，且 Real Server 节点如何返回数据给用户，是 IPVS 实现的重点技术。

知识巩固

1. 单选

（1）创建一个 2 核、4GB 内存、20GB 硬盘的云主机需要的步骤是（　　）。

A. 选择云主机规格为 2 核、4GB 内存、20GB 硬盘

B. 选择操作系统

C. 选择网络类型

D. 以上所有选项都正确

（2）在备份和恢复快照时，需要特别注意的因素是（　　）。

A. 带宽和存储空间　　B. 数据完整性和可靠性

C. 暂停系统运行　　D. 以上所有选项都正确

（3）在 OpenStack 中，使用（　　）命令可以创建一个卷的快照。

A. openstack snapshot create　　B. openstack image create

C. openstack volume snapshot create　　D. openstack server snapshot create

（4）常见的负载均衡器类型不包括（　　）。

A. 硬件负载均衡器　　B. 软件负载均衡器

C. 云负载均衡器　　D. 手动负载均衡器

（5）HAProxy 支持的协议是（　　）。

A. TCP/UDP　　B. HTTP/HTTPS

C. SMTP/POP3　　D. 以上所有协议

2. 填空

（1）HAProxy 常用的调度算法包括轮询、______、IP 哈希、URL 哈希和源地址哈希。

（2）Keepalived 软件主要是通过______协议实现高可用功能的。

（3）负载均衡系统通过______来将大量的请求分发到不同的服务器上，以达到分担服务器负载的目的。

3. 简答

（1）云应用系统迁移需要考虑哪些因素？

（2）如何构建高可用架构？

拓展任务

熟悉 LVS 3 种负载均衡技术的实现方案

通过阅读其他参考文献，熟悉 LVS 实现负载均衡技术的 3 种方案，并对比其应用场景和优缺点。